Applied Mathematical Sciences

EDITORIAL STATEMENT

The mathematization of all sciences, the fading of traditional scientific boundaries, the impact of computer technology, the growing importance of mathematical-computer modelling and the necessity of scientific planning all create the need both in education and research for books that are introductory to and abreast of these developments.

The purpose of this series is to provide such books, suitable for the user of mathematics, the mathematician interested in applications, and the student scientist. In particular, this series will provide an outlet for material less formally presented and more anticipatory of needs than finished texts or monographs, yet of immediate interest because of the novelty of its treatment of an application or of mathematics being applied or lying close to applications.

The aim of the series is, through rapid publication in an attractive but inexpensive format, to make material of current interest widely accessible. This implies the absence of excessive generality and abstraction, and unrealistic idealization, but with quality of exposition as a goal.

Many of the books will originate out of and will stimulate the development of new undergraduate and graduate courses in the applications of mathematics. Some of the books will present introductions to new areas of research, new applications and act as signposts for new directions in the mathematical sciences. This series will often serve as an intermediate stage of the publication of material which, through exposure here, will be further developed and refined. These will appear in conventional format and in hard cover.

MANUSCRIPTS

The Editors welcome all inquiries regarding the submission of manuscripts for the series. Final preparation of all manuscripts will take place in the editorial offices of the series in the Division of Applied Mathematics, Brown University, Providence, Rhode Island.

SPRINGER-VERLAG NEW YORK INC., 175 Fifth Avenue, New York, N.Y. 10010

Applied Mathematical Sciences | Volume 72

Applied Mathematical Sciences

1. *John:* Partial Differential Equations, 4th ed.
2. *Sirovich:* Techniques of Asymptotic Analysis.
3. *Hale:* Theory of Functional Differential Equations, 2nd ed.
4. *Percus:* Combinatorial Methods.
5. *von Mises/Friedrichs:* Fluid Dynamics.
6. *Freiberger/Grenander:* A Short Course in Computational Probability and Statistics.
7. *Pipkin:* Lectures on Viscoelasticity Theory.
9. *Friedrichs:* Spectral Theory of Operators in Hilbert Space.
11. *Wolovich:* Linear Multivariable Systems.
12. *Berkovitz:* Optimal Control Theory.
13. *Bluman/Cole:* Similarity Methods for Differential Equations.
14. *Yoshizawa:* Stability Theory and the Existence of Periodic Solution and Almost Periodic Solutions.
15. *Braun:* Differential Equations and Their Applications, 3rd ed.
16. *Lefschetz:* Applications of Algebraic Topology.
17. *Collatz/Wetterling:* Optimization Problems.
18. *Grenander:* Pattern Synthesis: Lectures in Pattern Theory, Vol I.
20. *Driver:* Ordinary and Delay Differential Equations.
21. *Courant/Friedrichs:* Supersonic Flow and Shock Waves.
22. *Rouche/Habets/Laloy:* Stability Theory by Liapunov's Direct Method.
23. *Lamperti:* Stochastic Processes: A Survey of the Mathematical Theory.
24. *Grenander:* Pattern Analysis: Lectures in Pattern Theory, Vol. II.
25. *Davies:* Integral Transforms and Their Applications, 2nd ed.
26. *Kushner/Clark:* Stochastic Approximation Methods for Constrained and Unconstrained Systems
27. *de Boor:* A Practical Guide to Splines.
28. *Keilson:* Markov Chain Models—Rarity and Exponentiality.
29. *de Veubeke:* A Course in Elasticity.
30. *Sniatycki:* Geometric Quantization and Quantum Mechanics.
31. *Reid:* Sturmian Theory for Ordinary Differential Equations.
32. *Meis/Markowitz?* Numerical Solution of Partial Differential Equations.
33. *Grenander:* Regular Structures: Lectures in Pattern Theory, Vol. III.
34. *Kevorkian/Cole:* Perturbation methods in Applied Mathematics.
35. *Carr:* Applications of Centre Manifold Theory.
36. *Bengtsson/Ghil/Källén:* Dynamic Meteorology: Data Assimilation Methods.
37. *Saperstone:* Semidynamical Systems in Infinite Dimensional Spaces.
38. *Lichtenberg/Lieberman:* Regular and Stochastic Motion.
39. *Piccini/Stampacchia/Vidossich:* Ordinary Differential Equations in R^n.
40. *Naylor/Sell:* Linear Operator Theory in Engineering and Science.
41. *Sparrow:* The Lorenz Equations: Bifurcations, Chaos, and Strange Attractors.
42. *Guckenheimer/Holmes:* Nonlinear Oscillations, Dynamical Systems and Bifurcations of Vector Fields.
43. *Ockendon/Tayler:* Inviscid Fluid Flows.
44. *Pazy:* Semigroups of Linear Operators and Applications to Partial Differential Equations.
45. *Glashoff/Gustafson:* Linear Optimization and Approximation: An Introduction to the Theoretical Analysis and Numerical Treatment of Semi-Infinite Programs.
46. *Wilcox:* Scattering Theory for Diffraction Gratings.
47. *Hale et al.:* An Introduction to Infinite Dimensional Dynamical Systems—Geometric Theory.
48. *Murray:* Asymptotic Analysis.
49. *Ladyzhenskaya:* The Boundary-Value Problems of Mathematical Physics.
50. *Wilcox:* Sound Propagation in Stratified Fluids.
51. *Golubitsky/Schaeffer:* Bifurcation and Groups in Bifurcation Theory, Vol. I.
52. *Chipot:* Variational Inequalities and Flow in Porous Media.
53. *Majda:* Compressible Fluid Flow and Systems of Conservation Laws in Several Space Variables.
54. *Wasow:* Linear Turning Point Theory.

(continued on inside back cover)

P. Lochak C. Meunier

Multiphase Averaging for Classical Systems

With Applications to Adiabatic Theorems

Translated by H.S. Dumas

With 60 Illustrations

Springer-Verlag
New York Berlin Heidelberg
London Paris Tokyo

Pierre Lochak
Centre de Mathematiques
Ecole Normale Superieure
F-75230 Paris Cedex 05
France

Claude Meunier
Centre de Physique Theorique
Ecole Polytechnique
F-91128 Palaiseau Cedex
France

Translator

H. Scott Dumas
Department of Mathematics
University of New Mexico
Albuquerque, NM 87131
USA

Mathematics Subject Classification (1980): 34XX

Library of Congress Cataloging-in-Publication Data
Lochak, P. (Pierre)
Multiphase averaging for classical systems : with applications to adiabatic theorems / P. Lochak, C. Meunier : translated by H.S. Dumas.
p. cm.—(Applied mathematical sciences : v. 72)
Bibliography: p.
ISBN 0-387-96778-8
1. Averaging method (Differential equations) 2. Differential equations—Asymptotic theory. 3. Adiabatic invariants.
I. Meunier, C. (Claude, 1957– . II. Title. III. Series: Applied Mathematical sciences (Springer-Verlag New York Inc.) ; v. 72.
QA1.A647 vol. 72
[QA372]
510 s—dc19
[515.3'5] 88-20020

Camera ready text prepared by the authors.
Printed and bound by R.R. Donnelley & Sons, Harrisonburg, Virginia.
Printed in the United States of America.

9 8 7 6 5 4 3 2 1

ISBN 0-387-96778-8 Springer-Verlag New York Berlin Heidelberg
ISBN 3-540-96778-8 Springer-Verlag Berlin Heidelberg New York

FOREWORD

In many physical situations, the various degrees of freedom do not play the same role; some are macroscopic or evolve slowly, while others vary on smaller spatial or temporal scales. A number of physical examples displaying this disparity come to mind: Work and heat in thermodynamics, hydrodynamical and molecular dynamical quantities in fluid mechanics; the reader can no doubt add to this list. In such situations, one often seeks to eliminate the fast variables and to determine the gross behavior of the slow variables by constructing "macroscopic" evolution equations which bring into play only certain average characteristics of the small-scale motion, and which give a good approximation to the true evolution of the slow variables on a certain time interval. This is the aim of "homogenization methods" (deterministic or stochastic) and of the method of averaging, in which we are interested here. These methods are by nature asymptotic, as they always suppose a large separation of the scales involved (fast time t and slow time $\tau = \varepsilon t$ in the case of averaging). If such a separation does not exist in the strict sense, the derivation of the macroscopic equations becomes problematic, as shown, for example, by the difficulties encountered in attempts to describe hydrodynamic turbulence on a large scale.

From a mathematical point of view, the principal reference on averaging for several decades was, and to some extent still is, the book by Bogoliubov and Mitropolski, in which certain relatively elementary techniques are discussed and justified mathematically, then applied to a number of very interesting problems, most notably in electronics (cf. [Bog] in the bibliography at the end, as well as the bibliographical notes). In the last twenty-five years however, more sophisticated or general *multiple phase* averaging theorems have been proved, again mostly by Soviet mathematicians (D. Anosov, V. Arnold, A. Neistadt), which in particular allow for rigorous statements (and proofs) of adiabatic theorems in general cases. These results, though admittedly simple, seem to have suffered a lack of recognition as a consequence of the great popularity of works connected with the KAM theorem. Yet they offer beautiful examples of the use of perturbation techniques and, apart from their intrinsic interest, constitute in that alone a good introduction to perturbation theory.

In physics articles, these averaging theorems are often "invoked" without reference to a specific mathematical statement; in fact, the whole theory is sometimes compressed into the single statement: "To obtain the evolution of the slow variables, one eliminates the fast variables by averaging along the trajectories of the unperturbed motion". Such overly general slogans cannot always be applied with impunity, as we shall see.

It is our aim in the present work to clarify and render accessible the main results in averaging for ODE's since the appearance of Bogoliubov and Mitropolski's book, and to give at the same time a survey of the adiabatic theorems which more or less readily follow from those results. We should point out that concision is a distinctive trait of the original articles, many of which remain untranslated from the Russian. Often appearing in the form of "Notes" in which the proofs are only sketched, these articles are full of important remarks which the uninitiated reader may have trouble deciphering. The sequel is therefore more of an explication than a résumé of the literature, including what we believe are the first complete proofs of some results.

The material is organized as follows.

Chapter One presents an overview of the averaging methods and introduces the basic notation to be used throughout the text.

Chapter Two is devoted to a very general averaging theorem which was proved by Anosov and relies heavily on the notion of ergodicity. The result is stated in Section 2.1 and an outline of the proof is given in Section 2.2. The later sections, which contain the details of the proof, may be skipped on first reading.

In Chapter Three we briefly review the current status of the theory of averaging for one frequency systems (which is not the main concern of the present book); we then start our study of multiple phase averaging with an elementary result due to Arnold for two frequency systems.

The detailed proof of Neistadt's optimal averaging theorem for two frequency systems is given in Chapter Four. The basic ideas are quite elementary but their application involves some technicalities which may appear tedious at first reading; this chapter aims mainly at giving a very concrete description of the resonances in a nontrivial particular case.

In Chapter Five we show how an averaging theorem for n frequency systems can be established via Anosov's method. Since the result obtained in this way is not optimal, this chapter may be considered of a somewhat pedagogical nature.

The optimal theorem for the same class of systems is proved in Chapter Six via another technique.

Chapter Seven is devoted to Hamiltonian systems, the specific features of which lead to a very powerful result embodied in Nekhoroshev's theorem. We also review the content of the celebrated KAM theorem.

Chapter Eight introduces adiabatic invariants. Most of the known results for one degree of freedom Hamiltonian systems are briefly discussed.

In Chapter Nine we give the proofs of the classical adiabatic theorems for general Hamiltonian systems with n degrees of freedom.

The proof of the quantum adiabatic theorem is given in Chapter Ten, where we also examine the connections between this theorem and classical ones.

Basic technical notions are recalled in the appendices which should help make this book (almost) self-contained, keeping the prerequisites at less than graduate level.

Following the bibliography, which is by no means exhaustive, we have included two brief notes on Bogoliubov and Mitropolski's book and Freidlin and Wentzell's book which may help to put the present work in perspective.

It is a pleasure for one of the authors (P. L.) to thank the "Milano-Padova group" (G. Benettin, L. Galgain, A. Giorgilli) who introduced him to part of the material contained in this book; the proof of Nekhoroshev's theorem in Chapter 7 together with the organization of Appendix 8 closely follow their work ([Ben4] and [Ben5]).

CONTENTS

CHAPTER 1: INTRODUCTION AND NOTATION

1.1 Introduction

The typical averaging problem may be defined as follows: One considers an unperturbed problem in which the slow variables remain fixed. Upon perturbation, a slow drift appears in these variables which one would like to approximate independently of the fast variables. This situation may be described with the aid of a fiber bundle in which the base represents the slow variables and the fibers represent the fast variables. This may seem unnecessarily pedantic, but in fact it provides a convenient and precise language which supports the intuitive image of the "mixing" that occurs among the fast variables, as opposed to the drift of the slow variables on the base. For these reasons, we shall sometimes employ this geometrical language when appropriate but we will not make use of any nontrivial (i.e. global) property of fiber bundles in this work.

The unperturbed motion is thus described by a vertical vectorfield, in other words a vectorfield whose projection onto the base (the first components) is zero: The (slow) action variables remain fixed; the (fast) phase variables rotate rapidly. If the vectorfield is perturbed, its vertical component changes, but more importantly it acquires a horizontal component. One then seeks to describe the motion in the framework of a *first order* perturbation theory, that is, on time scales of order $1/\varepsilon$, by introducing a vectorfield on the base, called the averaged system, such that its trajectories are close to the projection onto the base of the trajectories of the unperturbed system (except perhaps for a small set of initial conditions).

Following this very general outline, which we hope underscores the natural setting of the theory, let us proceed to a more concrete - and necessarily more modest - description. We shall mainly consider so-called "standard" systems of the form:

(1) $\quad dI/dt = \varepsilon f(I, \varphi, \varepsilon), \quad d\varphi/dt = \omega(I, \varphi) + \varepsilon g(I, \varphi, \varepsilon),$

where $I \in \mathbb{R}^m$ (resp. $\varphi \in T^n$ or any compact smooth manifold) denotes the slow (resp. fast) variables and ε is a small parameter; the form of the functions f, ω and g will be specified in each of the cases to be examined. The above notation is used to suggest the important (but by no means ubiquitous) Hamiltonian case in which one perturbs an integrable system; the variables (I, φ) are then action-angle variables.

Let us first examine how the averaged vectorfield arises as the natural means of describing the approximate evolution of the slow variables. There are clearly two timescales in this problem: t for the unperturbed motion on the fiber (the fast scale), and $\tau = \varepsilon t$ for the motion on the base (the slow scale). We proceed to treat the problem by the method of multiple scales, beginning with expansions of the form:

$$(2) \qquad I = I_0(\tau) + \varepsilon I_1(\tau, t) + \ldots; \qquad \varphi = \varphi_0(t) + \varepsilon\varphi_1(\tau, t) + \ldots$$

and the corresponding system:

$$(3) \qquad \begin{aligned} dI/dt + \varepsilon dI/d\tau &= \varepsilon f(I, \varphi, 0) + O(\varepsilon^2) \\ d\varphi/dt + \varepsilon d\varphi/d\tau &= \omega(I, \varphi) + O(\varepsilon). \end{aligned}$$

To leading order, we obtain:

$$(4) \qquad dI_0/dt = 0, \quad d\varphi_0/dt = \omega(I_0, \varphi),$$

which describes the unperturbed motion. At the next order we have:

$$(5) \qquad dI_1/dt + dI_0/d\tau = f(I_0, \varphi_0, 0),$$

which describes the slow evolution of I_0. It follows that:

$$(6) \qquad I_1(t, \tau) = h(\tau) + \int_0^t f(I_0(\tau), \varphi_0(u), 0)\, du - t.dI_0/d\tau.$$

Demanding the absence of secularity, we thus obtain:

$$(7) \qquad dI_0/d\tau = \lim_{t\to\infty} 1/t.\int_0^t f(I_0(\tau), \varphi_0(u), 0)\, du.$$

The solutions of this equation are not necessarily integral curves of a vectorfield on the base, as they may depend on the initial phases. In other words, one does not obtain in this way a closed system describing the evolution of the slow variables.

However, this difficulty is removed if the unperturbed motion is *ergodic on the fiber*, in which case one obtains, for the slow motion on the base, the *averaged system*:

$$(8) \qquad dI/dt = \varepsilon\langle f\rangle(I), \quad \langle f\rangle(I) =_{def} \int f(I, \varphi, 0)\, d\mu_\varphi,$$

where $d\mu_\varphi$ designates the invariant measure associated with unperturbed motion on the fiber. The problem thus consists of understanding in what sense the system (8) approximates the evolution of the slow variables of (1) for small ε. In the sequel, we will study cases where ergodicity is realized on almost every fiber for the unperturbed system, which, as we just saw, is a necessary condition for the system (8) to make sense.

For more than one frequency ($n > 1$) the two important cases, or at least the two that have been studied in detail, are the following:

i) <u>Quasiperiodic systems with n frequencies</u>:

Here the fast variables evolve on a torus of dimension n and the unperturbed motion is associated with a constant field on the torus: $\varphi \in T^n$ and $\omega = \omega(I)$ is independent of φ. The invariant measure is of course ordinary Lebesgue measure on the torus. This is the direct (non-Hamiltonian) generalization of the problem of perturbation of an integrable Hamiltonian. Its importance lies in the fact that one can define resonances (which will be amply treated later in this text) and thus make use of the powerful tool of Fourier series.

ii) Systems which are ergodic on the energy surface:

We begin here with a system which is Hamiltonian and ergodic on almost every energy surface, for a certain energy interval. Each surface is assumed to be bounded and without boundary, and the invariant measure is Liouville measure. This case is especially important in connection with statistical thermodynamics (cf. Chapter 9).

Schematically, it may be said that the separation on the base between solutions of the perturbed and averaged systems will be small if the exact trajectory spends most of its time in regions where the unperturbed flow is ergodic and where the "ergodization" of the functions (the approximation of the spatial average by the time average along a given trajectory) takes place rapidly. This is nicely demonstrated in Anosov's work which will be discussed in Chapter 2. In fact, Chapter 2 is the only place where we consider multiple phase averaging for unperturbed systems that are not quasiperiodic (ω does depend on φ), and it contains about all that is currently known on the subject. In Chapter 5 we explain the equivalence of resonance in quasiperiodic systems and "bad" ergodization of functions.

In summary, the aim of the averaging method is most succinctly expressed by the following

Averaging-type Theorem:

Let Γ be an adequate (to be defined) set of initial conditions for the system (1). For any initial condition in Γ (not belonging to a certain set of small measure to be defined), the solution of the averaged system (8) remains close (in a way to be made precise) on the time interval $[0, 1/\varepsilon]$ to the projection on the base of the solution of the exact system (1) with the same initial condition for the slow variable (and any initial phase).

We recall that historically, such problems were first encountered in celestial mechanics in the study of perturbations of planetary motion, for example in the absence of secularity in the variations of the major axes of orbits, in the rotation of perihelia, by Lagrange, Laplace, and Gauss.

The following table indicates the principal averaging theorems for quasiperiodic systems to be treated in Chapters 3 through 6. Definitions of the conditions A and N for two-phase systems will be given in due time.

number of phases	order of approximation	measure of the set of initial conditions to be excluded	author	chapter
1 without resonance	ε	0	Arnold	3
2 condition A	$\sqrt{\varepsilon}\,\log^2\varepsilon$	0	Arnold	3
2 condition A optimal result	$\sqrt{\varepsilon}$	0	Neistadt	4
2 condition N optimal result	$\sqrt{\varepsilon}\,\log(1/\varepsilon)$	$\sqrt{\varepsilon}$	Neistadt	4
n	$\rho(\varepsilon)$ arbitrary	$\sqrt{\varepsilon}/\rho(\varepsilon)^{3/2}$	Neistadt	5
n optimal result	$\rho(\varepsilon)$ arbitrary	$\sqrt{\varepsilon}/\rho(\varepsilon)$	Neistadt	6

Table 1.1: Main results of the averaging method are listed in this table. The first column indicates the systems to which the result stated in the following two columns applies. The principal author is cited in column four. For further details, the reader should consult the Chapter indicated in the last column.

It is perhaps appropriate here to make some further remarks about the nature of averaging. In its general form, it is a *first order method*, that is to say valid on time intervals of order $1/\varepsilon$, where ε is the small parameter in the problem. In general, nothing further may be said about longer time intervals, and the averaged equation may in fact cease to be a good approximation to the exact system on such intervals. In this sense averaging is the most rudimentary of perturbation theories, and thus constitutes a good introduction to them.

One may then proceed to the classical methods of fixed finite order, with an error of order ε^p $(p > 1)$ on the time scale $1/\varepsilon$, or with a larger error but valid on timescales of order ε^{-q} $(q > 1)$. However, more details require much more stringent hypotheses: That there be only one fast variable ($n = 1$, see Chapter 3) or that the system be Hamiltonian (see Chapters 7 and 8), or, as in Bogoliubov and Mitropolski's book, that the unperturbed system have fixed frequencies; in other words one then perturbs a linear system with constant coefficients, or in physical language, a set of noninteracting harmonic oscillators (cf. Chapters 3, 5 and 6 and the note on Bogoliubov and Mitropolski's book).

Going to even longer timescales one finds more sophisticated methods of finite but variable order in ε, leading, for Hamiltonian systems only, to approximations on exponentially long time intervals of order $\exp[1/\varepsilon^r]$ $(r \leqslant 1$; cf. Nekhoroshev's theorem in Chapter 7), and finally, to problems with stability on infinite time intervals (KAM-type theorems; see again Chapter 7). Notice that in KAM-type theorems, one no longer tries to describe the behavior of the system on all of phase space, nor even on a regular part of it.

Averaging theorems are global theorems, obtained by a reduction to normal form through first order: Nonresonant normal forms on the nonresonant domain, resonant normal forms in the neighborhood of resonances (the reader will find a very short and elementary introduction to the theory of normal forms in Appendix 5). In addition, one makes use of an "ultraviolet cutoff" by taking into account only a finite number of low order resonances (which are defined in Appendix 3), and in turn only finitely many terms are eliminated in the reduction to normal form. To justify this truncation, one must impose minimal differentiability conditions, mainly on the phase variables, to ensure that the contribution from higher harmonics is minimal. In general it is not necessary to demand analyticity, although it simplifies the estimates considerably and is often assumed.

The first part of the book (Chapters 2 through 7) is devoted to averaging, with a final chapter on Hamiltonian systems for which we prove strong specific results. Single-phase averaging is reviewed in Chapter 3 but we do not dwell on this long standing subject to which Sanders and Verhulst devoted a detailed study ([Sand3]).

The second part of the work (Chapters 8 through 10) deals with adiabatic invariance, an important physical phenomenon leading to the statement of certain mathematical theorems which are direct consequences of averaging theorems, or at least are proved by entirely similar techniques. It seemed important to include results on adiabaticity, as the subject is often treated as folk wisdom and the proofs in the literature are generally inaccessible or lack rigor, a lack that is not simply one of academic detail; for example, the classical theorem for integrable systems in more than one dimension is generally stated erroneously.

We briefly indicate here the three adiabatic theorems to be proved in this book:

i) The first theorem concerns a family of integrable Hamiltonians depending on a small parameter which is made to vary "infinitely" slowly. The historical example is of course the isochronous pendulum whose length is varied on a timescale much longer than its period; this problem was discussed at the 1911 Solvay Congress in the context of the "old quantum theory" in Einstein's response to a question by Lorenz as they were trying to understand the mystery of the existence of Planck's constant [Ei1]. As is well known to physicists, the adiabatic invariant in this case is E/ω (E is the energy, ω the frequency), which represents the action of the unperturbed system and should be compared to the quantization formula $E = (n + 1/2)\hbar\omega$ for the harmonic oscillator, as pointing to a problematic "adiabatic stability of Planck's constant".

Historical examples are often peculiar; here in two respects:

- It is first of all a one degree of freedom system, and all such systems are integrable as well as ergodic, which underscores their nongenericity. Their properties of adiabatic invariance are much *stronger* than in the generic case, as we will show in Chapter 8. It may be of interest to point out that as far back as 1917, again in the context of the old quantum theory, Einstein devoted an important article ([Ei2]) to the problem of quantization of integrable systems with many degrees of freedom, in other words to the intrinsic character of the Bohr-Sommerfeld rules in that case. In most works, the case of one degree of freedom is the only one treated, which has no doubt led to the false statements of analogous results in higher dimensions.
- The pendulum example is also peculiar in that it is a linear system, from which follows the isochronism of small oscillations and the possibility of parametric resonance and we shall see that adiabatic invariance results are by contrast *weaker* for these systems than for generic (nonlinear) systems of one degree of fredom.

We will also be interested in adiabatic invariance on *infinite* time intervals for generic Hamiltonians with a single degree of freedom; results in this vein were deduced from the KAM theorem in a degenerate case by Arnold and Neistadt. We will study the connections between averaging and adiabatic invariance and show how the adiabatic invariance of action for integrable Hamiltonians is deduced from Neistadt's averaging theorem for systems with n frequencies (modulo a few minor changes).

ii) The second adiabatic theorem concerns families of *ergodic* Hamiltonians with slowly varying parameters. This is the case from which the phenomenon of adiabaticity derives its name, and we will briefly develop its connections with statistical thermodynamics in Chapter 9. To our knowledge, the only correct proof in print is Kasuga's ([Kas]), of which we give a slightly clarified version; the method is very similar to the one used by Neistadt to establish optimal results on averaging for systems with n phases (historically, it was Neistadt who took up Kasuga's idea, as he says himself).

iii) The final adiabatic theorem we deal with is the quantum theorem, mathematically simpler - as is often the case - because of the underlying linearity of the quantum theory. Chapter 10 is devoted to this theorem and to various related results which clarify among other things its connection to the classical theorems.

1.2 Notation

For convenience, we give here a partial list of the notations used in this book; most of them are standard or at least widely used:

- As usual, $C^p(D)$ designates the Banach space of scalar or vector valued functions on a domain D whose differentials exist and are continuous up to order p. It is endowed with the norm $\| f \|_p$ defined as the supremum of the pointwise norm of the differentials of order less than or equal to p. We shall often omit the subscript when it is 0, i.e., when we are dealing with the sup norm on the space of continuous functions which we denote by $C(D)$. As we shall not make use of either the L^p norms (except rarely, in which case we say so) nor the Sobolev norms, this notation should not be confusing.

- C^{p+} designates the subspace of C^p spanned by the functions whose p^{th} differential is Lipschitz continous; we abreviate C^{0+} as C^+, the space of Lipschitz continuous functions.

- Order symbols are used in the standard way; that is, if f and g are two scalar functions of the real variable x, $g > 0$, we write $f = O(g)$ at x_0 ($x_0 \in \mathbb{R} \cup \{\pm\infty\}$) if the ratio $|f|/g$ is bounded as x approaches x_0. We write $f = o(g)$ if that same ratio goes to 0 and $f \sim g$ if f/g goes to 1.

- It will sometimes be convenient to use various scales to compare functions; we shall say that $\delta(\varepsilon)$ is a modulus of continuity (at $\varepsilon = 0$) if it is defined and positive for ε positive small enough and satisfies $\delta(0) = 0$.

- Averages will obviously play an important role in this book and according to context we employ the two notations that are commonly in use. Namely the average or mean value of a function f with respect to phase variable(s) will be designated as f_0 or $<f>$. The first notation refers to the average being the 0th Fourier coefficient of a function $f(I, \varphi)$ which is 2π-periodic in its phase variables φ_i ($\varphi \in T^n = (\mathbb{R}/2\pi\mathbb{Z})^n$, $I \in \mathbb{R}^m$):

$$f_0(I) = (2\pi)^{-n} \oint f(I, \varphi)\, d\varphi.$$

$<f>$ will generally refer to a more general type of average for nonperiodic functions (see especially Chapter 3). The fluctuating or oscillating part $f - f_0 = f - <f>$ of f will be denoted by $\tilde{f}$.

- For functions $f(I, t)$ periodic in the scalar variable t, with say period 2π, $\Phi[f]$ designates the integral of the fluctuating part with zero mean value; that is, it satisfies $f = \partial/\partial t\, (\Phi[f]) + <f>$ and has zero average. Explicitely:

$$\Phi[f](I, t) = \int_0^t \tilde{f}(I, s)\, ds - \int_0^1 d\sigma \int_0^\sigma \tilde{f}(I, s)\, ds$$

or, using Fourier coefficients:

$$\Phi[f](I, t) = \sum_{k \neq 0} -i/k\, f_k(I) e^{ikt}.$$

The reader will find some basic facts about Fourier series in Appendix 1.

- For a vector $k \in \mathbb{Z}^n$ with integral components k_i, $| k |$ will denote one of the three equivalent norms $\sup_i | k_i |$, $\sum_i | k_i |$ or $(\sum_i | k_i |^2)^{1/2}$; we will specify which one.

- For real x, $[x]$ will denote its integer part.

- Concerning the domains of definition of the various functions, we often use $(I, \varphi) \in K \times T^n$, where K is a subset of $\mathbb{R}^m$, open or closed, compact or not; we shall only rarely specify all the requirements that K should meet but it will be plain in each case what "reasonable" domains will do.

- For $\rho > 0$, we will denote by $K - \rho$ the set of centers of balls of radius ρ contained in K.

We will also use complex extensions of the domain $K \times T^n$; for this purpose we denote by $K_{\rho,\sigma}$ ($\rho > 0$, $\sigma > 0$) the set:

$$K_{\rho,\sigma} = \{(I, \varphi) \in \mathbb{C}^m \times \mathbb{C}^n,\ \mathrm{Re}\, I \in K + \rho\ ,\ |\mathrm{Im}\ I| < \rho\ ,\ |\mathrm{Im}\ \varphi| < \sigma\},$$

where $|\mathrm{Im}\ I| < \rho$ means $|\mathrm{Im}\ I_j| < \rho$ for all j and the functions under consideration will be multiply periodic in φ in the complex plane, by analytic continuation.

- Lastly, there is a proliferation of constants in the text, and we have employed a notation for them which is a compromise between precision and readability: For isolated or less important constants we simply use a generic c; for those crucial to the statements or proofs of theorems, or to which repeated references are made, we use c_k , where the numbering is in order of appearance and is specific to each chapter.

CHAPTER 2: ERGODICITY

2.1 Anosov's result

Our aim in this chapter is to examine the role of ergodicity in the method of averaging from a general standpoint. To this end, we shall follow Anosov's article [Ano], and see how divergence between exact and averaged systems takes place when the fast variables belong to a fiber on which the unperturbed system is not ergodic, or does not permit sufficiently fast "mixing" (mixing is not meant in the technical sense here). In the case of quasiperiodic unperturbed systems (to be discussed in Chapters 3 through 6), these fibers correspond respectively to exact resonances and to the "resonant zones" around them, a distinction to which we will return at the beginning of Chapter 5.

Anosov's paper laid the groundwork for many of the latest results in averaging, and the present chapter seeks to give an understanding of the general principles in that article. However, for the reader desiring an acquaintance with particular averaging theorems, this material is not essential for an understanding of what follows and may thus be (re)read after the more specific results in later chapters.

We begin by considering the system:

(1) $\qquad dI/dt = \varepsilon f(I, \varphi, \varepsilon); \quad d\varphi/dt = \omega(I, \varphi) + \varepsilon g(I, \varphi, \varepsilon),$

where $I = (I_1, ..., I_m)$ and $\varphi = (\varphi_1, ..., \varphi_n)$ are the fast and slow variables belonging, respectively, to a "base" of dimension m and a compact "fiber" of dimension n. Here the spaces are equipped with the norm $\| x \| = \sum_i | x_i |$.

As it is, the system is in a very general form, for we have assumed neither that the fiber is diffeomorphic to a torus, nor, if it is, that the unperturbed motion is quasiperiodic on the torus, as would be the case, for example, for Hamiltonian perturbations of integrable Hamiltonians expressed in action-angle variables. For a discussion of the latter case we refer the reader to Chapter 7; we are presently concerned with systems where the unperturbed motion on the fiber $I = I_0$ is governed by the equation $d\varphi/dt = \omega(I_0, \varphi)$.

We suppose that:

i) The various functions are C^1 in all their arguments

ii) They satisfy the following estimates:

$$(2) \qquad \|f\| \leq 1, \quad \|g\| \leq 1, \quad \sum_{i,k} |\partial f_i/\partial I_k| \leq 1, \quad \sum_{i,k} |\partial f_i/\partial \varphi_k| \leq 1,$$
$$\sum_{i,k} |\partial g_i/\partial I_k| \leq 1, \quad \sum_{i,k} |\partial g_i/\partial \varphi_k| \leq 1, \quad \sum_{i,k} |\partial \omega_i/\partial I_k| \leq 1,$$
$$\sum_{i,k} |\partial \omega_i/\partial \varphi_k| \leq 1,$$

which may always be obtained by a suitable change of scale.

iii) The unperturbed motion is ergodic on (almost) each fiber $I = I_0$. We will write $<\psi>(I_0)$ to designate the time average of a continuous function ψ on the fiber $I = I_0$. This time average coincides with the space average, by the definition of ergodicity:

$$(3) \qquad <\psi>(I_0) =_{def} \lim_{T\to\infty} 1/T.\int_0^T \psi(I_0, \varphi^*(t))\, dt,$$

where $\varphi^*(t)$ is the solution of the unperturbed system on the fiber above I_0.

iv) Although this is not necessary we assume for simplicity that the invariant measure $d\mu_\varphi(I_0)$ on the fiber is simply $d\varphi$, that is, ω is divergence free as a function of φ, and we write $\mu = dI d\varphi$, the ordinary measure on the whole space. For a set A: $\mu(A) = \int_A dI d\varphi$.

Let T_ε^t be the flow of the exact system. It is obvious that

$$(4) \qquad e^{-\varepsilon Mt}\mu(A) \leq \mu(T_\varepsilon^t(A)) \leq e^{\varepsilon Mt}\mu(A)$$

for a certain constant M because the divergence of the vector field in (1) is of order ε.

In what follows we will confine ourselves to a compact regular subset K of the base and to its corresponding fibers; i.e., to the domain $\Gamma = \pi^{-1}(K)$, where π is the projection onto the base (the I component). Here, we will not concern ourselves too much with the domain problems which arise when a trajectory approaches the boundary of K; we will see in later chapters how simple "bootstrap" arguments take care of such details. Let us only say that the set $D(\varepsilon)$ of initial conditions to be considered will be such that for all $t \in [0, 1/\varepsilon]$, the exact solution $(I(t), \varphi(t))$ remains in Γ and the solution $J(t)$ of the averaged system, to be defined shortly, remains in K.

During the perturbed motion, the base point is displaced, and its slow evolution is described by the first set of equations (1): $dI_k/dt = \varepsilon f_k(I, \varphi, \varepsilon)$ $(k = 1,\ldots, m)$. As explained in the introduction, it is natural to introduce the averaged system:

$$(5) \qquad dJ/dt = \varepsilon <f>(J)$$

where $<f>$ is obtained by integrating $f(I, \varphi, 0)$ with respect to $d\varphi$, which agrees with (3) almost everywhere.

We assume the following, which again can be obtained by a change of scale:

(6) $\quad | \partial\langle f\rangle_i / \partial I_k | \leq 1.$

In this setting, Anosov shows that the averaged system is a good approximation to the perturbed system in the following sense:

Theorem (Anosov):

For each fixed $\rho > 0$, the measure of the set of initial conditions in Γ such that, for $t \in [0, 1/\varepsilon]$:

i) The exact trajectory $(I(t), \varphi(t))$ remains in Γ,

ii) The averaged trajectory $J(t)$ remains in K,

iii) $\sup_{t\in[0, 1/\varepsilon]} \| I(t) - J(t) \| > \rho$,

tends to 0 with ε.

Let us introduce some necessary notation. As we said above, $D(\varepsilon)$ is the set of initial conditions in Γ such that, for $t \in [0, 1/\varepsilon]$, the exact trajectory remains in Γ and the averaged trajectory $J(t)$ remains in K; $D(\rho, \varepsilon)$ will be the subset of initial conditions in $D(\varepsilon)$ for which the separation between the exact and averaged trajectories does not exceed ρ. The theorem can then be reexpressed as:

(7) $\quad \lim_{\varepsilon\to 0} \mu[D(\varepsilon) - D(\rho, \varepsilon)] = 0.$

We now define certain subsets of Γ related with the rate of "ergodization" for the unperturbed motion of a given continuous function on Γ.

$E(T_0, \delta, \psi)$ (E for ergodization) is the set of points (I_0, φ_0) such that on the fiber $I = I_0$, the ergodization of the (vector) function ψ is realized to within δ in a time less than T_0. In other words, for (I_0, φ_0) in $E(T_0, \delta, \psi)$ and for all T strictly greater than T_0, the T-time average of ψ over the unperturbed trajectory with initial condition (I_0, φ_0) differs by less than δ from the spatial average of ψ:

(8) $\quad \| \langle\psi\rangle(I_0) - 1/T.\int_0^T \psi(I_0, \varphi^*(t))\, dt \| < \delta.$

From the definition of the averaged system, we see that it is the rate of ergodization of the function $f(I, \varphi, 0)$ which must be considered.

We next introduce sets of initial conditions defined in terms of the time spent by the exact trajectories in certain measurable subsets of Γ. In what follows we apply this notation to regions in Γ of the form $\Gamma - E(T_0, \delta, \psi)$ where the ergodization of ψ is slow.

For each trajectory with initial condition (I_0, φ_0) and for each measurable subset A

of Γ, we define the subset $S(\varepsilon, t, I_0, \varphi_0, A)$ (S for "spent") of $[0, t]$ consisting of the intervals of time spent by the trajectory in $\Gamma - A$: $S(\varepsilon, t, I_0, \varphi_0, A)$ is the set of τ in $[0, t]$ such that $T_\varepsilon^\tau(I_0, \varphi_0)$ does not belong to A. We will shorten $S(\varepsilon, 1/\varepsilon, I_0, \varphi_0, A)$ to $S(\varepsilon, I_0, \varphi_0, A)$.

Finally, we define the set $N(\varepsilon, T_0, \delta, \psi, T_1)$ of "nice" initial conditions such that, on the time interval $[0, 1/\varepsilon]$, their associated trajectories spend a time less than T_1 in the zone $\Gamma - E(T_0, \delta, \psi)$. $N(\varepsilon, T_0, \delta, \psi, T_1)$ is thus the subset of $D(\varepsilon)$ consisting of points (I_0, φ_0) which satisfy $\mathrm{mes}[S(\varepsilon, I_0, \varphi_0, E(T_0, \delta, \psi))] < T_1$ (ordinary measure on $\mathbb{R}$). We shall be particularly interested in the set:

(9) $$N(\varepsilon, T_0, \delta, T_1) = N(\varepsilon, T_0, \delta, f(I, \varphi, 0), T_1)$$

of initial conditions such that the corresponding trajectories spend a relatively short time in zones where the perturbative term $f(I, \varphi, 0)$ is badly ergodized.

We trust that the reader will bear with this seemingly cumbersome notation, which, as we shall see, is unfortunately necessary.

2.2 Method of proof

To prove Anosov's result, we will make use of two lemmas from which the theorem will follow almost immediately.

Lemma 1:

For each fixed T_0 and each $\rho > 0$, there exist $\zeta(\rho)$, $\delta'(\rho)$ and $\varepsilon'(T_0, \zeta(\rho), \rho)$ such that for $T_1 \leqslant \zeta/\varepsilon$, $\delta \leqslant \delta'$ and $\varepsilon \leqslant \varepsilon'$, $N(\varepsilon, T_0, \delta, T_1)$ is contained in $D(\varepsilon, \rho)$.

This lemma states that exact and averaged trajectories with the same initial condition can only separate significantly (more than a fixed ρ) if the exact trajectory spends a long time in regions where the ergodization of the perturbative term $f(I, \varphi, 0)$ is badly realized.

The second lemma may be stated as

Lemma 2:

For each function ψ and for all real ζ, α and δ, there exists a T_0 such that

(1) $$\mu[D(\varepsilon) - N(\varepsilon, T_0, \delta, \psi, \zeta/\varepsilon)] < \alpha.$$

We shall see that this lemma follows readily from the ergodicity of the perturbed system on almost every level surface. It affirms that few trajectories spend a time longer than ζ/ε (ζ given) in regions of Γ where the ergodicity of a given function is realized very slowly ("slowly" being measured by T_0). It is possible to make the set of such trajectories arbitrarily small, even for ζ arbitrarily small, by choosing T_0 large enough.

The reader will find it useful to reflect on the content of these simple lemmas, and to translate them in terms of resonances and resonant zones as they are discussed in Chapter 5, for these ideas are the basic ingredients necessary to the justification of averaging. Before proving these lemmas however, we first deduce the theorem from them.

Fix $\rho > 0$ and let α be any positive number. Take $\psi = f(I, \varphi, 0)$. Let $\zeta'(\rho)$ and $\delta'(\rho)$ be such that Lemma 1 is satisfied for all T_0, where of course the maximal value ε_0 may take is adjusted with T_0. Lemma 2 permits us to choose $T_0(\zeta', \alpha, \delta', f)$ such that $\mu[D(\varepsilon) - N(\varepsilon, T_0, \delta', \zeta'/\varepsilon)] < \alpha$. This T_0 depends only on α, ρ and the function f. For this value of T_0, Lemma 1 permits us to choose ε' in such a way that whenever $\varepsilon \leqslant \varepsilon'$ $N(\varepsilon, T_0, \delta', \zeta'/\varepsilon)$ is contained in $D(\varepsilon, \rho)$. ε' depends only on α and ρ. We thus have:

(2) $\qquad N(\varepsilon, T_0, \delta', \zeta'/\varepsilon) \subset D(\varepsilon, \rho)$

and

(3) $\qquad \mu[D(\varepsilon) - N(\varepsilon, T_0, \delta', \zeta'/\varepsilon)] < \alpha.$

Consequently $\mu[D(\varepsilon) - D(\varepsilon, \rho)] \leqslant \alpha$, which proves the theorem. ■

2.3 Proof of Lemma 1

We will establish two preliminary lemmas which estimate the separation between an exact trajectory with initial condition in $N(\varepsilon, T_0, \delta, T_1)$ and the averaged trajectory with the same initial condition. The exact trajectory may be written:

(1) $\qquad I(t) = I(0) + \varepsilon\int_0^t f(I(\tau), \varphi(\tau))\, d\tau,$

whereas the averaged trajectory with the same initial condition is expressed as:

(2) $\qquad J(t) = I(0) + \varepsilon\int_0^t \langle f\rangle(J(\tau))\, d\tau.$

The first preliminary result is stated as

Lemma 3:

Any exact trajectory with initial condition in $N(\varepsilon, T_0, \delta, T_1)$ may be written:

$$(3) \qquad I(t) = I(0) + \varepsilon\int_0^t \langle f\rangle(I(\tau))\, d\tau + \xi(t)$$

where the function ξ is bounded in norm on $[0, 1/\varepsilon]$ by:

$$(4) \qquad \beta(\varepsilon, T_0, T_1, \delta) = \gamma(\varepsilon) + 4\varepsilon T_1 + 2\varepsilon T_0 + 2\delta + \\ + \varepsilon^2(1+T_1/T_0+1/(\varepsilon T_0))(8T_0{}^2+3e^{2T_0})$$

where $\gamma(\varepsilon) = \sup_\Gamma \| f(I, \varphi, \varepsilon) - f(I, \varphi, 0) \|$.

Lemma 3 is almost equivalent to Lemma 1 because of what we state as

Lemma 4:

Let $\rho > 0$ be given. If an exact trajectory can be written in the form (3) with ξ bounded by $\rho/4$, then the separation between the exact trajectory and the averaged trajectory with the same initial condition is less than ρ on the time interval $[0, 1/\varepsilon]$:

$$\| I(t) - J(t) \| \leqslant \rho.$$

This lemma is a straightforward consequence of (2), (3), (4), Gronwall's lemma and assumption (6) in Section (1) (the reader will find a statement and discussion of the elementary differential inequality known as Gronwall's lemma in the first section of Chapter 3). ▪

Lemma 1 follows from Lemma 4 and the following choice. In fact, for fixed ρ and T_0, we suppose that in Lemma 3, $\varepsilon \leqslant \varepsilon'$, $\delta \leqslant \delta'$, and $T_1 \leqslant \zeta/\varepsilon$, with $\zeta(\rho)$, $\delta'(\rho)$ and $\varepsilon'(T_0, \zeta(\rho), \rho)$ chosen in such a way that $\beta(\varepsilon, T_0, T_1, \delta) \leqslant \rho/4$ for $\varepsilon \leqslant \varepsilon'$, $\delta \leqslant \delta'$ and $T_1 \leqslant \zeta/\varepsilon$; this choice is possible since $\gamma(\varepsilon)$ tends to 0 with ε.

We have thus reduced the proof of Lemma 1 to the proof of Lemma 3, which is the most technical, and which we break up into several propositions. Though the following estimates appear laborious, they show, in the most concrete way, how one controls the various possible separations between the exact and averaged systems.

We begin with

Proposition 1:

If the function v is continuous on $[0, T]$, vanishes at 0, and satisfies the inequality $v(t) \leqslant \int_0^t v(\tau)\, d\tau + \varepsilon t + \varepsilon t^2/2$, then $v(t) < 2\varepsilon e^t$.

Let the function $w_\delta(t)$ be defined by $w_\delta(0) = \delta > 0$ and $w_\delta(t) = \int_0^t w_\delta(\tau)\, d\tau + \varepsilon t + \varepsilon t^2/2$. It can be evaluated as $(\delta + 2\varepsilon)\, e^t - \varepsilon t - 2\varepsilon$. Now $v(t) < w_\delta(t)$ on $[0, T]$, and thus $v(t) < (\delta + 2\varepsilon)e^t$ for all δ, from which the proposition follows. This is again a particular case of the integral Gronwall's lemma, to be stated in the next chapter. ▪

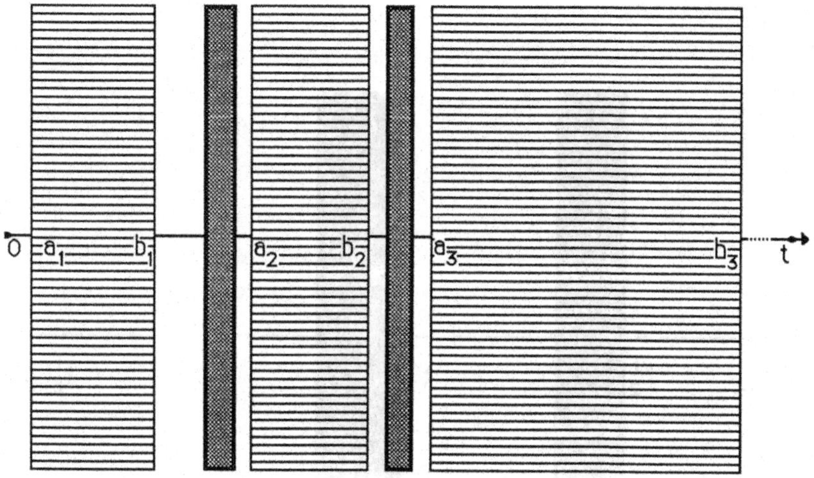

Figure 2.1: Schematic representation of the subsets $S(t)$ and $C(t)$ of $[0, t]$. S is composed of open intervals; some of them are longer than T_0 and are denoted by $[a_i, b_i[$ (hatched blocks) while the others are smaller than T_0 (shaded blocks). $C(t)$ is the complement of long intervals; this subset is the union of the complement of S and of the short intervals which belong to S.

The following is immediate:

Proposition 2:

$$(5) \qquad I(t) = I(0) + \varepsilon \int_0^t f(I(\tau), \varphi(\tau), 0)\, d\tau + \xi_1(t)$$

where $\| \xi_1(t) \| \leqslant \gamma(\varepsilon)$.

Let us now consider the subset $S(t) = S(\varepsilon, t, I(0), \varphi(0), E(T_0, \delta, f(I, \varphi, 0)))$ of $[0, t]$ which the trajectory - with initial condition in $N(\varepsilon, T_0, \delta, T_1)$ - spends in regions of Γ where the ergodization of $f(I, \varphi, 0)$ is slow. Since $E(T_0, \delta, f(I, \varphi, 0))$ is closed, this set is composed of open intervals. These intervals have total length less than $\min\{T_1, t\}$, and at most $[T_1/T_0]$ of them have a length greater than or equal to T_0. We will denote these longer intervals by $]a_i, b_i[$, $1 \leq i \leq p$ (p depends on t; see Figure 2.1).

We adopt the notation $C(t)$ for the complement of $\cup_{1 \leq i \leq p}]a_i, b_i[$ in $[0, t]$. When studying systems with n frequencies, we will also need to distinguish between long and short crossings of resonant zones and we will see that short crossings are of little importance.

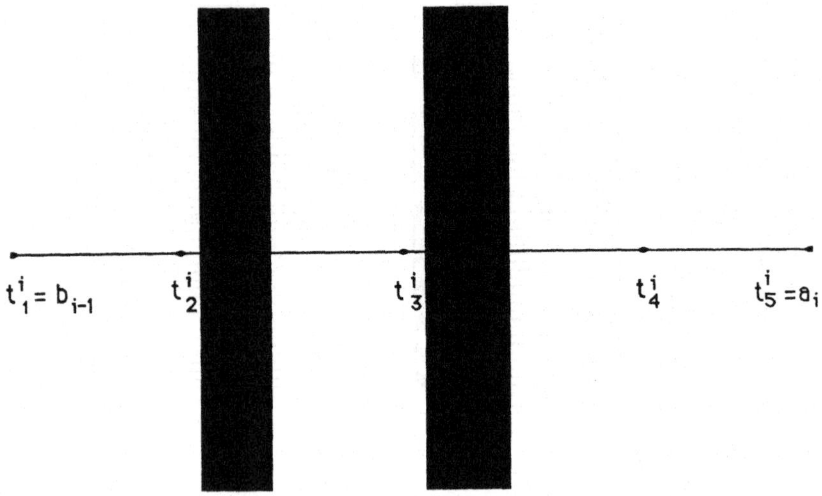

Figure 2.2: Schematic representation for the partition of a segment $]b_{i-1}, a_i[$. The points t^i_k are in the complement of $S(t)$ (this latter set appears on the figure as black blocks). The distance between successive points lies between T_0 and $2T_0$, except possibly for the terminal segment.

Since $\| f \| \leq 1$, the following proposition is immediate:

<u>Proposition 3</u>:

$$(6) \qquad \int_0^t f(I(\tau), \varphi(\tau), 0)\, d\tau = \int_{C(t)} f(I(\tau), \varphi(\tau), 0)\, d\tau + \xi_2(t)$$

where $\| \xi_2(t) \| \leqslant T_1$.

$C(t)$ is composed of segments $[b_{i-1}, a_i]$; if necessary we set $b_0 = 0$ and $a_{p+1} = t$. We proceed to partition each of these segments by writing $b_{i-1} = t_1^{\,i} < t_2^{\,i} < \ldots < t_{mi}^{\,i} = a_i$, where the $t_k^{\,i}$ are assumed to lie in the complement of $S(t)$; we will see the reason for this requirement shortly. We also assume that $T_0 \leqslant t_{k+1}^{\,i} - t_k^{\,i} \leqslant 2T_0$, except for terminal segments containing the endpoints a_i , which may possibly have length less than T_0 (see Figure 2.2).

This partition is constructed as follows (see Figure 2.3):

- if $a_i - b_{i-1} \leqslant 2T_0$, the whole segment is retained, with no subdivisions ($t_1^{\,i} = b_{i-1}$, $t_2^{\,i} = a_i$).

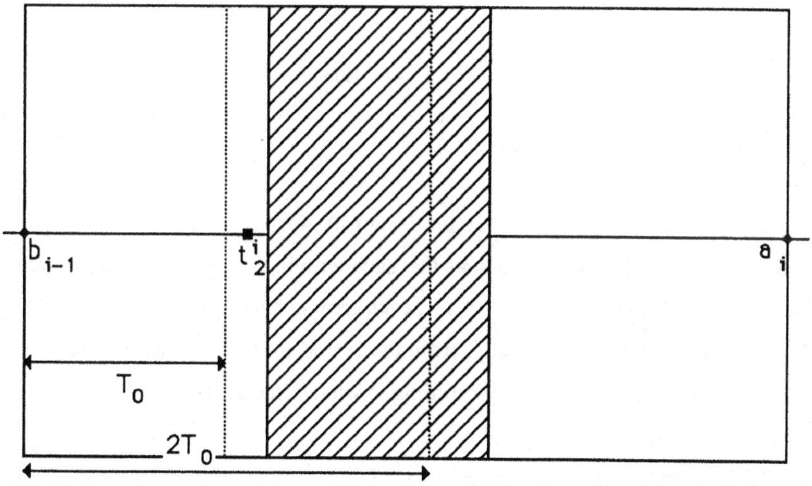

Figure 2.3: This explains how a segment $]b_{i-1}, a_i[$ must be subdivided when the point $b_{i-1} + 2T_0$ belongs to an interval of $S(t)$ (symbolized by a hatched rectangle). Some point t'^{i}_2 is chosen outside $S(t)$ and at a correct distance from a_i .

- if $a_i - b_{i-1} > 2T_0$, we distinguish two cases:

a) If $b_{i-1} + 2T_0$ does not belong to $S(t)$ we choose $t_2{}^i = b_{i-1} + 2T_0$; the length of the segment $[t_1{}^i, t_2{}^i]$ is then $2T_0$, and the construction is continued by proceeding to the subdivision of $[t_2{}^i, a_i]$.

b) If $b_{i-1} + 2T_0$ belongs to $S(t)$, then there are points in $[b_{i-1} + T_0, b_{i-1} + 2T_0]$ which do not, by definition of the b_i. We set $t_2{}^i$ equal to one of these points and continue the construction by proceeding to the subdivision of $[t_2{}^i, a_i]$.We will adopt the notation $t_{k+1}{}^i - t_k{}^i = h_k{}^i$ and $s(i, k) = [t_k{}^i, t_{k+1}{}^i]$.

Let us now estimate the number of subsegments $s(i, k)$ in the partition. From its definition, $C(t)$ comprises at most $[T_1/T_0] + 1$ segments $[b_{i-1}, a_i]$. $C(t)$ thus contains at most $[T_1/T_0] + 1$ terminal subsegments of length less than T_0. Since all other segments have length greater than T_0, and since the measure of $N(t)$ is less than $1/\varepsilon$, the number of these segments is less than $[(\varepsilon T_0)^{-1}]$, and the total number of subsegments $s(i, k)$ is bounded by $[(\varepsilon T_0)^{-1}+T_1/T_0] + 1$. We will now prove

<u>Proposition 4</u>:

$$(7)\qquad \int_{C(t)} f(I(\tau), \varphi(\tau), 0)\, d\tau = \sum_{i,k} \int_{s(i,k)} f(I(t_k{}^i), \varphi(\tau), 0) d\tau + \xi_3(t)$$

where $\| \xi_3(t) \| \leqslant 4\varepsilon T_0{}^2(1 + T_1/T_0 + (\varepsilon T_0)^{-1})$.

(7) is obviously satisfied with:

$$(8)\qquad \xi_3(t) = \sum_{i,k} \int_{s(i,k)} [f(I(\tau), \varphi(\tau), 0) - f(I(t_k{}^i), \varphi(\tau), 0)]\, d\tau.$$

Since $\sum_{i,k} | \partial f_i/\partial I_k | \leqslant 1$, it follows that each term in the series is bounded in norm by $\int_{s(i,k)} \| I(\tau) - I(t_k{}^i) \|\, d\tau$. Since $\| f \| \leqslant 1$, each term is bounded by $\varepsilon\int_{s(i,k)} (\tau - t_k{}^i)\, d\tau = \varepsilon(h_k{}^i)^2$, and thus by $4\varepsilon T_0{}^2$. Finally, from the bound on the number of subsegments $s(i, k)$, we see that $\xi_3(t)$ is bounded according to what is stated above. ▪

At this point, we have established that:

$$(9)\qquad \| I(t) - I(0) - \sum_{i,k} \varepsilon\int_{s(i,k)} f(I(t_k{}^i),\varphi(\tau),0)\, d\tau \| \leqslant \gamma(\varepsilon) + \varepsilon T_1 + $$
$$+ 4\varepsilon^2 T_0{}^2(1+T_1/T_0+(\varepsilon T_0)^{-1}).$$

For each subsegment $s(i, k)$, we now introduce the unperturbed solution $\varphi^*{}_k{}^i(t)$ with initial condition $\varphi^*{}_k{}^i(t_k{}^i) = \varphi(t_k{}^i)$ on the fiber $I = I(t_k{}^i)$. We then establish the following

<u>Proposition 5</u>:

(10) $$\sum_{i,k} \varepsilon \int_{s(i,k)} f(I(t_k^{\,i}),\varphi(\tau),0)d\tau = \sum_{i,k} \varepsilon \int_{s(i,k)} f(I(t_k^{\,i}),\varphi^{*}{}_k^{\,i}(\tau),0)\, d\tau + \xi_4(t),$$

where $\| \xi_4(t) \| \leqslant 2\varepsilon e^{2T_0}[1 + T_1/T_0 + 1/(\varepsilon T_0)]$.

In fact, since $\sum_{i,k} | \partial f_i/\partial\varphi_k | \leqslant 1$, we have:

(11) $$\| \xi_4(t) \| \leqslant \sum_{i,k} \int_{s(i,k)} \| \varphi^{*}{}_k^{\,i}(\tau) - \varphi(\tau) \|\, d\tau.$$

Consider an arbitrary term in the series. For t in $s(i, k)$:

(12) $$\int \| \varphi^{*}{}_k^{\,i}(\tau) - \varphi(t) \|\, d\tau \leqslant \varepsilon\int \| g(\varphi(\tau), I(\tau), 0) \|\, d\tau + \int \| \omega(\varphi(\tau), I(\tau)) - \omega(\varphi^{*}{}_k^{\,i}(\tau), I(t_k^{\,i})) \|\, d\tau,$$

where the integral is calculated between $t_k^{\,i}$ and t. Consequently,

(13) $$\begin{aligned}\int \|\varphi^{*}{}_k^{\,i}(\tau) - \varphi(\tau)\|\, d\tau &\leqslant \varepsilon(t - t_k^{\,i}) + \int\| \varphi(\tau) - \varphi^{*}{}_k^{\,i}(\tau) \|\, d\tau + \int\| I(\tau) - I(t_k^{\,i})\|\, d\tau \\ &\leqslant \int \| \varphi(\tau) - \varphi^{*}{}_k^{\,i}(\tau) \|\, d\tau + \varepsilon(t - t_k^{\,i}) + \varepsilon(t - t_k^{\,i})^2/2\end{aligned}$$

since $\| f \| \leqslant 1$, $\sum_{i,k} | \partial\omega_i /I_k | \leqslant 1$ and $\sum_{i,k} | \partial\omega_i/\varphi_k | \leqslant 1$.

Using Proposition 1, we obtain:

(14) $$\begin{aligned}\| \xi_4(t) \| &\leqslant \sum_{i,k} 2\varepsilon\int_{s(i,k)} e^{\tau}\, d\tau \leqslant 2\varepsilon \sum_{i,k} \exp(h_k^{\,i}) \\ &\leqslant 2\varepsilon\, e^{2T_0}(1 + T_1/T_0 + 1/(\varepsilon T_0)). \quad \blacksquare\end{aligned}$$

We will now compare $\int_{s(i,k)} f(\varphi^{*}{}_k^{\,i}(\tau), I(t_k^{\,i}), 0)\, d\tau$ with the average value $\langle f\rangle(I(t_k^{\,i}))$ of $f(I, \varphi, 0)$ on the fiber $I = I(t_k^{\,i})$ containing the unperturbed trajectory $\varphi^{*}{}_k^{\,i}$.

<u>Proposition 6</u>:

(15) $$\sum_{i,k} \int_{s(i,k)} f(\varphi^{*}{}_k^{\,i}(\tau), I(t_k^{\,i}), 0)\, d\tau = \sum_{i,k} h_k^{\,i} \langle f\rangle(I(t_k^{\,i})) + \xi_5(t)$$

where $\| \xi_5(t) \| \leqslant 2\delta/\varepsilon + 2(T_0+T_1)$.

To prove this proposition, we write the left hand side as two sums $\sum^{*}$ and $\sum^{**}$ corresponding, respectively, to segments $s(i, k)$ of length greater than or equal to T_0, and to terminal segments of length less than T_0.

Consider first an arbitrary term in the sum $\sum^{*}$. Since $h_k^{\,i} \geqslant T_0$, and since $t_k^{\,i}$ does not belong to $S(t)$ (it is here that this assumption enters in), $f(\varphi^{*}{}_k^{\,i}(\tau), I(t_k^{\,i}), 0)$ is "ergodized" to within δ in the time interval h_k^i:

$$(16)\qquad \| \int_{s(i,k)} f(\varphi^*{}_k{}^i(\tau), I(t_k{}^i), 0)\, d\tau \,/\, h_k{}^i - \langle f\rangle(I(t_k{}^i)) \| \leqslant \delta.$$

Thus:

$$(17)\qquad \| \textstyle\sum^*{}_{i,k} [\int_{s(i,k)} f(\varphi^*{}_k{}^i(\tau), I(t_k{}^i), 0)\, d\tau - h_k{}^i \langle f\rangle(I(t_k{}^i))] \| \leqslant \delta \sum^*{}_{i,k} h_k{}^i \leqslant 2\delta/\varepsilon.$$

Now consider $\sum^{**}$. Since $\| f \| \leqslant 1$, we have the bound:

$$(18)\qquad \| \int_{s(i,k)} f(\varphi^*{}_k{}^i(\tau), I(t_k{}^i), 0)\, d\tau - h_k{}^i \langle f\rangle(I(t_k{}^i)) \| \leqslant 2h_k{}^i \leqslant 2T_0 .$$

Consequently,

$$(19)\qquad \| \textstyle\sum^{**}{}_{i,k} [\int_{s(i,k)} f(\varphi^*{}_k{}^i(\tau), I(t_k{}^i), 0) d\tau - h_k{}^i \langle f\rangle(I(t_k{}^i))] \| \leqslant 2(T_0 + T_1),$$

since the number of terminal segments of length less than T_0 is bounded by $1 + T_1/T_0$. The proposition now follows from the estimates (17) and (19). ▪

The term $\sum_{i,k} h_k{}^i \langle f\rangle(I(t_k{}^i))$ is a Riemann sum for the integral $\int_{C(t)} \langle f\rangle(I(\tau))\, d\tau$, which allows us to establish the following:

<u>Proposition 7</u>:

$$(20)\qquad \textstyle\sum_{i,k} h_k{}^i \langle f\rangle(I(t_k{}^i)) = \int_{C(t)} \langle f\rangle(I(\tau))\, d\tau + \xi_6(t),$$

where $\| \xi_6(t) \| \leqslant \varepsilon(4T_0{}^2)(1 + T_1/T_0 + 1/(\varepsilon T_0))$.

In fact, $\| \xi_6(t) \|$ is bounded by $\sum_{i,k} h_k{}^i \, \mathrm{Var}(\langle f\rangle, s(i, k))$, where $\mathrm{Var}(\langle f\rangle, s(i, k))$ is the variation of the function $\langle f\rangle$ on the subsegment $s(i, k)$. Because $\sum_{i,k} | \partial f_i/\partial I_k | \leqslant 1$ and $\| f \| \leqslant 1$, this variation is bounded by $\varepsilon h_k{}^i$, and it follows that:

$$(21)\qquad \| \xi_6(t) \| \leqslant \varepsilon \textstyle\sum_{i,k} (h_k{}^i)^2 \leqslant 4\varepsilon T_0{}^2(1 + T_1/T_0 + 1/(\varepsilon T_0)).$$ ▪

Lastly:

<u>Proposition 8</u>:

$$(22)\qquad \int_0^t \langle f\rangle(I(\tau))\, d\tau = \int_{C(t)} \langle f\rangle(I(\tau))\, d\tau + \xi_7(t)$$

where $\| \xi_7(t) \| \leqslant T_1$.

Because $\| f \| \leqslant 1$ and the measure of $[0, t] - C(t)$ is less than T_1. ▪

Gathering together the various estimates established thus far, we obtain the estimates (3) and (4) of Lemma 3. ∎

It remains only to prove Lemma 2 to establish the validity of Anosov's theorem.

2.4 Proof of Lemma 2

We recall the result to be proved: for any function ψ and for all real numbers ζ, α, and δ, there exists a T_0 such that $\mu[D(\varepsilon) - N(\varepsilon, T_0, \delta, \psi, \zeta/\varepsilon)] < \alpha$.

We will work on the space $\Gamma \times \mathbb{R}$, with the ordinary measure $d\mu' = d\mu\, dt = dI d\varphi dt$. We define the suspension map T'_ε from $D(\varepsilon) \times [0, 1/\varepsilon]$ to $\Gamma \times [0, 1/\varepsilon]$ by $T'_\varepsilon(I, \varphi, \tau) = (T_\varepsilon^{\tau}(I, \varphi), \tau)$, and we define the set $B(\varepsilon, T_0, \delta, \psi)$ consisting of initial conditions in $D(\varepsilon)$ whose associated trajectories visit the region $\Gamma - E(T_0, \delta, \psi)$ of slow ergodization of ψ at least once in the time interval $[0, 1/\varepsilon]$. More precisely,

$$B(\varepsilon, T_0, \delta, \psi) = (T'_\varepsilon)^{-1}((\Gamma - E(T_0, \delta, \psi)) \times [0, 1/\varepsilon]),$$

or equivalently:

$$B(1/\varepsilon, T_0, \delta, \psi) = \{ (I_0, \varphi_0) \in D(\varepsilon): \text{there exists } \tau \in [0, 1/\varepsilon],\ T_\varepsilon^{\tau}(I_0, \varphi_0) \in \Gamma - E(T_0, \delta, \psi) \}.$$

For every measurable set A in Γ, we have:

(1) $$\mu'[T'^{-1}_\varepsilon(A \times [0, 1/\varepsilon])] = \int_0^{1/\varepsilon} \mu[(T_\varepsilon^{\tau})^{-1}(A \cap T_\varepsilon^{\tau}(D(\varepsilon))]\, d\tau \leq \int_0^{1/\varepsilon} e^M \mu(A)\, d\tau \leq e^M \mu(A)/\varepsilon.$$

Applying this result to $\Gamma - E(T_0, \delta, \psi)$, we obtain:

(2) $$\mu'[B(\varepsilon, T_0, \delta, \psi)] \leq e^M \mu[\Gamma - E(T_0, \delta, \psi)]/\varepsilon.$$

For each initial condition $(I(0), \varphi(0))$ in $D(\varepsilon)$, we define the set $S' = S'(I(0), \varphi(0), \varepsilon, E(T_0, \delta, \psi))$ of points in time, less than $1/\varepsilon$, where the exact trajectory lies in $\Gamma - E(T_0, \delta, \psi)$. That is,

$$S' = \{ \tau \in [0, 1/\varepsilon]: (I(\tau), \varphi(\tau)) \in \Gamma - E(T_0, \delta, \psi) \}.$$

We have:

(3) $$\mu'[B(\varepsilon, T_0, \delta, \psi)] = \int_{D(\varepsilon)} \mu(S')\, dt.$$

It follows by inverting the integrals that:

(4) $$\int_{D(\varepsilon)} \text{mes}(S')\, d\mu \leq e^M \mu[\Gamma - E(T_0, \delta, \psi)]/\varepsilon.$$

By bounding the integral in (4) from below by the same integral taken on the smaller domain $D(\varepsilon) - N(\varepsilon, T_0, \delta, \psi, \zeta/\varepsilon)$, which corresponds to trajectories spending a time greater than ζ/ε in the slow ergodization zone $\Gamma - E(T_0, \delta, \psi)$), we obtain:

$$(5) \qquad \zeta\, \mu[D(\varepsilon) - N(\varepsilon, T_0, \delta, \psi, \zeta/\varepsilon)] \leqslant e^M \mu[\Gamma - E(T_0, \delta, \psi)].$$

Since $\mu[\Gamma - E(T_0, \delta, \psi)]$ tends to zero with increasing T_0, because the unperturbed motion is ergodic on the fiber, the proof of Lemma 2 is complete. ■

CHAPTER 3: ONE FREQUENCY SYSTEMS AND FIRST RESULT FOR TWO FREQUENCY SYSTEMS

3.1 One frequency; introduction and first order estimates

After the rather abstract result of the last chapter, we now come back to a more concrete viewpoint in a much more restricted setting. In this chapter we first examine the case of one frequency systems when no resonances are involved and then start our study of the resonance phenomena with a result due to Arnold [Ar7] which relies on a rather unrealistic hypothesis but provides a clear introduction to some of the techniques we shall meet again and again from Chapter 4 onwards. One frequency averaging was explored in [Bog], where the original papers of the Russian school are summarized; it also forms the main subject of [Sand3], these two books containing interesting examples, many of practical interest. Once and for all we thus refer to them for a more detailed exposition of the material in this chapter; the latter book also contains references to the recent original papers and a short history of the subject starting with Lagrange.

As already mentioned, averaging can be seen, admittedly from a somewhat abstract point of view, as the "art" of putting a certain class of systems into normal form (cf. Appendix 5). This does not usually require sophisticated mathematical techniques once the possibility of resonance has been disposed with. It is thus a good starting point in the study of perturbation theory, especially since, as it is perhaps worth pointing out, most of the formal manipulations can be turned into valid results once the necessary (simple) evaluations have been traced through. Also, no complicated geometric arguments are involved since again these arise as a consequence of the resonance phenomena. Here we shall confine ourselves to introducing the relevant concepts with the help of the most significant results in a version which does not claim to be optimal. Let us add that, from a combinatorial point of view, there is no hope of exhausting the variations of new (or old) theorems, many of which are in fact "rediscovered" or rewritten again and again in the literature. Given a specific equation or problem, it is certainly more useful to have in mind the general ideas that will enable one to prove an estimate or a theorem tailored to one's needs.

Let us start with the case of one frequency "integrable" systems, i.e., systems where $\varphi \in S^1$ is a phase variable and ω is a function of $I \in \mathbb{R}^m$ alone. This furnishes a simple illustration of the use of the linearized equation. For such systems, which may be written in local coordinates as:

$$(1) \qquad dI/dt = \varepsilon f(I, \varphi, \varepsilon), \quad d\varphi/dt = \omega(I) + \varepsilon g(I, \varphi, \varepsilon)$$

the only possibility for a resonance is $\omega(I) = 0$ and this is ruled out in this section. In fact we assume that I belongs to a domain K in $\mathbb{R}^n$ (which here needs not be compact nor even bounded) on which ω is bounded from above and below ($0 < m < |\omega(I)| < M$). f and g are defined for small enough ε and have period 2π in φ; f and ω are of class C^1 (in all their arguments) whereas g is at least Lipschitz continuous. Moreover f, $\nabla_I f$, g, ω and $\nabla\omega$ are all uniformly bounded on $K \times S^1$ for ε small enough. The averaged system is:

$$(2) \qquad dJ/dt = \varepsilon f_0(J), \quad f_0(J) =_{def} 1/2\pi \int_0^{2\pi} f(J, \varphi, 0)d\varphi.$$

Throughout this section η will denote a small fixed positive constant and K' will then be the set of initial conditions for the averaged system whose associated trajectories remain in $K - \eta$ for at least time τ^*/ε, τ^* some positive constant. In fact, if τ^* is small enough and if K is a reasonably regular domain, K' will be nonempty and independent of ε, since it is determined by the time τ^* of the vector field f_0. These sets arise naturally, because if the averaged trajectory approaches the boundary of K, the exact trajectory, which will be shown to stay close to the averaged one, can cross over the boundary and the system will no longer be defined.

The above gives the rigorous meaning of "working on the time scale $1/\varepsilon$". It is important to at least bear in mind that averaging theorems also provide *existence* theorems: The local existence for (1) is garanteed by the general Cauchy-Lipschitz existence result and it can be extended to time intervals of order $1/\varepsilon$ once we know that I and J remain close; we spell out the bootstrap argument below but will not always do so in the future.

The theorem can be stated as follows:

<u>Theorem 1</u>:

Under the above hypothesis there exists a positive number ε_0 such that for all $\varepsilon < \varepsilon_0$ the solution of (1) exists for $t \in [0, \tau^*/\varepsilon]$ and for every initial condition in $K' \times S^1$ and:

$$(3) \qquad \sup_{t\in[0,\, \tau^*/\varepsilon]} \| I(t) - J(t) \| = O(\varepsilon).$$

Moreover this estimate cannot be improved in general.

The proof of the theorem is almost easier than its statement but it is instructive. As we indicated, the idea consists in reducing the exact system to the averaged system up to order ε^2 by a near identity change of variables, which is none other than a very simple version of the method of normal forms:

$$(4) \qquad I \to P = I + \varepsilon S(I, \varphi).$$

S is determined as follows. Along a trajectory, we have:

$$(5) \qquad dP/dt = \varepsilon[f(I, \varphi, 0) + \omega(I)\partial S/\partial\varphi] + R(I, \varphi, \varepsilon)$$

with the remainder R given below and estimated as $O(\varepsilon^2)$. Since S must be periodic in φ, it may not be possible to choose it so as to eliminate the term in square brackets entirely, but it is certainly possible to keep only the *secularity* $f_0(I)$; in fact if we set:

$$(6) \qquad S(I,\varphi) = -\ 1/\omega(I) \int^{\varphi} [f(I, \varphi, 0) - f_0(I)]\, d\varphi$$

where the lower bound is arbitrary (it can be set equal to 0), then the term in square brackets reduces to $f_0(I)$. Note that S is of class C^p if ω and f are, and so is the application $I \to P$. Now R reads:

$$(7) \qquad R(I, \varphi, \varepsilon) = \varepsilon^2 \partial S/\partial I . f(I, \varphi, \varepsilon) + \varepsilon[f(I, \varphi, \varepsilon) - f(I, \varphi, 0)] + \\ + \varepsilon^2 g(I,\varphi,\varepsilon).\partial S/\partial\varphi$$

and is checked to be of order ε^2 so that P satisfies:

$$(8) \qquad dP/dt = \varepsilon f_0(I) + R(I, \varphi, \varepsilon) = \varepsilon f_0(P) + \varepsilon[f_0(I) - f_0(P)] + O(\varepsilon^2) = \\ = \varepsilon f_0(P) + O(\varepsilon^2)$$

which to first order is identical with the averaged equation.

To bound the separation between P and J (the solution of the averaged equation), we make use of a simple inequality which appears very frequently in this kind of situation and which we have already used in the last chapter. We first state it in an integral form that allows a minimal regularity:

Proposition: (Gronwall's Lemma)

Let $x(t)$, $a(t)$ and $\alpha(t)$ be positive continuous functions on $[0, t]$; assume α is of class C^1 and the following holds:

$$(9) \qquad x(t) \leqslant \alpha(t) + \int_0^t a(s)x(s)\, ds$$

then:

$$(10) \qquad x(t) \leqslant \alpha(0)\exp[\int_0^t a(s)\, ds] + \int_0^t \alpha'(s) \exp[\int_s^t a(\sigma)\, d\sigma]\, ds.$$

To prove this, set $y(t) = \int_0^t a(s)x(s)\, ds$ so that $y' - ay \leqslant a\alpha$; solve this differential inequality and a few manipulations (integration by parts) will produce the result.

In particular, let $x(t)$ be of class C^1 and such that $|x(0)| = c$, $|dx/dt| \leqslant a|x| + b(t)$

where $a \geq 0$ and $b(t) \geq 0$; then for all t:

(11) $|x(t)| \leq e^{at} [c + \int_0^t b(u)e^{-au} du]$.

We now return to the problem taking initial conditions in $K' \times S^1$ and we set $x(t) = P(t) - J(t)$, where $J(t)$ is the trajectory for the averaged system and $(P(t), \varphi(t))$ is the exact trajectory in the transformed variables (with initial conditions $(P\{I(0), \varphi(0)\}, \varphi(0))$. Then:

(12) $\| dx/dt \| \leq \varepsilon \|\nabla f_0\|.\| x \| + O(\varepsilon^2)$

by (9) so that using (11) we get:

(13) $\| P(t) - J(t) \| \leq e^{\varepsilon ct} [\| P(0) - J(0) \| + c\varepsilon^2 t] < c\varepsilon e^{\varepsilon ct}$

for $t = O(1/\varepsilon)$ since $\| P(0) - J(0) \| = \| P(0) - I(0) \| = O(\varepsilon)$.

This result is a priori valid only for $t < \tau$ which is defined as the instant when $P(t)$ exits from K; it remains to show that τ is of order $1/\varepsilon$ at least, using a bootstrap argument. For small enough ε_0, (13) implies that $\| P(\tau) - J(\tau) \| < \eta/2$. Thus $P(\tau)$ lies in $K - \eta/2$ for $t \in [0, \tau^*/\varepsilon]$ (τ^* entering the definition of K'; see above) which implies that $I(t)$ also stays in K for the same interval of time, at least for small enough ε. Therefore $J(t)$, $P(t)$, and $I(t)$ all remain in K for times of order $1/\varepsilon$, whereupon we immediately obtain:

(14) $\| I(t) - J(t) \| \leq \| I(t) - P(t) \| + \| P(t) - J(t) \| \leq c\varepsilon(1 + e^{c\varepsilon t})$

which finishes the proof of the estimate in the statement of the theorem. Notice that this estimate remains unchanged if instead of taking $J(0) = I(0)$, one allows a $O(\varepsilon)$ discrepancy in the initial conditions, setting $\| J(0) - I(0) \| = O(\varepsilon)$.

It is not hard to find examples showing that this result is optimal. The trivial system:

(15) $dI/dt = \varepsilon \sin\varphi \quad (I \in \mathbb{R}), \quad d\varphi/dt = 1$

has the corresponding averaged equation $dJ/dt = 0$, and the error estimate $| I(t) - J(t) | = \varepsilon | \cos(\varphi(0)+t) - 1 |$ takes the value 2ε for all initial conditions. This completes the proof of the theorem.

This theorem gives a rough idea of what the first order theory looks like and how simple it really is, but even without going to higher orders or trying to extend the timescale, it lends itself to many qualifications, improvements and generalizations.

1 - There is first of all the question of putting a given set of equations in the so-called standard form (1). This is in essence a practical matter that arises in all perturbation schemes; at one point or another one has to actually integrate the supposedly integrable

system that is being perturbed. This may be very simple when for instance the unperturbed system is linear, a case which is often considered for illustration: the reader will find several examples in [Sand3] of the form $x'' + x = \varepsilon f(x, x', t)$ which includes e.g. the Van der Pol equation ($f = (1-x^2)x'$). It may of course also become more technical when one needs to use special functions for the integration, like the elliptic functions if the harmonic oscillator is replaced by the pendulum ($x'' + \sin x = 0$) in the above example.

In [Sand3], the reader will find examples which illustrate why the system has to be put into the standard form *before* averaging and why averaging on the equation as it is given ("crude" averaging as they call it) may lead to wrong results. A striking example is the weakly periodically damped oscillator: $x'' + 2\varepsilon(\cos^2 t)x' + x = 0$; in that case, direct averaging leading to a constant damping ($= \varepsilon$) does lead to a wrong asymptotic estimate on the time scale $1/\varepsilon$.

2 - A trivial remark is that the above theorem includes the often considered case of a periodic vector field, that is systems of the form $dI/dt = \varepsilon f(I, t, \varepsilon)$ with a time-periodic function f; it amounts to taking $g = 0$ and ω a constant in (1). This provides however an opportunity to discuss the regularity that is really needed. In fact, in that case, the averaging theorem is true if f is Lipschitz continuous in I, t and ε (Lipschitz continuity in t serves only to ensure local solvability). To see this, suppose for simplicity that f is independent of ε (this makes no difference to order 1 because f is C^+ as a function of ε) and write $z = I - J$. Then:

$$(16) \qquad z(t) \leqslant c\varepsilon\int_0^t z(s)\,ds + \varepsilon\int_0^t \tilde{f}\,(I(s), s)\,ds.$$

The result then follows from Gronwall's inequality (10) and the following almost trivial lemma applied to $F = \tilde{f}$:

<u>Lemma 1</u>: (see [Bes], [Sand3])

Let I solve $dI/dt = \varepsilon f(I, t)$, let f be Lipschitz in I, continuous and periodic in t (with period 1). If $F(I, t)$ is of the same regularity and has zero average, one has:

$$(17) \qquad \int_0^T F(I(s), s)\,ds = O(1)$$

for $T = O(1/\varepsilon)$.

$T = O(1/\varepsilon)$ really means $T < \tau^*/\varepsilon$, the maximum existence time for I; we shall often use such slight abuses of language in the future. The proof is straightforward; if

$N = [1/\varepsilon]$:

(18) $$\left| \int_0^N F(I(s), s)\, ds \right| = \left| \sum_{k=1}^N \int_{k-1}^k [F(I(s), s) - F(I(k-1), s)]\, ds \right|$$
$$\leqslant c \sum_{k=1}^N \int_{k-1}^k \| I(s) - I(k-1) \|\, ds$$

and $\| I(s) - I(k-1) \| = O(\varepsilon)$ since $dI/dt = O(\varepsilon)$.

From the standpoint of regularity, this illustrates how uneconomical the normal forms are, that were used in the proof of the theorem; one superfluous derivative is usually required. The same improvement works for the standard system (we have not found this written in the literature) where one should only require that f, g and ω be Lipschitz continuous in all their arguments (simple continuity in φ would be enough, except again for local solvability). The reader may check this using the same proof as above by direct evaluation and the following generalization of the above lemma:

Lemma 1':

Let (I, φ) solve the standard system (1), let $f, g, \omega \in C^+$ and let $F(I, \varphi)$ be Lipschitz in I, continuous and periodic in φ (of period 1) with zero average ($\oint F(I, \varphi)\, d\varphi = 0$). Then the estimate (17) holds, again for $T = O(1/\varepsilon)$.

This is a slight generalization of Lemma 1; if I^* is fixed, it is easy to check that:

(19) $$\int_0^{1/\omega(I^*)} F(I^*, s)\, ds = O(\varepsilon)$$

by performing the monotone change of variable $t \to \varphi$. Since F is Lipschitz, this is still true if we change I^* to $I(s)$ in the integrand because $1/\omega(I^*) = O(1)$ (ω is bounded from below). Now choose $t_0 = 0 < t_1 < \ldots < t_k < \ldots < t_N < \tau^*/\varepsilon$ such that $t_{k+1} = t_k + 1/\omega(I(t_k))$ ($k = 0, \ldots, N-1$) and N is maximal. Then $N = N(\varepsilon) = O(1/\varepsilon)$ because ω is bounded from above, and the result follows by using (19) on each interval with $I^* = I(t_k)$ (the last interval $[t_N, \tau^*/\varepsilon]$ is $O(1)$).

3 - Under the conditions first stated above (C^1 regularity), the theorem can be extended to the case where $\omega = \omega(I, \varphi)$ depends on the phase variable, yielding the same estimate with the same normal form technique, but for a different averaged system, as we will shortly see. In fact starting as in the proof of Theorem 1, one encounters the linearized conjugacy equation, much as in (5), under the form:

(20) $$f(I, \varphi, 0) + \omega(I, \varphi).\partial S/\partial \varphi = \langle f \rangle.$$

This equation is to be solved for S, keeping the solution bounded, while making the still undefined r.h.s. as simple as possible; this again stresses the fact that the average arises - and in fact is defined - as an unvoidable secularity. Here the reader may check that one must set:

$$(21) \qquad \langle f \rangle = [\int_{S^1} f/\omega \; d\varphi] \, / \, [\int_{S^1} 1/\omega \; d\varphi]$$

which reduces to the previous f_0 when ω is φ independent. The averaged system takes the form (2) with f_0 replaced by $\langle f \rangle$ and S is given by:

$$(22) \qquad S(I, \varphi) = - \int^{\varphi} 1/\omega(I, \varphi) \, [f(I, \varphi, 0) - \langle f(I) \rangle] \, d\varphi.$$

4 - We now come to a nontrivial generalization of the periodic case, which originated long ago in the work of the Russian school. We consider systems of the form $dI/dt = \varepsilon f(I, t)$, with f a vector field which allows for the definition of a general type of average. Since we are working at a first order level, we drop the ε dependence of f, which would only obscure the notation. Let us first define the *local average* $\langle f \rangle_T$ of a continuous function f from $K \times \mathbb{R}$ into $\mathbb{R}^m$ ($K \subset \mathbb{R}^m$) as:

$$(23) \qquad \langle f \rangle_T (I, t) =_{def} 1/T \int_0^T f(I, t+\tau) \, d\tau$$

following [Sand3] such a vector valued function f which is Lipschitz continuous in I and continuous in t will be called a KBM vector field (after Krylov, Bogoliubov and Mitropolski) if the generalized average:

$$(24) \qquad \langle f \rangle (I) =_{def} \lim_{t\to\infty} \langle f \rangle_T (I, t) = \lim_{t\to\infty} 1/T \int_0^T f(I, t+\tau) \, d\tau$$

exists. Note that the average $\langle f \rangle$ is automatically independent of t. This definition is obviously reminiscent of ergodicity, but differs in that no compactness is involved a priori (t is a real number). In fact Anosov's work (cf. Chapter 2) originated in this way and presents a different kind of generalization where the averaging mechanisms are clarified. Here we shall only briefly review the results pertaining to KBM fields, referring to [Sand3] for the proofs which resemble those of Chapter 2, but are simpler and more explicit. One also uses a version of Besjes' lemma (Lemma 1 above) adapted to KBM fields.

Before stating the main results, it is important to note the generality gained by considering such functions; there are in fact perhaps two important types of KBM functions (or fields):

- The first type comprises periodic, quasiperiodic and almost periodic functions; here t acts as a phase in the "physical" meaning of that word. The case of *periodic* functions is clear; a *quasiperiodic* function is by definition one that may be expressed as a periodic function $f(\varphi_1, \ldots, \varphi_n)$ with the same period in all its arguments by setting $\varphi_i = \omega_i t$ with rationaly independent ω_i. In other words f can be factored through a linear flow on a torus. Because of this, the case of quasiperiodic fields is formally included in Anosov's theorem, although better estimates can be obtained by direct methods. The quasiperiodic case may also be regarded as a multifrequency system with n frequencies obeying the equations $d\varphi_i/dt = \omega_i$ (see below Section 2 and Chapters 5 and 6 where we shall return to this briefly).

Going to the "normal modes" coordinates this allows the case of weakly interacting harmonic oscillators to be treated, i.e., systems of the form $x' = Ax + \varepsilon h(x)$ where x is a $2n$ vector (the coordinates and velocities of the oscillators), and A is a matrix with purely imaginary eigenvalues ($= \pm i\omega_i$) and with the normal modes as eigenvectors. Systems of this kind are found in [Bog] and are of course quite important in engineering; a periodic or quasiperiodic time dependence can also be added to the perturbation h.

We shall not go into defining *almost periodic* functions, of which quasiperiodic functions are a particular case. Let us only mention that these functions possess generalized Fourier series, and direct the reader to the classical book by Bohr ([Boh]) or to the more recent one by Fink ([Fi]) where almost periodic differential equations are discussed at length.

- The second class of KBM functions often considered in applications are simply those functions that approach a constant value at infinity, possibly undergoing relatively mild oscillations which allow the integral in (24) to converge. For instance, functions like $f(t) = (1+t)^{-s}$ ($s > 0$) or $f(t) = e^{-t}$ obviously have zero average in that sense. Here t can hardly be termed a "phase".

Sanders and Verhulst give a simple and enlightening example using this type of function; consider again a weakly damped oscillator: $x'' + \varepsilon(2-f(t))x' + x = 0$, where $f(0) = 1$, $f(+\infty) = 0$ and f decreases monotonically. Now by means of the theorem to be stated below, it is easy to show that on the time scale $1/\varepsilon$, $x(t)$ behaves as if there were a constant damping 2ε (the asymptotic value) with an accuracy which depends on the exact shape of f. It is worth pointing out that this result is hardly intuitive.

We now state the main results for a system $dI/dt = \varepsilon f(I, t)$ ($I \in K \subset \mathbb{R}^m$) with f

KBM. The averaged system is of course:

$$(25) \qquad dJ/dt = \varepsilon\langle f\rangle(J).$$

We define K' as above, by supposing that a trajectory of (25) starting in K' stays well inside K until the time τ^*/ε (see the discussion below (2) for more precision). Then we have:

Theorem 2:

For any initial condition in K'

$$(26) \qquad \sup_{t\in[0,\,\tau^*/\varepsilon]} \| I(t) - J(t) \| < c\delta^{1/2}(\varepsilon)$$

where $J(t)$ satisfies (25), $J(0) = I(0)$ and $\delta(\varepsilon)$ is given by:

$$(27) \qquad \delta(\varepsilon) =_{def} \sup_{I\in K,\, t\in[0,\,\tau^*/\varepsilon]} \varepsilon\| \int_0^t [f(I, s) - \langle f\rangle(I)]\, ds \|.$$

Because the definition of the average involves a limit process, the evaluation of the remainders is not nearly as simple as in the periodic case, which is only natural considering what complicated objects almost periodic functions are. In particular this theorem cannot be proved by a normal form technique as Theorem 1 above, even if f is assumed to be smooth (say C^1), because the remainder $R(I, t, \varepsilon)$ arising in (5) cannot be controlled well enough. Instead one uses simple lemmas which avoid the use of differentiation (cf. [Sand3]). It is interesting to notice that in order to compare I and J, each is first compared to the solution of (25) with $\langle f\rangle$ replaced by $\langle f\rangle_T$ and $T = T(\varepsilon)$ large but $\varepsilon T = o(1)$; this is precisely what was done in Chapter 2 in the proof of Anosov's theorem. Moreover in the periodic case, one has $\langle f\rangle_T = \langle f\rangle$ when T is chosen to be the period.

Observe that when f is time periodic, $\delta(\varepsilon)$ is order ε; thus we do *not* directly recover in this way the theorem in the periodic case. There are other simple cases where $\delta(\varepsilon)$ is of order ε: In particular, this is trivially true when f is a finite sum of periodic functions (it is then quasiperiodic); it is also true if f can be written as an absolutely convergent sum:

$$(28) \qquad f(I, t) = \langle f\rangle(I) + \sum_{n\neq 0} c_n(I) \exp[i\lambda_n t]$$

with λ_n bounded away from zero ($|\lambda_n| > \alpha > 0$). This is because we are dealing with almost periodic functions with almost periodic integrals (once the average has been removed). This is not a general property of a.p. functions, as shown by the example of $f(t) = \sum_{n\geq 1} 2^{-n} \sin(2^{-n}t)$, or any function of that type which has very many low frequency harmonics; it in fact easy to prove that the integral of f is not almost periodic, because it is not bounded (boundedness and almost periodicity are equivalent for the integral of an almost periodic function) and that, consequently, the associated remainder for the trivial system $dI/dt = \varepsilon f(t)$ is not of order ε.

In the second important category of KBM functions we considered, it is much easier to find examples where $\delta(\varepsilon)$ is not of order ε, for instance the functions:

$$f(t) = (1+t)^{-s} \quad (0 < s \leqslant 1).$$

Assuming more regularity allows one to find an $O(\delta(\varepsilon))$ estimate for the remainder. Define:

$$(29) \qquad S(I, t) = \varepsilon\int_0^t [f(I, s) - \langle f\rangle(I)]\, ds$$

which is of order $\delta(\varepsilon)$ on a timescale $1/\varepsilon$. With minor notational changes this is of course the quantity considered in the proof of Theorem 1 and which naturally arises in the problem. Suppose f is of class C^1 and define as in [Sand3] $f_1 =_{def} \nabla f.S - \langle f\rangle.\nabla S$ (the dot products are indicated with periods). This quantity shows up when one performs the near identity change of variables $P \rightarrow I = P + S(P, t)$. In fact in the proof of Theorem 1, we explicitly considered the inverse mapping $I \rightarrow P$ (with S changed into $-S$) but to first order all these things are equivalent; we shall say more about this in the next section. A very similar quantity then appears in the evaluation of the remainder (see equations (7) and (8)) which is trivially dealt with in that case.

Now if f is KBM of class C^1 and f_1 is KBM the difference in (26) is of order $\delta(\varepsilon)$. In fact in that case, Sanders and Verhulst proceed to prove a second order result which contains more information. Define $\delta_1(\varepsilon)$ as in (27) with f_1 in place of f and consider the second order averaged system:

$$(30) \qquad dJ/dt = \varepsilon\langle f\rangle(J) + \varepsilon\langle f_1\rangle(J), \quad J(0) = I(0)$$

where the second term on the r.h.s. is in effect smaller than the first. Then one has:

$$(31) \qquad \sup_{t\in[0,\, \tau^*/\varepsilon]} \| I(t) - (J(t)+S(I, t)) \| < c\delta(\varepsilon)[\delta(\varepsilon) + \delta_1^{1/2}(\varepsilon)].$$

We have mentioned this in anticipation of the next section because although the proof is not particularly difficult, it seems to be the only existing higher order result for general KBM vector fields and we shall not come back to these in the sequel (see however the next section for more on quasiperiodic fields). We now turn to the available higher order results for the standard system (1) and the particular case of time periodic fields, still keeping a timescale $1/\varepsilon$.

3.2 Increasing the precision; higher order results

Before we turn to the exposition of the results, a somewhat trivial remark may help the reader to understand more clearly why extension of the timescale is not possible without special assumptions. Let $x(t)$ and $y(t)$ be real functions satisfying:

(1) $$dx/dt = f(x, \varepsilon), \quad dy/dt = f(y, \varepsilon) + g(y, \varepsilon)$$

$f \in C^+$ with Lipschitz constant $\delta_1(\varepsilon)$ ($|f(x, \varepsilon) - f(y, \varepsilon)| \leqslant \delta_1(\varepsilon)|x - y|$), g of order $\delta_2(\varepsilon)$ as ε goes to 0, $\delta_2(\varepsilon) = o(\delta_1(\varepsilon))$. Typically, one can think of $\delta_1(\varepsilon) = \varepsilon$, $\delta_2(\varepsilon) = \varepsilon^n$ ($n > 1$). Then, if we set $z = x - y$ and $z(0) = \delta_3(\varepsilon)$, it is easy to check that Gronwall's estimate produces:

(2) $$|z(t)| \leqslant (\delta_2/\delta_1 + \delta_3)\exp[\delta_1 t] - \delta_2/\delta_1$$

and this cannot be improved in general, which means that one is generally confined to the timescale $1/\delta_1(\varepsilon)$; in our case $\delta_1(\varepsilon) = \varepsilon$ by the definition of ε as the ratio of the velocities of the slow and fast variables.

This method of estimating the difference between the solutions of two equations is certainly the simplest and most direct one and is sometimes called "secular perturbation theory", as opposed to the normal form (or coordinate transform) technique. In the next section we shall examine the nontrivial supplementary hypotheses which do allow an extension of the time-scale.

We have already mentioned a higher order result at the end of last section, namely a second order result for general KBM fields. An nth order result could certainly be stated and proved, as is (implicitly) done in [Perk] but, as the author himself points out, it imposes regularity conditions on very unnatural quantities (see above for the second order case) and it is probably best to turn now to the case of periodic fields before coming back to the general system in standard form and (briefly) to the case of quasiperiodic fields. For periodic systems, the following holds:

Theorem 1:

Let $f = f(I, t, \varepsilon)$ be defined on $K \times \mathbb{R} \times [0\ \ \varepsilon_0]$ ($\varepsilon_0 > 0$, K a sufficiently regular domain in $\mathbb{R}^m$), continuous in t and of class C^{n-1+} ($n \geqslant 1$) in (I, ε) and consider the system

$$dI/dt = \varepsilon f(I, t, \varepsilon).$$

Then there exists a nth order autonomous averaged system of the form:

(3) $$dJ/dt = \textstyle\sum_{k=1}^{n} \varepsilon^k X_k(J)$$

and a time $T^* = \tau^*/\varepsilon$ ($\tau^* > 0$) such that $I(t)$ exists until T^* and satisfies:

$$(4) \qquad I = J + \sum_{k=1}^{n-1} \varepsilon^k S_k(J, t) + O(\varepsilon^n)$$

where J solves (3). The X_k's and S_k's will be determined in the proof and $X_1 = f_0$.

This is certainly the main theorem of nth order averaging for periodic systems and it has been around for quite a long time. A very detailed proof, with somewhat different regularity assumptions can be found in [El] which we follow in part; a result which is essentially the same is proved by L.M. Perko in [Perk].

In the above statement τ^* is the same as in the first order result and is determined by the first order averaged system; this is fairly obvious since we are only trying to increase precision.

Because of the regularity assumptions for f, there exist functions f_k such that:

$$(5) \qquad f(I, t, \varepsilon) = \sum_{k=1}^{n} \varepsilon^{k-1} f_k(I, t) + \varepsilon^n r(I, t, \varepsilon)$$

where r is a remainder which is bounded as ε goes to 0, at least for I in a compact set, which will be enough for the proof, and $f_k \in C(\mathbb{R}_t, C^{n-k+}(K))$, i.e., is continuous in t and C^{n-k+} in the slow variable(s) I. The system may be written:

$$(6) \qquad dI/dt = \sum_{k=1}^{n} \varepsilon^k f_k(I, t) + \varepsilon^{n+1} r(I, t, \varepsilon)$$

a form we could have assumed from the start.

The proof consists in finding an intermediate variable P such that:

$$(7) \qquad P = J + \sum_{k=1}^{n-1} \varepsilon^k S_k(J, t)$$

and which satisfies:

$$(8) \qquad dP/dt = \sum_{k=1}^{n} \varepsilon^k f_k(P, t) + \varepsilon^n g(P, t) + \varepsilon^{n+1} h(P, t, \varepsilon)$$

where g and h are bounded and g has zero average. Suppose for a moment that the X_k's and S_k's have been determined; if $P(0) = I(0)$ Gronwall's lemma together with Lemma 1 of the last section show that $\| I(t) - P(t) \| = O(\varepsilon^n)$ for $t = O(1/\varepsilon)$. This proves (4). To determine $J(0)$, it is enough to invert (4) at $t = 0$, working within $O(\varepsilon^n)$, which produces:

$$(9) \qquad J(0) = I(0) + \sum_{k=1}^{n-1} \varepsilon^k Q_k(I(0), 0)$$

where the Q_k's have universal expressions as functions of the S_k's; these are given in [Perk] and [El]. It then remains to solve the averaged system (3) to get an nth order approximation of I.

Before giving the construction of P, it may be worthwhile to point out that this scheme, which is used by Ellison et al. is not the only possible one. In fact, whereas to order one all schemes are equivalent, to higher order the conversion exists but is not straightforward. For instance, if we write (7) as $P = U(J, t)$ where $U = 1 + O(\varepsilon)$ is a transformation that is invertible order by order, one could use, as an alternative intermediate variable, $P' = U^{-1}(I, t)$, producing $\| P' - J \| = O(\varepsilon^n)$ and $P'(0) = J(0) = U^{-1}(I(0), 0)$.

We now come to the determination of X_k and S_k which are found recursively and satisfy $X_k \in C^{n-k+}(K)$ $(k = 1,\dots, n)$, $S_k \in C(\mathbb{R}_t, C^{n-k+}(K))$ $(k = 1,\dots, n-1)$. As is always the case the X_k's, that is, the coefficients of the averaged system, will arise as unavoidable secularities. Differentiating the tentative definition (7) and using (3) produces:

$$(10) \qquad dP/dt = \sum_k \varepsilon^k X_k(J) + \sum_k \varepsilon^k \nabla_J S_k(J, t) [\sum_j \varepsilon^j X_j(J)] + \sum_k \varepsilon^k \partial S_k/\partial t(J, t)$$

and this has to be compared with (8) which takes the form:

$$(11) \qquad dP/dt = \sum_k \varepsilon^k f_k(J+\sum_j \varepsilon^j S_j(J, t)) + \varepsilon^n g(J+\varepsilon S, t, \varepsilon) + \varepsilon^{n+1} h(J+\varepsilon S, t, \varepsilon)$$

where the arguments of g and h have not been spelled out and we abreviate $\sum_{j=1}^{n-1} \varepsilon^j S_j$ as εS. Equating like powers, we obtain at order one:

$$(12) \qquad X_1(J, t) + \partial S_1/\partial t(J, t) = f_1(J, t)$$

which entails:

$$(13) \qquad X_1 = \langle f_1 \rangle, \quad S_1 = \Phi[f_1] + S'_1(J)$$

The notation $\Phi[.]$ for the integral of the fluctuating part has been explained in Section 1.2; the S_k's are all determined within functions S'_k of the slow variables alone that need only satisfy $S'_k \in C^{n-k+}$ in order to preserve regularity.

Comparing again (10) and (11) at order $k = 2,\dots, n-1$, we have:

$$(14) \qquad X_k(J, t) + \partial S_k/\partial t(J, t) = f_k(J, t) + \sum_{j=1}^{k-1} 1/j!\, \partial/\partial \varepsilon^j\big|_{\varepsilon=0} \times [f_{k-j}(J+\varepsilon S, t)] \sum_{j=1}^{k-1} \nabla_J S_j(J, t).X_{k-j}(J, t) =_{def} F_k(J, t).$$

This formula also works for $k = 1$, in which case the sums are empty. It gives, much as in (13):

$$(15) \qquad X_k = \langle F_k \rangle, \quad S_k = \Phi[F_k] + S'_k(J).$$

An explicit but not very enlightening expression for F_k is available in [El], by working out the coefficients in the differentiation of the first sum on the r.h.s. of (14).

Finally, at order n, we get $X_n = \langle F_n \rangle$, $g = -\tilde{F}_n$ and the rest is lumped into the remainder h. It is easy to check recursively that X_k and S_k really have the claimed regularity. ∎

It is now to be noticed that, at a *formal* level, the scheme above may be literally repeated for quasiperiodic system. Namely suppose that:

$$(16) \qquad f = f(I, \varphi, \varepsilon) = \sum_{j=1}^{n-1} \varepsilon^{j-1} f_j(I, \varphi) + \varepsilon^n r(I, \varphi, \varepsilon)$$

is multiply periodic - say with period 2π - in the variables $\varphi_i = \omega_i t$ ($\varphi \in T^s$, $d\varphi/dt = \omega \in \mathbb{R}^s$); then one only has to read the above scheme as it is, replacing t by φ and the operator $\partial/\partial t$ by $\omega.\nabla_\varphi$ .The equations to be solved are:

$$(17) \qquad X_j(J) + \omega.\nabla_\varphi S_j(J, \varphi) = F_j(J, \varphi), \quad j = 1, \ldots, n-1$$

with F_j still defined according to formula (14) (we use j as the running index because k will denote the generic frequency vector). Averaging over the torus entails:

$$(18) \qquad X_j = \langle F_j \rangle = (2\pi)^{-s} \oint F_j(J, \varphi)\, d\varphi$$

and, still working formally, if $F_j^{(k)}$ denote the Fourier coefficients of F_j :

$$(19) \qquad F_j(J, \varphi) = \sum_{k \in \mathbb{Z}^s} F_j^{(k)}(J) e^{ik.\varphi}, \quad k.\varphi =_{def} \sum k_j \varphi_j$$

we get:

$$(20) \qquad S_j(J, \varphi) = -i \sum_{k \neq 0} (\omega.k)^{-1} F_j^{(k)}(J) e^{ik.\varphi}$$

up to an arbitrary (smooth) function of J alone. This is of course the multidimensional analogue of the Φ operator and our first encounter with the "small denominators" $(\omega.k)$. Since there will be many more in the subsequent chapters (and Appendices 3, 4, 5, 8) this is not the place to discuss their role at length. Suffice it to notice that here the formal scheme is not immediately convertible into the proof of a theorem and one has to add conditions that ensure convergence or rather the asymptotic character of the series at hand.

The accustomed reader will find no difficulty in stating such conditions and the corresponding theorems for this problem which is significantly simpler than those we shall meet in the sequel; he may find it useful to look in Appendices 4 and 8 and examine the following two cases:

- The functions $f_j(I, \varphi)$ have only a finite number of harmonics, i.e., they are polynomials in the variables $\exp[\pm i\varphi_i] = \exp[\pm i\omega_i t]$ ($i = 1, \ldots, s$) and $\omega.k \neq 0$, at least for $|k| \leqslant K$ where K depends on the degree of the polynomials and on the order n of the approximation being sought.
- The functions f_j are of a more general nature, but their Fourier coefficients decrease fast enough and ω is "very irrational", that is it satisfies $|\omega.k| \geqslant \psi(|k|)$ where ψ is some positive decreasing function (see Appendix 4 for the case of the so-called diophantine conditions). This is worked out in details in [Perk], thus providing an nth order averaging theorem in the non resonant quasi-periodic case.

In any case these notions will be detailed in the rest of this book, and on second reading the above should appear rather transparent. Let us only add that the quasiperiodic case also pertains to Chapters 5 and 6, as it may be viewed as a simple case of a multiphase system (with phase equations $d\varphi/dt = \omega$).

We now come to the higher order theory for the general standard system with one phase, which reads:

(21) $$dI/dt = \varepsilon f(I, \varphi), \quad d\varphi/dt = \omega(I) + \varepsilon g(I, \varphi); \quad (I \in \mathbb{R}^m, \varphi \in T^1).$$

For the sake of simplicity, we shall assume that f and g are smooth and independent of ε; in fact, if this parameter is not given a priori, it may be *defined* as the ratio of the rate of change of the slow and fast variables which ensures that f does not depend on ε. Here we shall content ourselves with providing the formal scheme at order two (for notational simplicity) and leave to the reader the easy but tedious tasks of:

- extending the formal scheme to order n, including a possible dependence of f and g on ε.
- finding the minimal regularity assumptions.
- stating and rigorously proving the resulting theorem.

We still assume that there are no resonances ($|\omega(I)| \geqslant \omega_0 > 0$) and we again work on the time interval τ^*/ε, where τ^* is determined as in Section 1 by the first order averaged system $dJ/d\tau = f_0(J)$.

At n^{th} order, one must keep track of the phase to order $n-1$; thus at first order, we had to know it within $O(1)$, i.e., not at all since $\varphi \in T^1$. The scheme we give below is not a generalization of the one we used in the periodic case, but rather the implementation of the remark below equation (9). The original system is (21); the averaged system will read:

(22) $$dJ/dt = \varepsilon X_1(J) + \varepsilon^2 X_2(J), \quad d\psi/dt = \omega(J) + \varepsilon Y_1(J)$$

and we shall look for a pair of intermediate variables (P, χ) such that:

(23) $$I = P + \varepsilon S_1(P, \chi) + \varepsilon^2 S_2(P, \chi), \quad \varphi = \chi + \varepsilon T_1(P, \chi)$$

and which solve a system of the form:

(24) $$dP/dt = \varepsilon X_1(P) + \varepsilon^2 X_2(P) + O(\varepsilon^3), \quad d\chi/dt = \omega(P) + \varepsilon Y_1(P) + O(\varepsilon^2).$$

To this end we simply differentiate the tentative definitions (23) and compare them with the original system (21):

$$(25)\qquad dI/dt = \varepsilon X_1 + \varepsilon^2 X_2 + \varepsilon^2 \nabla_P S_1 . X_1 + \varepsilon\omega . \partial S_1/\partial\chi + \varepsilon^2 Y_1 \partial S_1/\partial\chi + + \varepsilon^2 \omega . \partial S_2/\partial\chi + O(\varepsilon^3)$$

which is to be compared with:

$$(26)\qquad dI/dt = \varepsilon f(I, \varphi) = \varepsilon f(P, \varphi) + \varepsilon^2(\nabla_P f . S_1 + \partial f/\partial\chi . T_1) + O(\varepsilon^3)$$

As for the phase, the corresponding equations read:

$$(27)\qquad d\varphi/dt = \omega(P) + \varepsilon\omega\, \partial T_1/\partial\chi + \varepsilon Y_1(P) + O(\varepsilon^2)$$

$$(28)\qquad d\varphi/dt = \omega(P) + \varepsilon\nabla\omega . S_1 + \varepsilon g(P, \chi) + O(\varepsilon^2).$$

There result the equations for X, Y, S, and T:

$$(29)\qquad X_1(P) + \omega(P)\partial S_1/\partial\chi(P, \chi) = f(P, \chi) X_2(P) + \omega(P)\partial S_2/\partial\chi\,(P, \chi) = = [\nabla_P f . S_1 + \partial f/\partial\chi . T_1](P, \chi) - Y_1(P) . \partial S_1/\partial\chi(P, \chi)$$

and:

$$(30)\qquad Y_1(P) + \omega(P)\partial T_1/\partial\chi(P, \chi) = g(P, \chi) + \nabla\omega(P) . S_1(P, \chi).$$

One then first solves the first of (29) for (X_1 , S_1), to get:

$$(31)\qquad X_1 = \langle f\rangle, \quad S_1 = 1/\omega\ \Phi[f].$$

Again, S_1 is determined only up to the addition of a function of P, but the above choice has the advantage that $\langle S_1\rangle = 0$. Then (30) yields:

$$(32)\qquad Y_1 = \langle g\rangle\,, \quad T_1 = 1/\omega\ \Phi[g] + 1/\omega\ \nabla\omega . \Phi[S_1] = 1/\omega\ \Phi[g] + + 1/\omega^2\ \nabla\omega . \Phi^2[f]$$

where we have used the expression for S_1 and the fact that it has zero average. It then remains to solve the second of (29) to get (X_2 , S_2). By now it should be clear how the general n^{th} order scheme is constructed. Notice that however cumbersome the algorithm may be, the equations to be solved are always of the same type; they are what we call in Appendix 5 the linearized conjugacy equations or the "homological" equations, as Arnold puts it.

Once X, Y, S and T have been obtained, the procedure is simple: if $P(0) = J(0)$, there will be $\| J(t) - P(t) \| = O(\varepsilon^2)$ for $t = O(1/\varepsilon)$ and $| \psi - \chi | = O(\varepsilon)$ on the same time interval, by a straightforward application of Gronwall's lemma. This will relate I and J through (23) and in fact, we get:

$$(33)\qquad I = J + \varepsilon S_1(J, \psi) + O(\varepsilon^2), \quad \varphi = \psi + O(\varepsilon)$$

still for $t = O(1/\varepsilon)$. The initial conditions for the averaged system are determined by inverting these relations for $t = 0$ order by order, and within the necessary accuracy in ε; here we simply get:

$$(34)\qquad J(0) = I(0) - \varepsilon S_1(I(0), \varphi(0)), \quad \psi(0) = \varphi(0)$$

which completes the prescriptions to order two.

By now one may suspect that the computation of S_2 and T_1 was rather unnecessary, and this is no coincidence because we included them only to make it plain what the higher order scheme is; in fact we almost completed the *third* order computation. The information we did not use is given by Lemma 1' of last section which allows a gain of one order in the computation and correspondingly of one derivative in the regularity needed at a given order. We can dispense with the use of S_2 and T_1 by demanding that (P, χ) satisfy:

$$(35) \qquad dP/dt = \varepsilon X_1(P) + \varepsilon^2 X_2(P) + \varepsilon^2 X'_2(P, \chi) + O(\varepsilon^3)$$

$$d\chi/dt = \omega(P) + \varepsilon Y_1(P) + \varepsilon Y'_1(P, \chi) + O(\varepsilon^2)$$

instead of (24). Here X'_2 and Y'_1 are functions with zero averages. One then uses a combination of Gronwall's lemma and Lemma 1' of Section 1 (as was done in the periodic case) to estimate $\| J - P \|$ and $| \psi - \chi |$ with the same accuracy as above. The analogue of this of course works at any order.

In fact the scheme above can be pushed much further: Assuming the analyticity of f, g and ω, and working recursively, one can show that it is possible to take $n \sim 1/\varepsilon$, which leads to a remainder on the order of $e^{-c/\varepsilon}$ $(c > 0)$ instead of ε^n (n fixed). This is the simplest instance (due to the absence of resonances) of what can be called a Nekhoroshev type scheme, and nearly the only case in which it can be implemented in a noncanonical framework. As this will be detailed in the more complicated original setting in Chapter 7, it seems premature to discuss it at length at this point and we shall restrict ourselves to the statement of a theorem, the proof of which can be found in [Nei7] and consists of nothing but a much simplified and non-Hamiltonian version of what is called the analytic lemma in Chapter 7.

<u>Theorem 2</u>:

Let f, g and ω in (21) be defined and analytic on the complex extension $K_{\rho,\sigma}$ of $K \times T^1$ (see Notation, Section 1.2), and assume that they are bounded on the closure of this domain, with $| \omega |$ also bounded from below. Then, there exists a change of variables $(I, \varphi) \to (J, \psi)$ such that for small enough ε and $| t | \leqslant T^* = \tau^*/\varepsilon$, (J, ψ) satisfy the averaged system:

$$(36) \qquad dJ/dt = \varepsilon X(J, \varepsilon) + \alpha(J, \psi, \varepsilon), \quad d\psi/dt = \omega(J) + \varepsilon Y(J, \varepsilon) + \beta(J, \psi, \varepsilon)$$

where $X(J, 0) = \langle f \rangle$ and the following expected estimates hold true:

$$(37) \qquad | I - J | + | \varphi - \psi | = O(\varepsilon), \quad | X(J, \varepsilon) - \langle f \rangle | = O(\varepsilon), \quad | Y(J, \varepsilon) | = O(1).$$

Moreover, the remainders satisfy:

$$(38) \qquad |\alpha(J, \psi, \varepsilon)| + |\beta(J, \psi, \varepsilon)| \leq c_1 \exp(-c_2/\varepsilon) \quad (c_1 > 0, c_2 > 0).$$

So within exponentially small terms, the original analytic sytem can be reduced to an averaged (phase independent) system on the timescale $1/\varepsilon$ and, as far as precision is concerned, this is in fact the best one can hope for. Note that obviously we could have accomodated an (analytic) dependence of f and g on the parameter ε. We shall meet again with this nonresonant Nekhoroshev-type scheme in Chapter 8.

We end this section by mentioning that one can combine the quasiperiodic and the standard case above in a further generalization, considering a system like (21) with $\varphi \in T^s$ and $\omega \in \mathbb{R}^s$ a *constant* vector. This case certainly has more to do with the material of Chapters 5 and 6 but it can be understood in the present context; one needs only repeat the averaging scheme presented above, replacing $\partial/\partial\chi$ everywhere by ∇_χ and considering ω as a constant vector. The remarks about the small denominators arising in the quasiperiodic case can also be repeated here. As will become clearer after Chapters 5 and 6, this provides the most general possible system for which a higher order theory is available in the generic, non-Hamiltonian framework.

3.3 Extending the timescale; geometry enters

In this section we examine some important cases in which an extension of the timescale for the validity of the averaging method is possible and we also briefly discuss the problem of combining a greater precision with a longer timescale. As the title emphasizes, some of the hypotheses will be of a geometrical nature and in fact the subject of this section hinges upon the general geometric theory of dynamical systems, especially the invariant manifold theory. This combination of asymptotic analysis and geometry cannot be considered as having reached the form of a well-established theory and we shall have to content ourselves with partial, yet suggestive results.

We shall consider systems in standard form (see (1) of Section 3.1) as well as the particular case of time periodic systems and leave it to the reader to adapt some of the results to KBM type systems. As for the necessary regularity conditions, they are the same as in the last two sections and we will not bother with them any further.

At the beginning of the last section we saw how a naïve yet generically optimal Gronwall's estimate shows that one is generally confined to a timescale on the order of the inverse of size of the slow vector field. We now try to formulate less and less restrictive hypotheses which remove this obstacle. Note that in the case of averaging, the two systems to be compared (in (1) of the last section) are those for the original and the intermediate variables, as are (6) and (8) or (22) and (24) in last section.

The first simple case (often encountered in practice) where the timescale can be extended is where the $(r-1)^{th}$ order ($r \geq 1$) averaged system for the slow variables vanishes; this corresponds to the vanishing of f in (1) of the last section. According to the scheme presented at the end of the previous section the nth order averaged system for a system in standard form reads:

(1) $$dJ/dt = \textstyle\sum_{k=1}^{n} \varepsilon^k X_k(J); \quad d\psi/dt = \omega(J) + \textstyle\sum_{k=1}^{n-1} \varepsilon^k Y_k(J).$$

For time periodic systems, $\psi = t$ and only the first equation appears. Suppose $X_k = 0$ for $k = 1, \dots, r-1$ ($r < n$). Then:

(2) $$dJ/dt = \varepsilon^r X_r(J) + O(\varepsilon^{r+1})$$

and the same equation, with P in place of J, holds for the intermediate variable P, as seen from the obvious n^{th} order analogue of equation (24) in Section 3.2. This entails that $\| P - J \| = O(\varepsilon)$ for $t = O(\varepsilon^{-r})$ and thus also $\| I - J \| = O(\varepsilon)$ since I and P are ε-close. So if the $(r-1)^{th}$ order system for the slow variables vanishes, an $O(\varepsilon)$ approximation is achieved on the timescale $O(\varepsilon^{-r})$.

Now it is also obviously true that $\| P - J \| = O(\varepsilon^{r+1}T)$ ($P(0) = J(0)$) for $|t| \leq T = O(\varepsilon^{-r})$, so that in particular $\| P - J \| = O(\varepsilon^{r-p+1})$ for $t = O(\varepsilon^{-p})$, $p \leq r \leq n$. Also, (2) entails that $J(t) = J(0) + O(\varepsilon^r t)$ for $t = O(\varepsilon^{-r})$; more precisely, J may be determined by the equation:

(3) $$dJ/dt = \varepsilon^r X_r(J)$$

to within $O(\varepsilon^{r-p+1})$ on the time-scale $O(\varepsilon^{-p})$, $p \leq r$.

The transformation equations from the original variables (I, φ) to the intermediate ones (P, χ), are written as in (23) of the last section:

(4) $$I = P + \textstyle\sum_{k=1}^{n} \varepsilon^k S_k(P, \chi); \quad \varphi = \chi + \textstyle\sum_{k=1}^{n-1} \varepsilon^k T_k(P, \chi).$$

In the case of time periodic systems, $\varphi = \chi = t$ and the prescription to obtain a greater precision on a shorter time interval is particularly simple:

- Compute $J(0)$ by using:

(5) $$I(0) = J(0) + \textstyle\sum_{k=1}^{n} \varepsilon^k S_k(J(0), 0).$$

Inversion can be carried to any degree of accuracy.

- Use the estimate of $\| P - J \|$ and of $\| J - J(0) \|$ (or better still equation (3)) to produce an estimate of $I(t)$. In particular, making no use of (3):

(6) $$I(t) = J(0) + \sum_{k=1}^{r-p-1} \varepsilon^k S_k(J(0), t) + O(\varepsilon^{r-p})$$

for $t = O(\varepsilon^{-p})$, $p \leqslant r-1$.

In fact, using the nth order system ($n \geqslant r$), it is easy to derive an $O(\varepsilon^{n-r+1})$ approximation of I on the timescale ε^{-r}, just as this was shown in the last section, which corresponds to the generic case $r = 1$. The main phenomenon to be noticed here is simply that once the $(r-1)^{th}$ order averaged system vanishes, everything that could be done on the timescale ε^{-1} may be done on the timescale ε^{-r} which is the natural scale for the motion of the slow variables. We note that a more general case, where only part of the averaged system for the slow variables vanishes, is examined in [Pers] under further stability assumptions.

The case of general standard systems is only slightly more complicated and can be discussed in much the same way using equations (1) trough (4). The phase can be approximated once the amplitudes (slow variables) are known to within a certain degree of accuracy, as is plain from system (1).

An important generalization of the above arises when the $(r-1)^{th}$ order averaged system for the slow variables, though nonvanishing, posesses one or more first integrals. Let $\Phi(J, \varepsilon)$ be such a conserved quantity, so that:

(7) $$\nabla\Phi(J, \varepsilon).(\sum_{k=1}^{r-1} \varepsilon^k X_k(J)) = 0$$

where the dot indicates the ordinary scalar product. Then the same reasoning used above can be repeated to compare $\Phi(I, \varepsilon)$, $\Phi(J, \varepsilon)$ and $\Phi(P, \varepsilon)$. In particular

$$| \Phi(I, \varepsilon) - \Phi(J, \varepsilon) | = O(\varepsilon) \text{ for } t = O(\varepsilon^{-r})$$

which defines an approximate invariant manifold for I.

The case of vanishing higher order averaged systems arises in particular in the study of perturbed integrable Hamiltonian systems, whereas the use of first integrals is implicit in the proof of Nekhoroshev's theorem, using resonant normal forms. The integrals then correspond to projections transverse to the "plane of fast drift". We mention this only for the benefit of the reader who is familiar with the contents of Chapter 7.

Let us now turn to more geometric hypotheses concerning the averaged system. Again, in the case of standard systems, one needs only impose conditions on the amplitude (slow) part of the system. Here we shall assume that the slow part of the nth

order averaged system is *asymptotically stable*. We write the system as in (1):

(8) $$dJ/dt = \varepsilon X(J, \varepsilon); \quad d\psi/dt = \omega(J) + \varepsilon Y(J, \varepsilon)$$

with $X(J, \varepsilon) =_{def} \sum_{k=1}^{n} \varepsilon^k X_k(J)$, $Y(J, \varepsilon) =_{def} \sum_{k=1}^{n-1} \varepsilon^k Y_k(J)$. Let U_t denote the flow of the vector field εX, i.e., $U_t(J(0)) = J(t)$, when J satisfies the first of equations (8). We require that there exist a modulus of continuity $\delta(\varepsilon)$ (which means $\delta(\varepsilon) > 0$, $\lim_{\varepsilon \to 0} \delta(\varepsilon) = 0$; see Notation) and $\varepsilon_0 > 0$ such that for any J_0, J_0', $\| J_0 - J_0' \| \leqslant \varepsilon \leqslant \varepsilon_0$ implies:

(9) $$\| J(t) - J'(t) \| = \| U_t(J_0) - U_t(J_0') \| \leqslant \delta(\varepsilon), \quad 0 \leqslant t < \infty.$$

Note that $\delta(\varepsilon) \geqslant \varepsilon$ and that in fact we need only require the asymptotic stability of a given trajectory, i.e., for given $J_0 = J(0)$. We now evaluate $\| I(t) - J(t) \|$ for $t = O(\varepsilon^{-n})$ by using a kind of normal form technique.

Let π: $K \times S^1 \to K$ be the projection onto the base space (recall that $K \subset \mathbb{R}^m$ is the domain of the slow variables). The intermediate variables (P, χ) satisfy a system close to (8) (see end of last section) which we write as:

(10) $$dP/dt = \varepsilon X(P, \varepsilon) + \varepsilon^{n+1} X'(P, \chi, \varepsilon); \quad d\chi/dt = \omega(P) + \varepsilon Y(P, \varepsilon) + \\ + \varepsilon^n Y'(P, \chi, \varepsilon).$$

We let V_t denote the flow of these equations on $K \times S^1$. Lastly, let $\xi(0) = (P(0), \chi(0))$ and $z(t) = U_{-t} \circ \pi \circ V_t(\xi(0))$. We have:

(11) $$dz/dt = dU_{-t}/dt \circ \pi \circ V_t(\xi(0)) + dU_{-t}[\pi \circ V_t(\xi(0))].\pi \circ dV_t/dt(\xi(0)) \\ = -\varepsilon X(z(t)) + dU_{-t}(\pi \circ V_t(\xi(0)).[\varepsilon X(\pi \circ V_t(\xi(0)), \varepsilon) + \\ + \varepsilon^{n+1} X'(V_t(\xi(0)), \varepsilon)].$$

Here we have used the defining property of U_t : $dU_t . X(J, \varepsilon)) = X(U_t(J), \varepsilon)$. Applying this again, we find that:

(12) $$dz/dt = \varepsilon^{n+1} dU_{-t}(\pi \circ V_t(\xi(0))) . X'(V_t(\xi(0)), \varepsilon) = O(\varepsilon^{n+1})$$

which implies that $\| z(t) - z(0) \| = O(\varepsilon)$ for $t = O(\varepsilon^{-n})$. Because of the assumed stability of the averaged system, for small enough ε it follows that: $\| U_t(z(t)) - U_t(z(0) \| \leqslant \delta(c\varepsilon) =_{def} \delta'(\varepsilon)$ where c is a constant, δ' is a new modulus of continuity and the above holds on the same interval of time. Now $U_t(z(t)) = \pi \circ V_t(\xi(0)) = P(t)$ and $U_t(z(0)) = U_t(P(0)) = U_t(J(0)) = J(t)$ and we obtain $\| P(t) - J(t) \| = O(\delta'(\varepsilon))$ for $t = O(\varepsilon^{-n})$; since $\| I - P \| = O(\varepsilon)$, we at last find that $\| I(t) - J(t) \| = O(\delta'(\varepsilon))$ again on the same time interval. This somewhat abstract-looking result - which we have not found in the literature - serves to illustrate how two systems can be compared, not by directly evaluating their difference, but rather by pulling back the evolution of one of them using the flow of the other, which is indeed a normal form idea.

In most cases however, or at least in most tractable cases, asymptotic stability comes from contraction, and this in turn can often be determined from the linearized equation. We now restrict ourselves to this familiar situation, following in particular [Sand1] and [Sand3]. We first recall the classical Poincaré-Liapunov theorem which is easily proved in the present framework and asserts the exponential nature of the contraction for linearly stable systems.

Let $dx/dt = f(x, t)$ be a smooth differential system for $x \in \mathbb{R}^n$ and $x(t) = x_0(t)$ a particular trajectory. By changing x into $x - x_0$ we may assume that $x_0(t) = 0$ and write the equation as:

$$dx/dt = A(t)x + g(x, t) \tag{13}$$

where $A(t) \in M_n(\mathbb{R})$ is a ***bounded*** continuous family of real matrices and $g(x, t) = o(\| x \|)$ is the nonlinear part, which is in fact at most quadratic since f is smooth. In this situation we have the following

Theorem: (Poincaré-Liapunov) (see also e.g. [Hal] and [Sand3])

Assume that for any positive t:

i) The eigenvalues of $A(t)$ have negative real parts: $\text{spec } A(t) \subset \{ z \in \mathbb{C}, \text{Re } z < -\lambda, \lambda > 0\}$.

ii) $\| g(x,t) \| \leqslant c \| x \|^2$.

Then there exists a constant c_1 such that for any x_0 and δ, $\delta < \lambda/(2c_1)$, $\| x_0 \| \leqslant \delta/(2c_1)$, $x(0) = x_0$ implies:

$$\| x(t) \| \leqslant \delta e^{-\lambda t}, \quad 0 \leqslant t < \infty. \tag{14}$$

Thus, in the so-called Poincaré-Liapunov domain $\| x \| \leqslant \delta/(2c_1)$ (which may be much smaller than the basin of attraction of 0), the system is exponentially contracting. To prove this, let $U(t)$ be the fundamental matrix of the linear system $dx/dt = A(t)x$; under hypothesis i), $\| U(t) \| \leqslant c_1 e^{-\lambda t}$. Then use the variation of constants formula:

$$x(t) = U(t)x_0 + \int_0^t U(t-s)g(x(s),s)\, ds \tag{15}$$

and Gronwall's lemma to reach the conclusion. One may also evaluate $x(t)$ as:

$$\| x(t) \| \leqslant c_1 \| x_0 \| \exp[(c_1\delta-\lambda)t] \tag{16}$$

with a slower rate of contraction but explicit dependence on the initial condition.

An easy consequence of the proof is that trajectories starting in the Poincaré-Liapunov domain get close exponentially fast. Precisely, for two initial conditions $x_1(0)$

and $x_2(0)$, with $\| x_i(0) \| \leqslant \delta/(2c_1)$ $(i = 1, 2)$, we have:

$$(17) \qquad \| x_1(t) - x_2(t) \| \leqslant c_2 \| x_1(0) - x_2(0) \| \exp(c_3\delta - \lambda)t], \quad 0 \leqslant t < \infty$$

for some constants c_2 and c_3 $(c_3\delta < \lambda)$.

With this in mind, let us return to averaging, first for time periodic systems; in other words we consider the equation $dI/dt = \varepsilon f(I, t)$, f being t-periodic, and the associated averaged system $dJ/dt = \varepsilon f_0(J)$. Attraction then allows us to extend the approximation to all (forward) times as demonstrated by the following, which is taken from [Sand3] and summarizes previous work by various authors:

Theorem:

Suppose that 0 is a fixed point of f_0, which is *linearly asymptotically stable*, with a domain of attraction $D \subset K \subset \mathbb{R}^m$. Then if $I(0) \in D - \eta$ $(\eta > 0)$, $\| I(0) - J(0) \| = O(\varepsilon)$ and $J(t)$ satisfies the averaged system, one has:

$$(18) \qquad \| I(t) - J(t) \| = O(\varepsilon), \quad 0 \leqslant t < \infty.$$

Note that more generally, as was done for the Poincaré-Liapunov theorem, one can start from any periodic linearly stable trajectory of the averaged sytem and reach the same conclusion (with the same proof) for trajectories originating in the neighborhood of that periodic solution.

That the evaluation holds for $t = O(1/\varepsilon)$ is the content of the simplest averaging theorem (Theorem 1 of Section 3.1). Let us as usual introduce the rescaled time $\tau = \varepsilon t$. After a finite rescaled time interval τ^*, the trajectories starting in $D - \eta$ enter the Poincaré-Liapunov domain for the averaged equation $dJ/d\tau = f_0(J)$, where exponential contraction occurs. Then because of (17), there exists $T > 0$ such that for two trajectories $J_1(t)$ and $J_2(t)$ one has:

$$(19) \qquad \| J_1((\tau+T)/\varepsilon) - J_2((\tau+T)/\varepsilon) \| \leqslant k \| J_1(\tau/\varepsilon) - J_2(\tau/\varepsilon) \|$$

with $0 < k < 1$ and for any $\tau \geqslant \tau^*$.

To extend the validity of this approximation one now uses a summation trick introduced in this context in [Sanc]. Prior to $t = \tau^*/\varepsilon$, one uses the ordinary averaging theorem; then one divides the time axis into intervals of length T/ε. We let $I_n = [(\tau^*+nT)/\varepsilon, (\tau^*+(n+1)T)/\varepsilon]$ and for any vector valued $v(t)$, $\| v \|_{I_n} = \sup_{t \in I_n} \| v(t) \|$. We denote by $J_n(t)$ the solution of the averaged system such that

$J_n(nT/\varepsilon) = I(nT/\varepsilon)$. Then:

$$(20) \qquad \| I - J \|_{I_n} \leqslant \| I - J_n \|_{I_n} + \| J - J \|_{I_n} \leqslant c\varepsilon + k\| J - J_n \|_{I_{n-1}}$$

where we have used the ordinary averaging theorem and (19). Thus:

$$(21) \qquad \| I - J \|_{I_n} \leqslant c\varepsilon + k\| I - J \|_{I_{n-1}} + k\| I - J_n \|_{I_{n-1}} \leqslant (1+k)c\varepsilon + \\ + k\| I - J \|_{I_{n-1}}$$

and this recursion relation immediately implies that:

$$(22) \qquad \| I - J \|_{I_n} \leqslant c\varepsilon(1 + k)/(1 - k) = O(\varepsilon)$$

for any n, as was to be shown. ▪

We note that this also works for KBM fields, replacing the $O(\varepsilon)$ estimate by $\delta(\varepsilon)$ (see (24) and (27) of Section 3.1) in the approximation on the timescale $1/\varepsilon$. This changes $c\varepsilon$ into $\delta(\varepsilon)$ in (22), while everything else remains unaltered.

The above result is improved in [Sanc] in the often-encountered case where the averaged *and* the exact equations have a common fixed point. The author then proceeds to prove in the same way that the difference between I and J in fact decreases exponentially; for general KBM fields, one has: $\| I(t) - J(t) \| \leqslant \delta(\varepsilon)e^{-\gamma\varepsilon t}$ $(\gamma > 0)$ for any $t > 0$. However, in any of these results, it is important to notice that the *relative* error $\| I - J \|/\| J \|$ does not in general decrease with time; on the contrary it usually tends to infinity and in that respect $J(t)$ may not correctly represent the qualitative behavior of $I(t)$.

The case of the general standard system is hardly more complicated (see [Sand1]); in that case an attracting fixed point in the slow part of the averaged system corresponds to an attracting limit cycle of the full system. At first order, the averaged system still appears as above ($dJ/dt = \varepsilon f_0(J)$) and the result and proof do not differ from the above; in fact the statement of the theorem may be literally repeated and the only change needed in the proof is the introduction of solutions (J_n, ψ_n) such that

$$(J_n, \psi_n)(nT/\varepsilon) = (I, \varphi)(nT/\varepsilon);$$

one then works as above. At this level of approximation, one can take $\varphi = \chi = \psi$ (notation from the last section) so that the phase variables do not play any part. Notice however that to be consistent with the higher order theory one should introduce (J, ψ) where $d\psi/dt = \omega(J)$, in which case $\varphi \neq \psi$, but this is of no importance for our purposes.

Two remarks are in order which extend the scope of the above for standard systems and a fortiori in the particular case of time periodic systems:

1 - The existence of a linearly stable fixed point - or trajectory - is not essential for the extension of averaging to an infinite interval of time; it is only a way to ensure, via the

Poincaré-Liapunov theorem, the validity of a *contraction* property (as expressed by (19)). But as we mentioned earlier, contraction often cannot be proved in practice, except in that simple case.

2 - One can produce higher order estimates on an infinite time interval, also taking into account a possible vanishing of the lower order terms of the averaged system. To this end, one again considers the nth order averaged system:

$$(23) \qquad dJ/dt = \sum_{k=r}^{n} \varepsilon^k X_k(J); \quad d\psi/dt = \omega(J) + \sum_{k=1}^{n-1} \varepsilon^k Y_k(J).$$

As we saw above, the two important indices are n and r, X_r being the first nonvanishing vectorfield for the slow part. If the first of equations (23) is exponentially contracting, one obtains, analogously to (19):

$$(24) \qquad \| J_1(t+t_0) - J_2(t+t_0) \| \leq c_1 \exp(-c_2 \varepsilon^r t) \| J_1(t_0) - J_2(t_0) \|.$$

This holds in particular, if X_r posesses a linearly stable fixed point, for trajectories originating inside the domain of attraction and for t_0 large enough so that at time t_0 they have entered the Poincaré-Liapunov domain of exponential contraction. As we saw previously, a $O(\varepsilon^{n-r+1})$ approximation of I can be obtained on the timescale ε^{-r}, using (24) and the summation trick (see (20),(21)), this estimate can be carried over to all forward times. In view of the phase part of (23), one can then approximate φ with an error on the order of $t\varepsilon^{n-r+1}$, in particular to within $O(\varepsilon)$ on the time-scale ε^{n-r}; but, as is to be expected, this is a secular approximation, and the phase *cannot* in general be approximated at all times. For details of the proof, we refer the reader to [Sand1]; a different proof in the case of time periodic systems will be given below.

Having examined the consequences of contraction properties, it is natural to go to the less stringent requirement of hyperbolicity, i.e., the case where the spectrum of the linearized averaged field is only assumed to stay away from the imaginary axis. Here the subject merges with the general geometric theory of dynamical systems, especially the invariant manifold theory. We shall only touch on this not-yet-complete subject and we shall freely use some of the basic concepts of geometric theory, referring for background and more details to [Gu], [Hal] and [Sand2], among others. We restrict ourselves to the case of time periodic systems, which is the only one to be treated in the literature; this is *not* mere convenience but stems from the fact that the period map (or Poincaré map) turns out to be an almost indispensable tool in this context. Since it can only be defined assuming periodicity, the extension to general standard systems is far from obvious. Let us also point out that the considerations below of course apply in the particular case of stable (contracting) systems, and they yield new proofs of the results above, and address different, complementary questions. The material at the end of this section is mainly drawn from [Hal], [Mu] and [Ro].

We begin again with the two equations:

(25.1) $dz/dt = \sum_{k=r}^{n} \varepsilon^k X_k(z) = X(z, \varepsilon)$

(25.2) $dz/dt = X(z, \varepsilon) + \varepsilon^{n+1} X'(z, t, \varepsilon)$

with $z \in \mathbb{R}^m$ and $X'(z, t, \varepsilon)$ t-periodic (of period 1). All quantities are assumed to be smooth (C^∞). (25.1) is the nth order averaged equation governing the evolution of J; (25.2) is the equation for the intermediate variable P; I and P are related by the first of equations (4) with $\chi = t$, so that any approximation of P by J yields an approximation of I, with the same accuracy.

U_ε will denote the time 1 mapping of the flow defined by (25.1) (Poincaré map); similarly, U_ε' denotes the period map for (25.2); since X' is nonautonomous, this depends on the initial time but all resulting maps are conjugate with each other. We write:

(26.1) $U_\varepsilon(z) = z + \varepsilon^r u_r(z) + \ldots + \varepsilon^{n-1} u_{n-1}(z) + \varepsilon^n u_n(z, \varepsilon)$

(26.2) $U_\varepsilon'(z) = U_\varepsilon(z) + \varepsilon^{n+1} u'(z, \varepsilon)$.

Let z_0 be such that $u_r(z_0) = 0$ and $Du_r(z_0)$ is nonsingular (D denotes the differential map). Then by the implicit function theorem, for small enough ε, there exist points z_ε and z_ε' which depend smoothly on ε and such that $U_\varepsilon(z_\varepsilon) = z_\varepsilon$, $U_\varepsilon'(z_\varepsilon') = z_\varepsilon'$. They define periodic trajectories of (25.1) and (25.2) respectively and satisfy the estimates:

(27) $\| z_\varepsilon - z_0 \| = O(\varepsilon)$, $\| z_\varepsilon' - z_0 \| = O(\varepsilon)$, $\| z_\varepsilon' - z_\varepsilon \| = O(\varepsilon^{n+1-r})$.

Suppose now that $Du_r(z_0)$ is hyperbolic with p eigenvalues inside the unit circle (and $m - p$ outside). Then z_ε and z_ε' are also hyperbolic, of the same type, by the stability of hyperbolicity under perturbation (the type of a fixed point being of course defined by the type of the differential). Before we exploit this to study the dynamics around the fixed points, let us briefly comment on the above hypotheses.

First, if $Du_r(z_0)$ is singular, one is in the realm of bifurcation theory, and we shall not pursue this further. Second, if $Du_r(z_0)$ is nonsingular but nonhyperbolic, one can consider an important generalization of the above which also serves as a cautionary note on hyperbolicity. Let $r \leqslant s \leqslant n$ and $z_\varepsilon^{(s)}$ be a fixed point of the *truncated* application $U_\varepsilon^{(s)} = z + \sum_{k=r}^{s} \varepsilon^k u_k(z)$; if $s = n$, u_n depends on ε and $z_\varepsilon^{(n)} = z_\varepsilon$. Then if $z_\varepsilon^{(s)}$ is hyperbolic, is it true that z_ε and z'_ε are also hyperbolic, as is the case when $r = s$? The answer is *no* in general, when $s > r$, and a more stringent condition on $DU_\varepsilon^{(s)}(z_\varepsilon^{(s)})$ than simple hyperbolicity is needed. One such sufficient condition was introduced by J. Murdock and

C. Robinson in [Mu], which they call strong s-hyperbolicity (s refering to the index above); we shall not discuss it in detail, but the purpose is to ensure that $DU_\varepsilon^{(s)}(z_\varepsilon^{(s)}) + O(\varepsilon^{s+1})$ is hyperbolic (of the same type as $DU_\varepsilon^{(s)}(z_\varepsilon^{(s)})$), whatever the $O(\varepsilon^{s+1})$ perturbation may be, and in general this is *not* the case: hyperbolicity can be destroyed by a higher order term. When $s = r < n$ however, we have $DU_\varepsilon^{(s)}(z_\varepsilon^{(s)}) + O(\varepsilon^{r+1}) = \varepsilon^r\,[Du_r(z_0\,) + O(\varepsilon)]$ which is indeed hyperbolic, because the unperturbed operator is independent of ε. The other important case is $s = n$; then, as already mentioned, $z_\varepsilon^{(n)} = z_\varepsilon$ and one is looking at a periodic trajectory of the nth order averaged system which must be assumed strongly hyperbolic.

To summarize, strong hyperbolicity is always needed when $Du_r(z_0)$ is nonhyperbolic; for instance, if $r = s = n = 1$, which corresponds to the first order averaged system, $u_1(z, \varepsilon)$ depends on ε and one must assume either that $Du_1(z_0\,,\,0)$ is hyperbolic, or that $Du_1(z_\varepsilon\,,\,\varepsilon)$ is strongly hyperbolic. So in the sequel, we assume that $L_\varepsilon =_{\text{def}} DU_\varepsilon(z_\varepsilon)$ is strongly s-hyperbolic for some s, $r \leqslant s \leqslant n$. Because of the estimate on $U_\varepsilon^{(s)} - U_\varepsilon$ and $z_\varepsilon^{(s)} - z_\varepsilon$, this is the same thing as requiring that $DU_\varepsilon^{(s)}(z_\varepsilon^{(s)})$ be strongly s–hyperbolic for some s and may in fact be determined using only the $O(\varepsilon^s)$ principal part of L_ε. This hypothesis ensures that z_ε and z_ε' are hyperbolic of the same type, hence so are the periodic trajectories of the two systems (25.1) and (25.2). Of course, if $Du_r(z_0)$ is itself hyperbolic this simplifies, as explained above, but in any event it is useful to have a condition that concerns only the averaged system itself.

Having discussed the conditions for the existence of hyperbolic periodic orbits in equations (25), we come to a related

<u>Theorem</u>:

For ε small enough, there exists a neighborhood of z_ε defined by:

$$\Omega_\varepsilon = \{\, z \in \mathbb{R}^m : \| z - z_\varepsilon \| < c_1 \varepsilon^{s-r} \,\}$$

($c_1 > 0$) and a global Lipschitz homeomorphism $H_\varepsilon : \mathbb{R}^m \to \mathbb{R}^m$ satisfying:

$$(28) \qquad \| H_\varepsilon(z) - z \| = O(\varepsilon^{n+1-r})$$

such that for $z \in \Omega_\varepsilon$, one has the conjugacy relation:

$$(29) \qquad H_\varepsilon \circ U_\varepsilon(z) = U_\varepsilon' \circ H_\varepsilon(z).$$

This is proven in [Mu] (in the case $r = 1$) and is essentially a version of the Hartman-Grobman theorem on the stability of diffeomeophisms near a hyperbolic fixed point; there are many slightly different statements and proofs of the latter theorem, the modern ones yielding conjugacy on more complicated hyperbolic sets, and we refer the reader to [Hal] and [Gu], as well as references therein, such as [Pu]. Here we only indicate the reasons for the size of Ω_ε and that of $H_\varepsilon - 1$; note that the above is in fact a local statement, and one can take $H_\varepsilon = 1$ far from z_ε.

The first thing is to translate z_ε and z_ε' back to the origin, introducing:

(30) $\quad U_\varepsilon(y) = U_\varepsilon(y+z_\varepsilon) - z_\varepsilon; \quad U_\varepsilon'(y) = U_\varepsilon(y + z_\varepsilon') - z_\varepsilon'.$

This will change the coefficients $u_i(z)$, but because of the estimate (27) on $\| z_\varepsilon - z_\varepsilon' \|$, the part of order n will still be the same in $U_\varepsilon(y)$ and $U_\varepsilon'(y)$. We are thus brought back to the same situation (with different coefficients), and this allows us to assume that $z_\varepsilon = z_\varepsilon' = 0$. Using the fact that $L_\varepsilon = DU_\varepsilon(0)$, it is easily seen that one can write:

(31) $\quad U_\varepsilon(z) = L_\varepsilon z + V_\varepsilon(z); \quad U_\varepsilon'(z) = L_\varepsilon z + V_\varepsilon'(z)$

where for $\| z \| = O(\varepsilon^{s-r})$, V_ε and V_ε' satisfy:

(32.1) $\quad \| V_\varepsilon(z) \| = O(\varepsilon^{2s-r}), \quad \| V_\varepsilon'(z) \| = O(\varepsilon^{2s-r}),$

(32.2) $\quad \| DV_\varepsilon(z) \| = O(\varepsilon^{s}), \quad \| DV_\varepsilon'(z) \| = O(\varepsilon^{s}).$

On the other hand, by the very definition of hyperbolicity, there exists a p (resp. $m - p$) dimensional plane Π_ε^s (resp. Π_ε^u) on which L_ε (resp. L_ε^{-1}) is contracting, with a contraction constant at most $1 - c_2\varepsilon^s$ ($c_2 > 0$), because L_ε is strongly s-hyperbolic. Taking c_1 small enough in the definition of Ω_ε and using the estimates (32.2) ensures that U_ε and U_ε' are also contracting (resp. dilating) when restricted to Π_ε^s (resp. Π_ε^u), with a Lipschitz constant at most $1 - c_2\varepsilon^s/2$ (resp. the same estimate for the inverse maps). This is the reason for the choice of the size of Ω_ε, which permits the derivation of (32.2). Now the theorem is proved by the usual fixed point argument for contracting maps in a Banach space, which shows the existence of H_ε and yields the estimate:

(33) $\quad \| H_\varepsilon - 1 \| \leqslant (1 - \mathrm{Lip}(U_\varepsilon))^{-1} . \| V_\varepsilon' - V_\varepsilon \|$

where $\mathrm{Lip}(U)$ denotes the Lipschitz (contraction) constant of U. For $z \in \Omega_\varepsilon$, one has

$$\| V_\varepsilon'(z) - V_\varepsilon(z) \| = O(\varepsilon^{n+1}).O(\varepsilon^{s-r}) = O(\varepsilon^{n+1+s-r})$$

and this together with the estimate of $(1 - \mathrm{Lip}(U_\varepsilon)) = O(\varepsilon^s)$ proves (28). ▪

As a first comment, note that in fact U_ε is conjugate to the linear operator L_ε on Ω_ε (with a possibly smaller constant c_1). The same proof yields a homeomorphism K_ε such that $K_\varepsilon \circ U_\varepsilon = L_\varepsilon \circ K_\varepsilon$ on Ω_ε and it satisfies the estimate $\| K_\varepsilon - 1 \| = O(\varepsilon^{s+1-r})$ since here the size of the perturbation is $O(\varepsilon^s)$. This is the standard way of proving the existence of stable and unstable manifolds, denoted as $W_\varepsilon^i(z_\varepsilon)$ ($i = s, u$). These are characterized as:

$$W_\varepsilon^s(z_\varepsilon) = \{ z \in \mathbb{R}^m : \lim_{n\to\infty} U_\varepsilon^n(z) = z_\varepsilon \}$$

while n is changed into $-n$ for the unstable manifold. $W_\varepsilon^s(z_\varepsilon)$ (resp. $W_\varepsilon^u(z_\varepsilon)$) is tangent to Π_ε^s (resp. Π_ε^u) at z_ε (if the fixed point has not been translated back to zero, the planes are those containing it). Everything can be repeated for U_ε' to produce the existence of K_ε', $W_\varepsilon^s(z_\varepsilon')$ and $W_\varepsilon^u(z_\varepsilon')$. From the characterization of stable and unstable manifolds, one has:

$$(34) \qquad H_\varepsilon(W_\varepsilon^i(z_\varepsilon) \cap \Omega_\varepsilon) \subset W_\varepsilon^i(z_\varepsilon'), \quad i = s, u.$$

Suppose that $U_\varepsilon^n(z) \in \Omega_\varepsilon$ for any positive n, which is the case when $z \in W_\varepsilon^s(z_\varepsilon) \cap \Omega_\varepsilon'$, where Ω_ε' is slightly smaller than and contained in Ω_ε. Then:

$$(35) \qquad \| \dot{U}_\varepsilon'^n \circ H_\varepsilon(z) - U_\varepsilon^n(z) \| = \| H_\varepsilon \circ U_\varepsilon^n(z) - U_\varepsilon^n(z) \| = O(\varepsilon^{n+1-r}),$$

an estimate which the reader is invited to translate in terms of averaging. In particular, if U_ε is contracting ($p = m$), and thus so is U_ε', one can write:

$$\begin{aligned}(36) \qquad \| U_\varepsilon'^n(z) - U_\varepsilon^n(z) \| &\leqslant \| U_\varepsilon'^n \circ H_\varepsilon(z) - U_\varepsilon^n(z) \| + \\ &+ \| U_\varepsilon'^n \circ H_\varepsilon(z) - U_\varepsilon'^n(z) \| \\ &= \| H_\varepsilon \circ U_\varepsilon^n(z) - U_\varepsilon^n(z) \| + \\ &+ \| U_\varepsilon'^n \circ H_\varepsilon(z) - U_\varepsilon'^n(z) \| \\ &\leqslant \| H_\varepsilon \circ U_\varepsilon^n(z) - U_\varepsilon^n(z) \| + \| H_\varepsilon(z) - z \| \\ &= O(\varepsilon^{n+1-r}).\end{aligned}$$

This is valid for $z \in \Omega_\varepsilon'$ where Ω_ε' may have to be slightly reduced; notice that Ω_ε (and Ω_ε') are much larger than the distance $\| z_\varepsilon' - z_\varepsilon \|$ and thus contain both fixed points. In terms of averaging, the above is none other than the estimate $\| J - P \| = O(\varepsilon^{n+1-r})$ for any $t \geqslant 0$, the proof of which was sketched above, using different tools.

We end this section with indications of some directions of current research. One of the main goals is to find a more global picture: Except in the case where $s = r$ (one then requires $Du_r(u_0)$ hyperbolic), Ω_ε shrinks to the fixed point when ε tends to 0. What kind of conditions will ensure that essentially the same results hold on a larger neighborhood ? Tracing through the proof, it is easily seen that one should look for a domain on which $\| DU_\varepsilon(z) \|$ is hyperbolic (or contracting) and the proof then goes through. Now in the stable case, one may also inquire about the basins of attraction, which in general are much larger than the region on which attraction takes place: How does the basin of U_ε determine that of U_ε' ? This is a difficult question, and there are examples which show that the two basins can in general be very different. Finding reasonable and verifiable sufficient conditions which lead to a simple relation is a subject of current interest; further details may be found in [Mu].

In the hyperbolic case, a classical and difficult question concerns the global arrangement of the stable and unstable manifolds. In the two dimensional case and when there is an unperturbed homoclinic (or heteroclinic) point, the possible splitting of the separatrices in the perturbed system is studied with the help of the Melnikov integral (see e.g. [Gu]; we briefly introduce it in the next chapter). However, in the case of averaging, J.A. Sanders has pointed out a difficulty (cf. [Sand2]) which has not yet been overcome (and perhaps may never be). In fact one then wishes to compare the two systems:

(37) $dz/d\tau = X(z), \quad dz/d\tau = X(z) + X'(z, \tau/\varepsilon, \varepsilon), \quad \tau = \varepsilon t$

(in averaging, $z = J$ and $X = \langle f \rangle$). The trouble is that the perturbation X' has frequency $1/\varepsilon$, once the time and thus the unperturbed field have been rescaled. The Melnikov integral "often" turns out to be exponentially small. This makes the validity of the method itself dubious since in the usual case the integral gives only a first order estimate.

3.4 Resonance; a first encounter

We now suppose that the domain K contains resonant points I for which $\omega(I) = 0$; although it may seem strange to speak of "resonance", this refers to the case when ω represents a detuning frequency, i.e., the difference between the natural frequency of two motions. Generically, the points where ω vanishes form a surface. We give here an example due to Arnold which displays the effect of resonance in the simple case of a single fast variable. Consider the system:

(1) $dI/dt = \varepsilon(1 - \cos\varphi), \quad d\varphi/dt = I.$

The corresponding averaged system is $dJ/dt = \varepsilon$. Take the initial condition $(I = 0, \varphi = 0)$ on the resonance $I = 0$. The corresponding exact trajectory has projection $I(t) = 0$, whereas the averaged trajectory is $J(t) = \varepsilon t$. The two trajectories thus separate to order 1 during the time interval $[0, 1/\varepsilon]$, because the exact trajectory is "trapped" on the resonance. In the next section we will eliminate trajectories for which the exact and averaged systems differ markedly, and we will show that only "few" such trajectories need be removed.

Even in the absence of "capture into resonance," the resonances make their presence known. To see this, consider the system:

(2) $dI_1/dt = \varepsilon, \quad dI_2/dt = \varepsilon\cos\varphi, \quad d\varphi/dt = I_1.$

Choosing the initial condition $(I_1 = 0, I_2 = 1, \varphi = 0)$ on the resonance $I_1 = 0$, we find that the projection on the base of the exact trajectory is:

(3) $\quad I_1(t) = \varepsilon t, \quad I_2(t) = 1+\varepsilon\int_0^t \cos(\varepsilon u^2/2)\, du = 1+(2\varepsilon)^{1/2} C(t[\varepsilon/2]^{1/2})$

where $C(x)$ is the Fresnel integral $C(x) = \int_0^x \cos(u^2)\, du$ (see Figure 3.1).

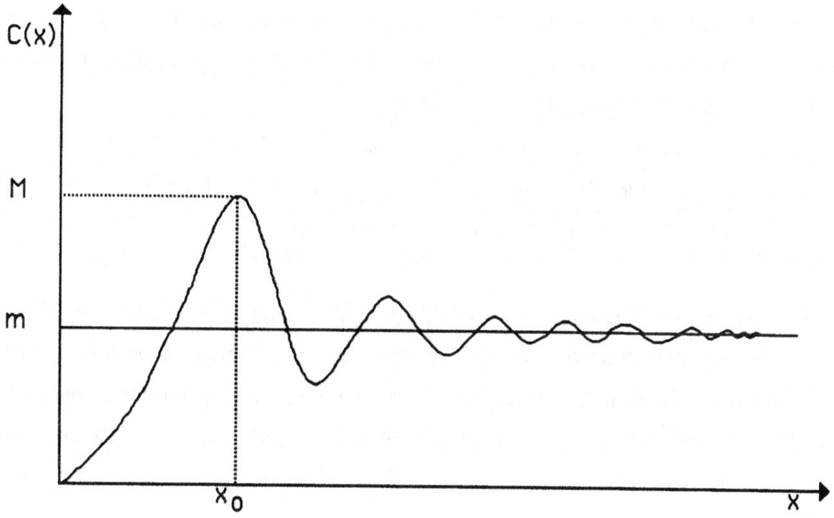

Figure 3.1: The Fresnel function $C(x)=\int_0^x \cos(u^2)\, du$ reaches its peak value M at $x_0=(\pi/2)^{1/2}$ and then tends asymptotically to the constant value $m = 1/2\, (\pi/2)^{1/2}$.

This trajectory is thus not captured by the resonance. The averaged system is:

(4) $\quad dJ_1/dt = \varepsilon, \quad dJ_2/dt = 0.$

The averaged trajectory with the same initial condition is therefore $J_1(t) = \varepsilon t$, $J_2(t) = 1$, and the separation in norm between the two trajectories is $(2\varepsilon)^{1/2} C(t[\varepsilon/2]^{1/2})$. Since $C(x)$ attains its maximum M at x_0, for $\varepsilon < 1/(2x_0^2)$ $(= 1/\pi)$, we obtain

$$\sup_{t\in[0,1/\varepsilon]} \| I(t) - J(t) \| = M(2\varepsilon)^{1/2}.$$

The separation is thus no longer of order ε as it was in the nonresonant case. There is a simple explanation for this order $\sqrt{\varepsilon}$ behavior. Consider a set of trajectories with initial conditions $I_1(0) = 0$, $I_2(0) = 1$ and $\varphi(0) = \varphi_0$ arbitrary, differing only in their initial phase. The generic trajectory may be written:

$$(5) \qquad I_1(t) = \varepsilon t$$

$$I_2(t) = 1 + (2\varepsilon)^{1/2}[\cos\varphi_0\, C(t[\varepsilon/2]^{1/2}) - \sin\varphi_0\, S(t[\varepsilon/2]^{1/2})]$$

where S is the Fresnel integral $S(x) = \int_0^x \sin(u^2)\, du$. We see that the separation between two exact solutions differing only in initial phase is of order $\sqrt{\varepsilon}$, so that the separation between exact and averaged solutions is necessarily also at least of this order. This *diffusion* of the trajectories is a general phenomenon, in as much as one cannot in general find an error estimate better than $\sqrt{\varepsilon}$, as we shall see.

This separation is attained in a time of order $1/\sqrt{\varepsilon}$, during which the trajectory travels a distance of order $\sqrt{\varepsilon}$ on the base. Consequently the size of the "resonant zones" is "naturally" of order $\sqrt{\varepsilon}$, a fact that will be used in Chapter 4.

There is not much point in establishing a general averaging result taking resonance into account for one frequency systems, because these systems may be seen as two frequency systems by adding another phase λ obeying $d\lambda/dt = 1$. It is better to study two frequency systems directly, noting that, conversely, these may be reduced to one frequency resonant systems, at least locally in space, by a change of timescale (see below).

3.5 Two frequency systems; Arnold's result

In this section we begin to examine averaging results for two frequency systems, following Arnold [Ar7] in a simplified, somewhat unrealistic case that nonetheless displays some of the structure of resonance. In Chapter 4 we consider a stronger - and in some sense optimal - result due to Neistadt, and we study the behavior of trajectories as they pass through resonances. As we shall see, the results of these two chapters do *not* apply to Hamiltonian systems.

We are interested here in systems of the form:

$$(1) \qquad \begin{aligned} dI/dt &= \varepsilon f(I, \varphi_1, \varphi_2, \varepsilon) \\ d\varphi_1/dt &= \omega_1(I) + \varepsilon g_1(I, \varphi_1, \varphi_2, \varepsilon) \\ d\varphi_2/dt &= \omega_2(I) + \varepsilon g_2(I, \varphi_1, \varphi_2, \varepsilon). \end{aligned}$$

Here I belongs to K, say a smooth compact subset of Euclidean $\mathbb{R}^m$. Each function is real, C^1 in I and ε, periodic with period 2π in the phase variables, and each possesses an analytic extension on $|\text{Im}\ \varphi_1| < \sigma$, $|\text{Im}\ \varphi_2| < \sigma$, $\sigma > 0$ (we write $|\text{Im}\ \varphi| < \sigma$ for short) which is continuous on the closure, the latter hypothesis being a simple means of controlling the Fourier coefficients (cf. Appendix 1). We will not need a complex extension in I in this section, and according to our conventions (see Section 1.2), we sometimes denote the extended domain as $K_{0,\sigma}$. In addition, as we remarked before, such a system may be reduced to a (resonant) single frequency system by eliminating the second frequency through a change of timescale. We will nevertheless retain the symmetric form (1) in the present chapter, saving the reduced form for Chapter 4.

The averaged system corresponding to (1) may be written:

$$(2) \qquad dJ/dt = \varepsilon f_0(J), \quad f_0(J) = (4\pi)^{-2} \oint f(J, \varphi, 0)\, d\varphi.$$

As in Section 1, we introduce K', the set of initial conditions (in J) such that the corresponding solutions of the average system remain well inside K (in $K - \eta$, $\eta > 0$) until at least time τ^*/ε ($\tau^* > 0$), and we work on this domain and its complex extension. From now on, we set $\tau^* = 1$ for notational simplicity.

As we saw in the last section, during a time of order $1/\varepsilon$, the separation between the exact and averaged trajectories may be of order 1 as the following example again shows:

$$(3) \qquad dI_1/dt = \varepsilon, \qquad dI_2/dt = 2\varepsilon \cos(\varphi_1 - \varphi_2)$$
$$d\varphi_1/dt = I_1, \qquad d\varphi_2/dt = I_2.$$

The averaged system is:

$$(4) \qquad dJ_1/dt = \varepsilon, \quad dJ_2/dt = 0.$$

As initial conditions, take $I_1(0) = I_2(0) = 1$, $\varphi_1(0) = \pi/3$, $\varphi_2(0) = 0$, situated on the resonance $\omega_1 = \omega_2$. The averaged solution is $J_1(t) = 1 + \varepsilon t$, $J_2(t) = 1$, whereas in the exact system $I_1(t) = 1 + \varepsilon t$ and $I_2(t) = 1 + \varepsilon t$ (the trajectory remains on the resonance).

As in the one frequency system, which is in fact equivalent, the large separation is due to the resonances. The simplest way of avoiding this behavior is to restrict attention to systems in which trajectories pass quickly through resonance and avoid capture. Arnold formulates this as:

<u>Condition A</u>:

One of the frequencies - say ω_2 - is assumed nonvanishing on K, and one sets

$\omega = \omega_1/\omega_2$. One then assumes that on every trajectory of the *exact* system with initial condition in $K' \times T^2$ the ratio ω of the frequencies varies strictly monotonically with time:

$$| d\omega/dt | > c_1\varepsilon.$$

This condition is incompatible with capture into resonance, defined by a rational ω, and it precludes trajectories from passing arbitrarily slowly through resonance or revisiting the same resonance several times. It is of course not satisfied in the preceding example, where $\omega = I_1/I_2$ and $d\omega/dt$ vanishes for $I_2 = 2I_1 \cos(\varphi_1-\varphi_2)$. In fact, condition A is quite restrictive, and we will eventually consider a larger class of systems satisfying Neistadt's condition N (cf. Chapter 4). Moreover, condition A cannot be satisfied by a Hamiltonian system, for which the averaged vectorfield vanishes.

For systems satisfying condition A, Arnold proves the following result:

Theorem 3:

For every initial condition in $K' \times T^2$, we have the bound:

(5) $$\sup_{t\in[0,\, \tau^*/\varepsilon]} \| I(t) - J(t) \| < c\sqrt{\varepsilon}(\log\varepsilon)^2.$$

This estimate is not optimal. The following example, inspired by the one frequency example, shows that the optimal estimate is not better than $\sqrt{\varepsilon}$ and we will obtain this result - for the average separation - in Chapter 4 . Consider the system:

(6) $$dI_1/dt = \varepsilon, \qquad dI_2/dt = \varepsilon \cos(\varphi_1-\varphi_2)$$

$$d\varphi_1/dt = I_1 + I_2, \quad d\varphi_2/dt = I_2.$$

For this system, condition A is satisfied on every compact subset of the open set $I_2 > I_1$, $I_2 > 0$. It is easily verified that for the initial condition $I_1 = 0, I_2 = 1, \varphi_1 = \varphi_2 = 0$, the separation is of order $\sqrt{\varepsilon}$.

3.6 Preliminary lemmas

We start implementing the geometric construction depicted in Appendix 3, which we shall meet again and again, by first decomposing K into resonant and nonresonant domains. For $N \in \mathbb{N}$ $(N > 1)$, we define the *nonresonant set:*

(1) $$K_N = \{ I \in K \text{ such that } \omega.k=\omega_1 k_1 + \omega_2 k_2 \neq 0 \text{ for every } k = (k_1 , k_2), |k_1| + |k_2| \leqslant N \}$$

which is the complement of the set of resonances of order less than N.

For $0 < d < 1$, each resonance k ($\omega(I) = -k_2/k_1$) is surrounded by a (closed) resonant zone $V_{d,k}$, defined by $|\omega(I) + k_2/k_1| \leq d$. We define the *resonant domain* $R_{d,N} = \cup_{|k|<N} V_{d,k}$, where $|k| = |k_1| + |k_2|$ is the order of the resonance (cf. Appendix 3). The *nonresonant domain* $K_{d,N} = K - R_{d,N}$ is the complement of $R_{d,N}$.

On $K_{d,N}$, we have:

(2) $\quad |\omega_1 k_1 + \omega_2 k_2| > cd, \quad c = \mathrm{Inf}_K |\omega_2|.$

Now consider a trajectory of the exact system with initial condition in $K' \times T^2$. We may associate a partition of $[0, 1/\varepsilon]$ to this trajectory corresponding to its successive passages from the resonant to the nonresonant domain and back. We define $r_{d,N}$ as the set of points in time for which $I(t) \in R_{d,N}$ (passage through the resonant domain), and $k_{d,N} = [0, 1/\varepsilon] - r_{d,N}$ as the set of times for which $I(t) \in K_{d,N}$ (passage through the nonresonant domain). The reader should recognize that this is the analogue, in this particular case, of the decompositions used in Anosov's theorem.

Lemma 1:

$r_{d,N}$ consists of at most N^2 segments.

A segment of $r_{d,N}$ corresponds to passage through a connected component of $R_{d,N}$. Since $\omega(I(t))$ is monotone by condition A, the trajectory may visit each resonance at most once. An arbitrary connected component of $R_{d,N}$ is the union of p resonant zones and is thus visited at most p times. The number of segments is thus less than the number of resonances of order less than N, and the number of such resonances is equal to the number of rationals in irreducible form such that the sum of the absolute values of the numerator and denominator is less than N. This number is less than N^2. ∎

Lemma 2:

Let $[\alpha, \beta]$ be a segment of $k_{d,N}$, and let x be a positive number less than $(\beta-\alpha)/2$. Then the segment $[\alpha + x, \beta - x]$ is included in $k_{d(x),N}$, where $d(x) = d + c_1\varepsilon x$.

This technical lemma follows from condition A and allows us to refine certain estimates, as we shall see. Suppose, for definiteness, that $d\omega(I(t))/dt$ is positive. We must show that for each resonance of order strictly less than N, the inequality $|\omega - k_1/k_2| > d + c_1\varepsilon x$ holds on $[\alpha + x, \beta - x]$. From condition A, we deduce that for

$y \in [\alpha + x, \beta - x]$:

(3) $\qquad \omega(\alpha) + c_1\varepsilon x \leqslant \omega(y) \leqslant \omega(\beta) - c_1\varepsilon x.$

Now $\omega(y) - k_1/k_2$ does not change sign on the nonresonant segment $[\alpha, \beta]$. Therefore, either $\omega(\alpha) - k_1/k_2 \geqslant d$, which implies $\omega(y) - k_1/k_2 \geqslant d + c_1\varepsilon x > 0$, or $\omega(\beta) - k_1/k_2 \leqslant -d$, which implies $\omega(y) - k_1/k_2 \leqslant -d - c_1\varepsilon x < 0$. In either case the lemma follows. ∎

The above lemma is simply another way of expressing the fact that by virtue of condition A resonant zones of width c are traversed in a time of order c/ε. The dependence of the trajectories of the averaged system on the initial conditions is controlled by the following lemma, where we denote the solution with initial condition $J(0) = J_0$ by $J(J_0, t)$. This solution remains in K for $t \in [0, t^*]$, t^* depending on the initial condition.

Lemma 3:

For t less than $t^*(J_0)$ and $t^*(J'_0)$, we have:

(4) $\qquad \| J(J_0, t) - J(J'_0, t) \| < \| J_0 - J'_0 \| e^{c\varepsilon t}.$

This is straightforward by Gronwall's lemma. ∎

We now undertake a change of variables in the nonresonant domain analogous to the one we used for one frequency systems.

Lemma 4:

There are constants ε_0, c_2, and c such that whenever $\varepsilon < \varepsilon_0$, $N \geqslant [c \log(1/\varepsilon)]$, and $d \geqslant c_2\sqrt{\varepsilon}$, there exists a C^1 function $S(I, \varphi_1, \varphi_2, \varepsilon)$, which is 2π periodic in its phase variables, is defined for $I \in K_{d,N}$, $|\operatorname{Im} \varphi| \leqslant \sigma/2$, and which, for $|\operatorname{Im} \varphi| \leqslant \sigma/2$ and $I \in K_{d,N} \cap (K-\eta)$ satisfies:

a) $\qquad \| S(I, \varphi_1, \varphi_2, \varepsilon) \| < c/d$

b) $\qquad \| dP/dt - \varepsilon f_0(P(t)) \| < c(\varepsilon/d)^2$, where $P = I + \varepsilon S(I, \varphi_1, \varphi_2, \varepsilon)$.

a) We may write f in the form $f = f_0(I) + \tilde{f}$ as usual. The Fourier series of f is convergent in norm for $|\operatorname{Im} \varphi| < \sigma/2$ and $I \in K_{d,N}$. S will be an *approximate* solution of the linearized equation:

(5) $\qquad \omega_1 \partial S/\partial\varphi_1 + \omega_2 \partial S/\partial\varphi_2 + \tilde{f} = 0.$

We define the truncated Fourier series f_N of $\tilde{f}$ by:

$$(6) \qquad f_N(I, \varphi_1, \varphi_2, \varepsilon) = \sum_{0<|k|<N} f_k(I, \varepsilon)e^{ik.\varphi}$$

and define S to be:

$$(7) \qquad S(I, \varphi_1, \varphi_2, \varepsilon) = i\sum_{0<|k|<N} f_k(I, \varepsilon)(\omega_1 k_1 + \omega_2 k_2)^{-1}e^{ik.\varphi}$$

where the f_k are the Fourier coefficients of f.

S is thus the solution of the equation obtained by replacing $\tilde{f}$ by f_N in (5). Using the exponential decrease of the Fourier coefficients (cf. Appendix 1), and the lower bound from Lemma 3 which excludes the small divisor problems, we easily obtain the desired bound.

b) Again using lemma 3 along with the exponential decrease of the Fourier coefficients of f we obtain the inequalities:

$$(8) \qquad \| \partial S_i/\partial I_j \| < c/d^2, \quad \| \partial S_i/\partial \varphi_j \| < c/d$$

for $I \in K_{d,N}$, $|\operatorname{Im} \varphi| < \sigma/2$. Let $R_N = \tilde{f} - f_N$ be the remainder of the Fourier series of $\tilde{f}$. For N large enough ($N \geq N_0 = [c \log(1/\varepsilon)]$), we have, using the exponential decrease of the Fourier coefficients: $\| R_N \| < c\varepsilon$ for $I \in K_{d,N}$, $|\operatorname{Im} \varphi| < \sigma/2$. We have:

$$(9) \qquad dP/dt - \varepsilon f_0(P) = \varepsilon[f_0(I)-f_0(P)] + \varepsilon[f_0(I, \varepsilon)-f_0(I)] + \varepsilon f_N(I, \varphi, \varepsilon) + \\ + \varepsilon R_N + \varepsilon^2 \partial S/\partial I.f + \varepsilon\omega(I).\partial S/\partial\varphi + \varepsilon^2[\partial S/\partial\varphi_1\, g_1 + \partial S/\partial\varphi_2\, g_2].$$

By the choice of S, the third and sixth terms combine to vanish; moreover, the sum of the other terms, with the exception of the first, is bounded in norm by $c(\varepsilon/d)^2$ for $N \geq N_0$. Now consider the first term. According to Lemma 4a, for c_2 large enough, the segment $[I, P]$ is contained in $K_{d/2,N}$ for I in $K_{d,N} \cap (K-\eta)$, which allows us to extract the bound $\| \varepsilon[f_0(I) - f_0(P)] \| \leq c(\varepsilon/d)^2$ without encountering domain problems. From these two bounds we deduce the desired inequality:

$$(10) \qquad \| dP/dt - \varepsilon f_0(P) \| < c\varepsilon^2/d^2. \quad \blacksquare$$

3.7 Proof of Arnold's theorem

We fix the values of N and d: $N = N_0$, $d = c_2\sqrt{\varepsilon}$. Then for ε_0 small enough, the resonant zones corresponding to resonances of order strictly less than N, of which there are a finite number, are pairwise disjoint. For if two resonant zones $V_{d,k}$ and $V_{d,k'}$ met, we would have, at a point I of K, $|\omega(I) + k_2/k_1| \leqslant d$ and $|\omega(I) + k'_2/k'_1| \leqslant d$, whence $|k_2/k_1 - k'_2/k'_1| \leqslant 2c_2\sqrt{\varepsilon}$. But $|k_2/k_1 - k'_2/k'_1| > 1/N^2 > 1/(c \log^2\varepsilon)$ contradicts the latter inequality for sufficiently small ε_0.

This implies that the length of a resonant segment of $r_{d,N}$ is less than $c/\sqrt{\varepsilon}$ since a segment corresponds to passage through a single resonant zone defined by $|\omega(I) + k_2/k_1| \leqslant 2c_2\sqrt{\varepsilon}$ and, according to condition A, $|d\omega(I(t))/dt| \geqslant c_1\varepsilon$.

Let $[t_1, t_2]$ be a segment of $r_{d,N}$ corresponding to passage through a resonance. It is easy to estimate the variation of the exact solution $I(t)$ on this time interval. Since $dI/dt = \varepsilon f(I, \varphi, \varepsilon)$ and $|t_2 - t_1| < c/\sqrt{\varepsilon}$, it follows that on the domain $I \in K$, $|\mathrm{Im}\varphi| < \sigma/2$, we have:

(1) $\quad \| I(t_2) - I(t_1) \| < c\| f \| \sqrt{\varepsilon} < c\sqrt{\varepsilon}.$

In the same way, for an averaged trajectory we have:

(2) $\quad | J(t_2) - J(t_1) \| < c\sqrt{\varepsilon}.$

The variation of a solution is thus small simply because the passage through resonance takes place quickly.

We now consider a segment $[t_1, t_2]$ of $k_{d,N}$ corresponding to passage through a nonresonant domain and we estimate the separation on the time interval $[t_1, t_2]$ between an exact trajectory $I(t)$ and an averaged trajectory $J(t)$ with the same initial condition at t_1. To do this, we of course make use of the change of variables for the nonresonant zone and we set $P(t) = P(I(t), \varphi, \varepsilon)$. We have:

(3) $\quad \| I(t) - J(t) \| \leqslant \| I(t) - P(t) \| + \| P(t) - J(t) \| \leqslant \varepsilon \| S \| + \| P(t) - J(t) \|.$

The first term is then bounded by $c\sqrt{\varepsilon}$ by Lemma 4a. To estimate the second term we use Gronwall's lemma which yields:

(4) $\quad \| P(t) - J(t) \| \leqslant e^{c\varepsilon(t_2 - t_1)} [\| P(t_1) - J(t_1) \| + \int_{t_1}^{t_2} b(u)\, du]$

where $b(t) = \| dP/dt - \varepsilon f_0(P) \|$. From Lemma 4a, we have $b(u) \leqslant c\varepsilon^2/d^2 \leqslant c\varepsilon$. Since, a priori, $t_2 - t_1$ may take the value $1/\varepsilon$ for certain initial conditions, this implies:

(5) $\quad \| P(t) - J(t) \| \leqslant c[\varepsilon \| S \| + c]$

which is too rough an estimate, as it leads to the bound on the separation in the nonresonant domain:

(6) $\quad \| I(t) - J(t) \| \leqslant c\varepsilon \| S \| + c$

which is of order 1.

To obtain a more refined estimate, we use Lemma 2. Let $x \leqslant (t_2 - t_1)/2$. For

$t \in [t_1 + x, t_2 - x]$, $I(t)$ belongs to $K_{d(x),N} \cap (K-\eta)$, where $d(x) = d + c_1 \varepsilon x$ by Lemma 2. Consequently, from Lemma 4b we have $b(t) \leqslant c\varepsilon^2/d(x)^2$. For $t \in [t_1\,,\,(t_1 + t_2)/2]$, we may choose $x = t - t_1$, which yields:

(7) $$b(t) \leqslant c\varepsilon^2/(d + c_1\varepsilon[t-t_1])^2.$$

In the same way, for t in $[(t_1 + t_2)/2,\, t_2]$, by taking $x = t_2 - t$ we obtain:

(8) $$b(t) \leqslant c\varepsilon^2/(d + c_1\varepsilon[t-t_1])^2$$

Using these bounds and the Gronwall lemma gives:

(9) $$\| P(t) - J(t) \| \leqslant c\varepsilon\| S \| + c\int_{t_1}^{(t_1+t_2)/2} c\varepsilon^2/(d + c_1\varepsilon[t-t_1])^2\, dt + $$
$$+ c\int_{(t_1+t_2)/2}^{t_2} c\varepsilon^2/(d + c_1\varepsilon[t_2-t])^2\, dt.$$

And, using Lemma 4a, we have:

(10) $$\| P(t) - J(t) \| \leqslant c\sqrt{\varepsilon}.$$

For the separation between the exact and averaged trajectories, we thus have:

(11) $$\| I(t) - J(t) \| \leqslant c\sqrt{\varepsilon}$$

which is a considerable improvement over the gross estimate (6), which did not take into account the fact that, by virtue of condition A, the exact trajectory quickly leaves the resonances.

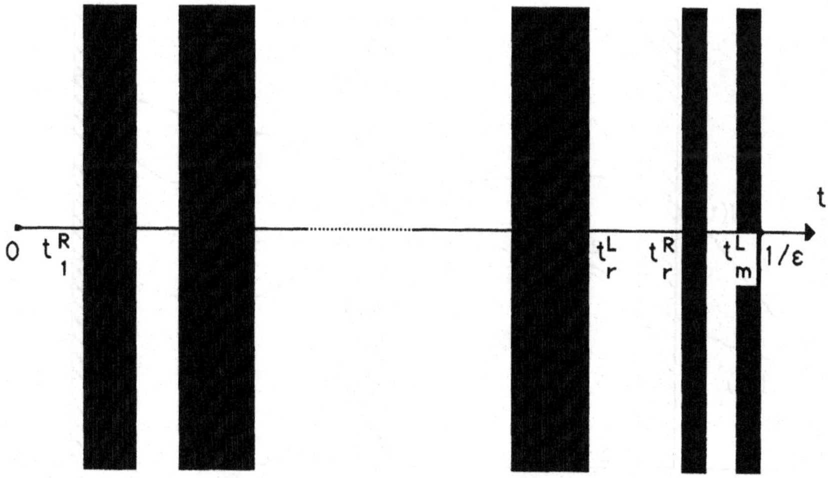

Figure 3.2: Schematic representation of the set $r_{d,N}$ (passages through resonances, depicted as black blocks) and its complement $k_{d,N}$ (passages trough the nonresonant domain). Segments of $k_{d,N}$ are denoted by $[t_r^R,\, t_r^L]$.

We now return to the structure of the time interval $[0, 1/\varepsilon]$ for a trajectory with initial condition in $K' \times T^2$. To simplify notation, we assume that 0 belongs to $k_{d,N}$ and $1/\varepsilon$ belongs to $r_{d,N}$. $k_{d,N}$ is composed of m segments denoted by $[t_r^R, t_r^L]$ (L for left, R for right) corresponding to successive passages through the nonresonant domain $K_{d,N}$. $r_{d,N}$ also comprises m segments corresponding to successive passages through (different) resonances (see Figure 3.2 above).

We set $I(t_r^L) = I_r^L$ and $I(t_r^R) = I_r^R$, and we introduce the m functions $J_r(t) = J(I_r^L, t_r^L, t)$, which are solutions of the averaged system with initial condition $J_r(t_r^L) = I_r^L$ (see Figure 3.3). In this section we assume that for all r, $J_r(t)$ remains in K for $t \in [t_r^L, 1/\varepsilon]$.

On $[t_r^L, t_{r+1}^L]$ we have the inequality:

(12) $\| I(t)-J(t) \| \leqslant \| I(t)-J_r(t) \| + \| J_r(t)-J_{r-1}(t) \| + \dots + \| J_2(t)-J_1(t) \|$

and on $[t_m^L, 1/\varepsilon]$ we have:

(13) $\| I(t)-J(t) \| \leqslant \| I(t)-J_m(t) \| + \dots + \| J_2(t)-J_1(t) \|$.

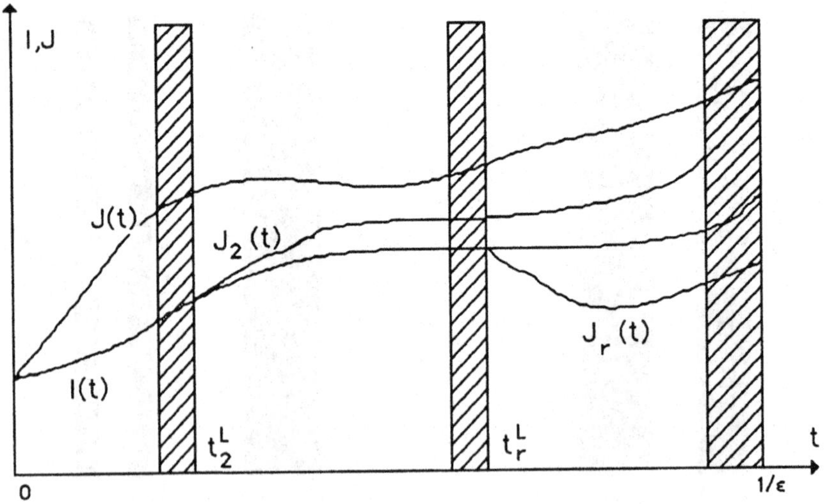

Figure 3.3: To compare I(t) and J(t) over the time interval one introduces the auxiliary functions $J_r(t)$ with initial condition $J_r(t_r^L) = I_r^L$.

These inequalities allow for an estimate of the separation between the exact and averaged trajectories on the time interval $[0, 1/\varepsilon]$. On $[t_{r+1}^{L}, 1/\varepsilon]$, we have, by Lemma 3:

$$(14)\qquad \|J_{r+1}(t) - J_r(t)\| \leqslant c\|J_{r+1}(t_{r+1}^{L}) - J_r(t_{r+1}^{L})\| \leqslant c[\|I_{r+1}^{L} - I_r^{R}\| + \|I_r^{R} - J_r(t_r^{R})\| + \|J_r(t_r^{R}) - J_r(t_{r+1}^{L})\|].$$

The first term in this decomposition represents the variation of $I(t)$ between the extremities of the resonant segment $[t_r^{R}, t_{r+1}^{L}]$, the third represents the variation of $J_r(t)$ on the same segment, and the second term arises from the separation, at the extremity t_r^{R} of the nonresonant segment $[t_r^{L}, t_r^{R}]$, between the exact solution $I(t)$ and the averaged solution which coincides with it at t_r^{L} (see Figure 3.4).

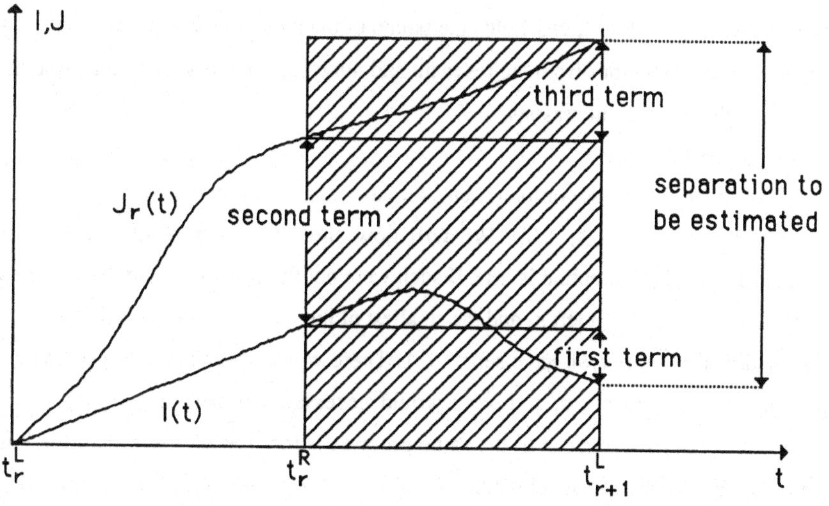

Figure 3.4: The separation between $I(t)$ and $J_r(t)$ at the end of the resonant segment $[t_r^{R}, t_{r+1}^{L}]$ is the sum of three terms as indicated.

We will apply the results of the previous subsection in order to bound this last term, whereas the other two will be bounded with the help of estimates (1) and (2). Proceeding in this way, on $[t_r^{L}, 1/\varepsilon]$ we obtain:

$$(15)\qquad \|J_{r+1}(t) - J_r(t)\| \leqslant c\sqrt{\varepsilon}.$$

Consequently, on $[t_r^L, t_{r+1}^L]$, we have:

(16) $$\| I(t) - J(t) \| \leqslant \| I(t) - J_r(t) \| + rc\sqrt{\varepsilon}.$$

It remains to bound $\| I(t) - J_r(t) \|$ on $[t_r^L, t_{r+1}^L]$. On the nonresonant segment $[t_r^L, t_r^R]$, this term is less than $c\sqrt{\varepsilon}$, by Lemma 4. On the resonant segment $[t_r^R, t_{r+1}^L]$, we have:

(17) $$\| I(t) - J_r(t) \| \leqslant \| I(t_r^R) - J_r(t_r^R) \| + c\varepsilon(t_{r+1}^L - t_r^R).$$

Since the length of a resonant segment is not greater than $c/\sqrt{\varepsilon}$:

(18) $$\| I(t) - J_r(t) \| \leqslant \| I(t_r^R) - J_r(t_r^R) \| + c\sqrt{\varepsilon}$$

and, using the estimate on the nonresonant segment $[t_r^L, t_r^R]$:

(19) $$\| I(t) - J_r(t) \| \leqslant c\sqrt{\varepsilon}.$$

On $[0, 1/\varepsilon]$, we therefore obtain the bound:

(20) $$\| I(t) - J(t) \| \leqslant \sqrt{\varepsilon}(c + mc).$$

Because $m \leqslant N^2 \leqslant c \log^2\varepsilon$, this yields the bound of the theorem for small enough ε_0. We note that the contributions from the resonant and nonresonant zones are of the same order $\sqrt{\varepsilon}$, although for different reasons.

The proof is almost complete, but we have assumed thus far that for all r, $J_r(t)$ remains in K for $t \in [t_r^L, 1/\varepsilon]$. We will now show by contradiction (a bootstrap argument) that this assumption is true. If such were not the case, we could define for each J_r the time t_r^* at which it left K, and the smallest of these times would be strictly less than $1/\varepsilon$, or rather τ^*/ε, but we have set $\tau^* = 1$. There would then exist a largest integer s such that $t_r^L < t^*$ for $r \leqslant s$, and $J_r(t)$ would necessarily remain in K on $[t_r^L, t^*]$ for $r \leqslant s$. We could then proceed to the same proof as before for all $r \leqslant s$ with the initial condition $I(t_r^L) = I_r^L$. We would obtain $\| I(t) - J_r(t) \| \leqslant c\sqrt{\varepsilon} \log^2\varepsilon$ for $t \in [t_r^L, t^*]$. For small enough ε_0, this would entail $c\sqrt{\varepsilon} \log^2\varepsilon < \eta$, so that $J_r(t)$, for all $r \leqslant s$, would remain in K until a time t' strictly greater than t^*, contradicting the definition of t^*. Therefore $t^* \geqslant 1/\varepsilon$. ∎

CHAPTER 4: TWO FREQUENCY SYSTEMS; NEISTADT'S RESULTS

4.1 Outline of the problem and results

In this chapter, following [Nei2], we give the most detailed and concrete account of the mechanism of passage through resonance, which may be described, for two frequency systems, by a nonlinear pendulum model (Section 4.3). By making use of more abstract techniques, more general results are obtained in the two subsequent chapters, but these results are not as strong when applied to the particular case of two frequency systems and give less insight into the details of the phenomena.

In the present chapter we will be interested in systems of the form:

$$
\begin{aligned}
&dI/ds = \varepsilon f_I(I, \varphi, \lambda, \varepsilon)\\
(1)\qquad &d\varphi/ds = \omega_\varphi(I) + \varepsilon f_\varphi(I, \varphi, \lambda, \varepsilon)\\
&d\lambda/ds = \omega_\lambda(I) + \varepsilon f_\lambda(I, \varphi, \lambda, \varepsilon)
\end{aligned}
$$

which are defined for $I \in K$, a regular compact subset of $\mathbb{R}^m$. We assume that all functions on the right hand side have period 2π in φ and λ, are analytic for $(I, \varphi, \lambda) \in K_{0,\sigma}$ $(\sigma > 0)$ and are C^1 in ε. Away from equilibrium points of the unperturbed system, one of the frequencies, say ω_λ, is nonvanishing. Since for small ε, λ is monotone on every trajectory, we may take λ to be the time by introducing:

$$(2)\qquad t = \int_0^s [\omega_\lambda(I(u)) + \varepsilon f_\lambda(I(u), \varphi(u), \lambda(u), \varepsilon)]\, du.$$

This new time, which is a "fast" time of the same order as s, depends on the trajectory as well as on ε, but this is not troublesome in the framework of a first order perturbation theory such as averaging. The system then takes the standard form , which could have been used in the preceding chapter:

$$(3)\qquad dI/dt = \varepsilon f(I, \varphi, \lambda, \varepsilon),\quad d\varphi/dt = \omega(I) + \varepsilon g(I , \varphi, \lambda, \varepsilon),\quad d\lambda/dt = 1.$$

We are thus dealing with a two frequency system in which one of the frequencies is unity and one of the phases is simply the time variable; we nonetheless retain the notation λ. The resonances are the points where ω takes rational values and it is very important to notice that this implies that *resonant surfaces in two frequency systems* do not intersect: They are the level surfaces of a function.

The associated averaging system is again expressed as:

$$(4) \qquad dJ/dt = \varepsilon f_0(J), \quad f_0(J) = (2\pi)^{-2} \oint f(I, \varphi, \lambda, 0)\, d\varphi d\lambda.$$

We now define the condition N, which will replace the condition A of the preceding chapter. For this purpose we set: $L(J) = f_0(J).\partial\omega/\partial J$ (scalar product in $\mathbb{R}^m$), and we express Neistadt's condition as

Condition N:

The function $L(J)$ does not vanish for $J \in K$, so that $|L(J)| > c_1 > 0$.

Troughout the chapter, we shall assume this condition is satisfied. It is fundamental, as we shall see throughout this exposition and in the proof of optimality of the result for n frequencies (cf. Chapter 6). It entails the absence of singular points in ω on the domain K, and forces each trajectory of the *averaged* system to cross the resonances transversly, with finite speed, and in such a way that a given resonance is visited at most once.

With respect to condition A, this condition has the dual advantage of being less restrictive (condition A entails condition N, as is seen by integrating the scalar product $f(I, \varphi, 0).\partial\omega/\partial I$ over the phases), and of dealing with the averaged system, making it simpler to test for a given system. For Hamiltonian systems though, the averaged vectorfield is zero and neither condition is satisfied.

As before, for $(I, \varphi, \lambda) \in K_{0,\sigma}$ we expand $f(I, \varphi, \lambda, 0)$ in its Fourier series, grouping together terms associated to the same resonant linear combination of frequencies:

$$(5) \qquad f(I, \varphi, \lambda, 0) = f_0(I) + \textstyle\sum_s f_s(I, s_1\varphi + s_2\lambda).$$

Here $s = (s_1, s_2)$ and the sum is limited to s_1 and s_2 relatively prime and to $s_1 > 0$. The *order* of the resonance is by definition equal to $s_1 + |s_2|$; f_s may thus be expressed as:

$$(6) \qquad f_s(I, \gamma) = \textstyle\sum_{r\neq 0} f_{rs}(I)e^{ir\gamma}$$

and has period 2π in γ. Because of the analyticity of $f(I, \varphi, \lambda, 0)$, its Fourier coefficients decrease exponentially fast (cf. Appendix 1), with decrement σ equal to the half-width of the strip of analyticity:

$$(7) \qquad \sup_{I\in K} \| f_{p,q}(I) \| \leq c(\sup_{K_{0,\sigma}} \| f(I, \varphi, \lambda) \|)e^{-\sigma(|p|+|q|)}.$$

Thus the remainder of the Fourier series:

(8) $$R_N(I, \varphi, \lambda) = \sum_{|p|+q>N} f_{p,q}(I)e^{i(p\varphi+q\lambda)}$$

is less than ε in norm for $N \geqslant N(\varepsilon)$, where $N(\varepsilon) = [c_2 \log(1/\varepsilon)]$.

In an analogous way, for $g(I, \varphi, \lambda, 0)$ we define $g_0(I)$, $g_s(I, \varphi, \lambda)$, and the remainder $S_N(I, \varphi, \lambda)$. Finally, for fixed s we define $\Phi_s(I, \gamma) = f_s(I, \gamma).\partial\omega/\partial I$.

We are also obliged to assume minor hypotheses connected with the possible degeneracy of certain functions. For this purpose, we introduce the set Θ_s' composed of points I in K associated to the resonance $\omega(I) = -s_2/s_1$ and such that:

(9) $$L(I) + \Phi_s(I, \gamma) = 0 \quad \text{and} \quad \partial\Phi_s(I, \gamma)/\partial\gamma = 0$$

for some γ with real part in $[0, 2\pi]$ and with imaginary part less than σ in absolute value. These points correspond to degeneracies of hyperbolic fixed-points of "pendulums" as we shall see later. Generically, Θ_s' is a finite union of analytic manifolds of dimension $m - 2$, since it is the projection onto $\mathbb{C}^m$ of an analytic manifold in the global space, defined by annihilation of the three functions $\Phi_s(I, \gamma) + L(I)$, $\partial\Phi_s(I, \gamma)/\partial\gamma$ and $\omega(I) + s_2/s_1$. In nongeneric cases, the dimension may of course equal $m - 1$ or m.

We define $\Theta_s = \Theta'_s \cap \mathbb{R}^m$. For $s = s_1 + |s_2| > N'$ (depending only on K and σ), Θ_s is empty. In fact, since f is analytic, we have:

(10) $$\| \Phi_s \| \leqslant c \| f \|.\|\partial\omega/\partial I\| e^{-\sigma|s|}.$$

For, $|s| \geqslant N'$, N' large enough, we have $\| \Phi_s \| \leqslant c_1/2$, and by condition N, it follows that

$$| L(I)+\Phi_s(I, \gamma) | > c_1/2.$$

Finally, we define $\Theta = \cup_{|s|<N'} \Theta_s$.

We point out that if the system satisfies condition A, it has no degeneracies. To see this, assume that $f(I, \varphi, \lambda, 0).\partial\omega/\partial I > c > 0$, and consider the resonance $s = (s_1, s_2)$. Since s_1 and s_2 are relatively prime, there exist m_1 and m_2 such that $m_1 s_1 + m_2 s_2 = 1$. We may define the second phase variable $\gamma' = -m_2\varphi + m_1\lambda$ and make use of the invertible coordinate change $(\varphi, \lambda) \to (\gamma, \gamma')$ on the torus T^2 (this coordinate change is in $SL_2(\mathbb{Z})$). Taking the average of $f.\partial\omega/\partial I$ in γ', with γ fixed, we find:

(11) $$(f_0 + f_s).\partial\omega/\partial I > c.$$

For ε sufficiently small it follows that:

(12) $$(f_0 + f_{s,N}).\partial\omega/\partial I > c$$

and:

$$(13) \qquad L(I) + \Phi_s(y) > c$$

which excludes the possibility of a degeneracy.

Before stating the precise results with which this chapter is concerned, we introduce some simplifying notation for certain subsets of the base.

- K_1 is a nonempty compact subset of $\mathbb{R}^m$ contained in the interior of $K - \Theta$. In the particular case $m = 1$, one can prove an averaging theorem (Theorem 2) without this domain restriction, even in nongeneric cases where Θ is nonempty, in which case $K_1 = K$.
- K_2 is the set of initial conditions of the averaged system such that the corresponding trajectories remain in $K_1 - \eta$ for $t \in (0, \tau^*/\varepsilon)$, $\tau^* > 0$. As usual, we set $\tau^* = 1$, which amounts to a scaling on ε.
- V is the set $K_2 \times T^2$.

In the general case, we prove the following theorem:

Theorem 1:

For ε sufficiently small, there exists a partition of V, $V = V' \cup V''$ such that:

a) $\operatorname{Sup}_{t \in [0,1/\varepsilon]} \| I(t) - J(t) \| < c\sqrt{\varepsilon}\log(1/\varepsilon)$ for $\{I(0), \varphi(0), \lambda(0)\}$ in V' that is, the averaged system is a good approximation to the exact system for initial conditions in V'.

b) $\mu(V'') < c\sqrt{\varepsilon}$ where μ is the ordinary measure on $\mathbb{R}^m \times T^2$ that is, most trajectories are well-approximated.

c) $\operatorname{Sup}_{t \in [0,\ 1/\varepsilon]} 1/\mu(V).\int_V \| I(t) - J(t) \| \, dI_0 d\varphi_0 d\lambda_0 < c\sqrt{\varepsilon}$ $((I_0, \varphi_0, \lambda_0) = (I(0), \varphi(0), \lambda(0)))$, that is, the average separation $M(\varepsilon)$ obtained by integrating over the initial conditions is at most of order $\sqrt{\varepsilon}$.

d) These estimates are optimal.

In the particular case where the base is one-dimensional and degeneracies exist, we establish the following weaker result:

Theorem 2:

For ε sufficiently small, there exists a partition of V, $V = V' \cup V''$ such that:

a) $\operatorname{Sup}_{t \in [0,\ 1/\varepsilon]} | I(t) - J(t) | < c\varepsilon^{1/4}$ (note the exponent change).

b) $\mu(V'') < c\sqrt{\varepsilon}$.

c) $M(\varepsilon) < c\sqrt{\varepsilon}$.

d) These estimates are optimal.

In the particular case where the *exact* system traverses the resonances with no trouble, i.e., the system satisfies condition A of Chapter 3, we prove the following result, which improves Arnold's estimate $\sqrt{\varepsilon}\,(\log\varepsilon)^2$ without recourse to further domain restrictions:

Theorem 3:

For a system satisfying condition A (any base dimension) and for ε sufficiently small:

a) $\mathrm{Sup}_{t\in[0,\,1/\varepsilon]} \| I(t) - J(t) \| < c\sqrt{\varepsilon}$ for every initial condition in V.

b) This estimate is optimal.

The proof of Theorem 1 takes up most of the remainder of this chapter, and the other two theorems follow easily from it. We add considerable detail to Neistadt's note [Nei2] in our scrutiny of the resonance phenomena.

4.2 Decomposition of the domain and resonant normal forms

As in the preceding chapter, for each resonance (p, q) of order less than the previously introduced $N = N(\varepsilon)$ ($|p| + |q| < N$), we define a resonant zone $V(-q, p, \delta)$ of width $O(\sqrt{\varepsilon})$ as the set of points in K such that $|\omega(I) - p/q| \leqslant c_3\delta(p, q, \varepsilon)$. δ is of the form $\delta(p, q, \varepsilon) = \sqrt{\varepsilon}h(p, q)$, where h is a positive cutoff function which decreases with increasing order of the resonance. The explicit choice:

$$(1) \qquad \delta(p, q, \varepsilon) = (\varepsilon/|q|)^{1/2} e^{-\sigma(|p|+|q|)/2}$$

will be justified later; with this definition, distinct resonant zones are necessarily disjoint for small enough ε, the proof being easy and identical to the one in Chapter 3.

The nonresonant zone is the complement of the union of the resonant zones of order less than N. On this domain, one may proceed with the usual change of variables (cf. Chapter 3) $P = I + \varepsilon S(I, \varphi, \lambda)$ in such a way that P satisfies the averaged equation to lowest order. S is again the solution to the approximated linearized equation which here writes $\omega.\partial S/\partial\varphi + \partial S/\partial\lambda = f_N$, where f_N is the truncated Fourier series of order N for the oscillating part of $f(I, \varphi, \lambda, 0)$. In the region of the nonresonant domain lying between two resonances $\omega = p'/q'$ and $\omega = p''/q''$, we have the estimates:

$$(2) \qquad \| S \| \leq c(1/\delta' + 1/\delta'')/c_3$$
$$\| dP/dt - \varepsilon f_0(P) \| \leq c\varepsilon^2(1/\delta'^2 + 1/\delta''^2)/c_3^{\,2}$$

where $c_3\delta'$ and $c_3\delta''$ are the half-widths of the resonant zones associated with the resonances under consideration.

To study the behavior of the system in a resonant zone $V(s, \delta)$ associated to the resonance $s = (-p, q)$, one reduces it to an appropriate first order *resonant* normal form by constructing a diffeomorphism $(I, \varphi, \lambda) \to (I^*, \varphi^*, \lambda)$. This diffeomorphism is defined on the product of the torus T^2 and a subset of $K_1 - \eta$ containing the resonant zone, namely the set of points satisfying $| \omega(I) - p/q | < 3c_3\delta(p, q, \varepsilon)$. This change of variables allows for elimination of the nonresonant oscillating terms through first order.

The remainder of this section is devoted to the construction of this variable change (which is reminiscent of what we did in Section 3.2), and to establishing estimates we will use in the sequel.

Though this is not entirely necessary, we will restrict ourselves to eliminating only the low-order Fourier components (of order less than N). In this way, the diffeomorphism will be defined by 2π-periodic trigonometric polynomials in φ^* and λ, with coefficients depending on I^*, v and w:

$$(3) \qquad I = I^* + \varepsilon v(I^*, \varphi^*, \lambda), \qquad \varphi = \varphi^* + \varepsilon w(I^*, \varphi^*, \lambda)$$

in such a way that:

$$(4) \qquad dI^*/dt = \varepsilon f_0(I^*) + \varepsilon f_{s,N}(I^*, \gamma^*) + o(\varepsilon)$$
$$d\varphi^*/dt = \omega(I^*) + \varepsilon g_0(I^*) + \varepsilon g_{s,N}(I^*, \gamma^*) + o(\varepsilon)$$

where γ^* is the resonant combination of phases $\gamma^* = p\lambda - q\varphi^*$. We note that here, as in Section 3.2, the *inverse* variable change $(I^*, \varphi^*) \to (I, \varphi)$ has been explicitly defined. Assuming invertibility, we have the following formal expressions, with implicit summation over k:

$$(5) \qquad dI_j/dt = \varepsilon f_j(I, \varphi, \lambda, \varepsilon)$$
$$= dI^*_j/dt + \varepsilon[dI^*_k/dt][\partial v_j/\partial I^*_k] + \varepsilon[d\varphi^*/dt][\partial v_j/\partial \varphi^*] + $$
$$+ \varepsilon \partial v_j/\partial \lambda$$
$$d\varphi/dt = \omega(I) + \varepsilon g(I, \varphi, \lambda, \varepsilon)$$
$$= d\varphi^*/dt + \varepsilon[\partial w/\partial I^*_k][dI^*_k/dt] + \varepsilon \partial w/\partial \varphi^* + \varepsilon \partial w/\partial \lambda.$$

To eliminate the nonresonant terms through first order, we require that v and w be solutions of the linearized equations:

(6) $$\omega(I^*).\partial v_j/\partial\varphi^* + \partial v_j/\partial\lambda = f^*_{s,N}(I^*, \varphi^*, \lambda)$$
$$\omega(I^*).\partial w/\partial\varphi^* + \partial w/\partial\lambda = g^*_{s,N}(I^*, \varphi^*, \lambda) + \sum_k v_k.\partial\omega(I^*)/\partial I^*_k$$

where $f^*_{s,N}$ and $g^*_{s,N}$ denote the truncated nonresonant oscillating parts of order N of f and g:

(7) $$f^*_{s,N} = \sum_{s'\neq s} f_{s',N}, \quad g^*_{s,N} = \sum_{s'\neq s} g_{s',N}.$$

We study the solubility of the first equation in (6) by considering the equation $\omega(I').\partial v/\partial\psi + \partial v/\partial\lambda = f^*_{s,N}(I', \psi, \lambda)$ on the subset of $K_1 \times T^2$ defined by

$$|\,\omega(I') - p/q\,| < 4c_3\delta(p, q, \varepsilon).$$

By writing v as its Fourier series:

(8) $$v = \sum v_{rs'}(I')\exp\{ir(s_1\psi+s_2\lambda)\}$$

where the sum is restricted to r and s satisfying $s' \neq s$ and $0 < |\,r\,|.|\,s'\,| \leqslant N$, we see that the equation is soluble and:

(9) $$v_{rs} = -if_{rs}[r(s_1\omega + s_2)]^{-1}.$$

$s_1\omega(I') + s_2$ does not vanish on the domain considered, because:

(10) $$|\,s_1\omega + s_2\,| > |\,\omega + s_2/s_1\,| > |\,(\omega-p/q) + (p/q+s_2/s_1)\,|.$$

Since $|\,p/q + s_2/s_1\,| > (|\,q\,|N)^{-1}$ and $|\,\omega - p/q\,| < 4c_3\delta$:

(11) $$|\,s_1\omega + s_2\,| > (|\,q\,|N)^{-1} - 4c_3\sqrt{\varepsilon} > (2c_2|\,q\,|\log(1/\varepsilon))^{-1}$$

for sufficiently small ε ($\varepsilon < \varepsilon_0$, where ε_0 does not depend on q), since we consider only resonances of order less than the ultraviolet cutoff ($|\,q\,| < N \leqslant c_2\log(1/\varepsilon)$).

We note that to prove the solubility of the equation, we could be content with the immediate lower bound on the small divisors $|\,s_1\omega + s_2\,| > 3c_3\sqrt{\varepsilon}$, which follows from the nonintersection of resonant zones for small enough ε. However, the refined estimates we use allow us to establish the bounds of Proposition 2.

We next consider the second equation of (6) on the same domain as before:

(12) $$\omega(I').\partial w/\partial\psi + \partial w/\partial\lambda = g^*_{s,N}(I', \psi, \lambda) + \sum_k v_k\,.\,\partial\omega(I')/\partial I'_k\,.$$

This equation is soluble *because* the solution v of the first equation has no resonant Fourier component and:

(13) $$w(I', \varphi, \lambda) = \sum_{r,s'} w_{rs'}(I')\exp\{ir(s_1\psi + s_2\lambda)\}$$

where the summation is restricted as in the definition of v and:

(14) $$w_{rs'} = -i(g_{rs'} + v_{rs'}\,.\,\partial\omega/\partial I')/r(s_1\omega + s_2)$$

We are now required to investigate the invertibility of the map:

(15) $I' \to I = I' + \varepsilon v(I', \psi, \lambda), \quad \psi \to \varphi = \psi + \varepsilon w(I', \psi, \lambda).$

Proposition 1:

For sufficiently small ε, the image of the map defined by equations (15) contains the points of $K_1 - \eta$ satisfying $|\omega(I) - p/q| < 3c_3\delta$.

The proof hinges on estimating the norms and derivatives of v and w on the domain $(\varphi, \lambda) \in T^2$, $I' \in K_1$, $|\omega(I') - p/q| < 4c_3\delta$. For the norm of v, we have:

(16) $\| v \| < \sum_{r,s'} \| f_{rs'} \| / \mathrm{Inf}_{I'} | s_1\omega(I') + s_2 |$

where the summation is restricted to the same terms as before. Using (11), along with the exponential decrease of the Fourier coefficients of f, we deduce:

(17) $\| v \| < c|q| \log(1/\varepsilon).$

Similarly, we obtain the bounds:

(18) $\| \partial v/\partial\psi \| < c|q| \log(1/\varepsilon), \quad \| \partial v/\partial\lambda \| < c|q| \log(1/\varepsilon).$

As for the partial derivatives of v with respect to the components of I', we find:

(19) $\partial v/\partial I'_j = \sum_{r,s'} [\partial f_{rs'}/\partial I'_j - s_1 \partial\omega/\partial I'_j \,.\, f_{rs'} (s_1\omega + s_2)^{-1}] \times$
$\times\, [ir(s_1\omega + s_2)]^{-1}.\exp\{ir(s_1\psi + s_2\lambda)\}$

where the summation is again restricted as in (8). We deduce as before that:

(20) $\| \partial v/\partial I'_j \| < cq^2 \log^2(\varepsilon).$

Turning now to w, we have:

(21) $\| w \| < \sum_{r,s'} (| g_{rs'} | + | v_{k,rs'} |.| \partial\omega/\partial I'_k |)/(|r|.| s_1\omega + s_2 |)$

where the summation over k is implicit and v_k is the kth component of v. Using the explicit form of $v_{k,rs'}$ and, as before, the lower bound (11) together with the analyticity of f, we obtain:

(22) $\| w \| < cq^2 \log^2(\varepsilon).$

In a similar way, we obtain the estimates:

(23) $\| \partial w/\partial\lambda \| < cq^2 \log^2(\varepsilon), \quad \| \partial w/\partial\psi \| < cq^2 \log^2(\varepsilon),$
$\| \partial w/\partial I_k \| < c|q|^3 \log^3(1/\varepsilon).$

From these various bounds, valid for $\varepsilon < \varepsilon_0$, where ε_0 is independent of q, the proposition follows easily and permits the change of variables $(I, \varphi, \lambda) \to (I^*, \varphi^*, \lambda)$ to be defined ▪

We now construct the equations in the variables $(I^*, \varphi^*, \lambda)$ describing motion in a resonant zone.

Proposition 2:

Consider the resonant combination of phase variables $\gamma^* = q\varphi^* - p\lambda$. For $\varepsilon < \varepsilon_0$, the motion in the resonant zone defined by $I \in K_1 - \eta$, $|\omega(I)-p/q| < 3c_3\delta(p, q, \varepsilon)$ satisfies the relations:

a)

$$(24) \qquad d^2\gamma^*/dt^2 = q\varepsilon[\Phi_s(I^*, \gamma^*) + L(I^*) + \sqrt{\varepsilon}\alpha(I^*, \varphi^*, \lambda, \varepsilon)].$$

b) $\qquad \| dI^*/dt \| < c\varepsilon$

where the scalar function α satisfies:

c) $\qquad \| \alpha \| < c|q|$

We have the additional estimates:

d) $\qquad |\omega(I^*) - p/q| < c\delta(p, q, \varepsilon)$.

e) $\qquad |d\gamma^*/dt - q[\omega(I)-p/q]| < cq^2\varepsilon \log(1/\varepsilon)$.

or: $\qquad |d\varphi^*/dt| < cq^2\varepsilon \log(1/\varepsilon)$.

The upshot of this proposition is to establish that the resonance phenomenon, in the adapted system of coordinates that has just been constructed, is governed by the generalized pendulum equation, to be studied in the next section. The reader may well prefer to skip the laborious proof on a first reading.

We begin by establishing the intermediate inequality:

$$(25) \qquad |d\varphi^*/dt - \omega(I)| < cq^2\varepsilon \log^2(\varepsilon).$$

From the definition of (I^*, φ^*):

$$(26) \qquad d\varphi/dt = (1+\varepsilon\partial w/\partial\varphi^*)d\varphi^*/dt + \varepsilon\partial w/\partial\lambda + \varepsilon\partial w/\partial I^*_k \cdot dI^*_k/dt$$

and

$$(27) \qquad dI_j/dt = (\delta_j^k+\varepsilon\partial v_j/\partial I^*_k)dI^*_k/dt + \varepsilon\partial v_j/\partial\varphi^*.d\varphi^*/dt + \varepsilon\partial v_j/\partial\lambda .$$

From this we deduce:

$$(28) \qquad d\varphi^*/dt = (\omega+\varepsilon g)[1+\varepsilon\partial w/\partial\varphi^* - \varepsilon^2\partial w/\partial I^*_j(\delta_j^k+\varepsilon\partial v_j/\partial I^*_k)\partial v_k/\partial\varphi^*]^{-1}$$
$$- [1+\varepsilon\partial w/\partial\varphi^* - \varepsilon^2\partial w/\partial I^*_j(\delta_j^k+\varepsilon\partial v_j/\partial I^*_k)\partial v_k/\partial\varphi^*] \times$$
$$\times [\varepsilon\partial w/\partial\lambda + \varepsilon^3\partial w/\partial I^*_j(\delta_j^k+\varepsilon\partial v_j/\partial I^*_k)(f_k - \partial v_k/\partial\lambda)]$$

where f_k is the kth component of f (*not* its Fourier coefficient) and the summations on j and k are implicit. Using estimates (17) through (23), we deduce (25) for fixed q and for ε small enough.

Note also that:

(29) $| \omega(I) - \omega(I^*) | < c| q |\epsilon \log(1/\epsilon),$

an inequality which follows from the definition of I^* and from the bound (16) and will be useful in the sequel.

We are now in position to establish inequality b) of Proposition 2. By (27):

(30) $dI^*_j/dt = (\delta_j^k + \epsilon \partial v_j/\partial I^*_k)^{-1} [dI_k/dt - \epsilon(\partial v_k/\partial \phi^*)(d\phi^*/dt) -$
                     $- \epsilon \partial v_k/\partial \lambda].$

The right hand side comprises three terms, the first of which is of order ϵ ($dI/dt = \epsilon f$), while the sum of the other two is a priori of order $\epsilon \log(1/\epsilon)$, which will not suffice. However, v is a solution of the linearized equation:

(31) $\partial v/\partial \lambda = - \omega(I^*)\partial v/\partial \phi^* + f^*_{s,N}(I^*, \phi^*, \lambda),$

and the last part of (30) may be rewritten as:

(32) $\epsilon(\delta_j^k + \epsilon \partial v_j/\partial I^*_k)^{-1} [(\partial v_k/\partial \phi^*)(d\phi^*/dt - \omega(I^*)) + f^*_{s,N}].$

To show inequality b), it will thus suffice to show that the term in brackets is bounded for fixed q and for small enough ϵ, which is easy using inequalities (25) and (29).

The bound e) is now easy to establish by starting from the identity:

(33) $d\gamma^*/dt - q [\omega(I) - p/q] = q(d\phi^*/dt - q\omega(I))$
                     $= \epsilon q[- (\partial w/\partial I^*_j)(dI^*_j/dt) -$
                     $- (\partial w/\partial \phi^*)(d\phi^*/dt) - \partial w/\partial \lambda + g(I, \phi, \lambda, \epsilon)].$

It will suffice to show that the sum of the first three terms is of order $q \log(1/\epsilon)$ for q fixed and ϵ small enough. In view of the bound on $\| \partial w/\partial I^*_j \|$ and inequality b), the first term is of order $\epsilon| q |^3 \log^3(1/\epsilon)$. We then use the fact that w solves (12) to rewrite the sum of the other two terms as:

(34) $\omega(I^*) \partial w/\partial \phi^* + \partial w/\partial \lambda = g^*_{s,N}(I^*, \phi^*, \lambda) + v.\partial \omega(I^*)/\partial I^*$
                     $= g^*_{s,N} + v.\partial \omega/\partial I^* + (\partial w/\partial \phi^*).\times$
              $\times (d\phi^*/dt - \omega).$

Together with the previously established estimates, this now yields e).

Inequality d) is now almost immediate, as may be seen by decomposing:

(35) $| \omega(I^*) - p/q | \leqslant | \omega(I) - p/q | + | \omega(I^*) - \omega(I) |.$

Since $|\omega(I) - p/q| < 3c_3\delta$ (the domain of definition), using (29) we find:

(36) $$|\omega(I^*) - p/q| < 3c_3\delta + c|q|\varepsilon \log(1/\varepsilon).$$

The desired inequality is then obtained by showing that $|q|\varepsilon \log(1/\varepsilon)$ is less than $\delta(p, q)$ by definition (1) of δ and by the fact that we consider only resonances of order less than the cutoff $N(\varepsilon) = [c_2 \log(1/\varepsilon)]$.

We must now show that γ^* is the solution to equation a), where α satisfies c). By definition of (I^*, φ^*):

(37) $$d\varphi^*/dt = \omega(I^*) + \beta(I^*, \varphi^*, \lambda, \varepsilon)$$
$$\beta(I^*, \varphi^*, \lambda, \varepsilon) = [\omega(I)/(1+\varepsilon\partial w/\partial\varphi^*) - \omega(I^*)] +$$
$$+ \varepsilon[g(I, \varphi, \lambda, \varepsilon)\partial w/\partial\lambda - (\partial w/\partial I^*_j)(dI^*_j/dt) -$$
$$- (\partial w/\partial\varphi^*)d\varphi^*/dt]/(1+\varepsilon\partial w/\partial\varphi^*).$$

We thus have:

(38) $$d^2\gamma^*/dt^2 = q\, d^2\varphi^*/dt^2$$
$$= q(\partial\omega/\partial I^*).(dI^*/dt) + q d\beta/dt.$$

Since $dI^*/dt = dI/dt - \varepsilon dv/dt$ and:

(39) $$dI/dt = \varepsilon[f_0(I^*) + f_s(I^*, \gamma^*)] + \varepsilon f^*_s(I^*, \varphi^*, \lambda) + \varepsilon R_N(I^*, \varphi^*, \lambda) +$$
$$+\varepsilon[f_{s,N}(I^*, \gamma^*) - f_s(I^*, \gamma^*)] + \varepsilon[f(I, \varphi, \lambda, \varepsilon) - f(I^*, \varphi^*, \lambda, 0)],$$

we see that γ^* satisfies an equation of type (24) with the explicit expression for α:

(40) $$\alpha(I^*, \varphi^*, \lambda, \varepsilon) = \partial\omega/\partial I^*.[f(I, \varphi, \lambda, \varepsilon) - f(I^*, \varphi^*, \lambda, 0)]/\sqrt{\varepsilon}$$
$$+ \partial\omega/\partial I^*.R_N/\sqrt{\varepsilon} + \partial\omega/\partial I^*.[f_{s,N}(I^*, \gamma^*) - f_s(I^*, \gamma^*)]/\sqrt{\varepsilon}$$
$$+ \partial\omega/\partial I^*.f^*_s(I^*, \varphi^*, \lambda)/\sqrt{\varepsilon} + d\beta/dt(I^*, \varphi^*, \lambda, \varepsilon)/\varepsilon^{3/2}.$$

We will not write down the details necessary for the bound on α; this follows by using the estimates already established and from the fact that v and w are solutions of the linearized equations (6). The following (easily obtained) inequalities are also required:

(41) $$\| \partial^2 w/\partial\varphi^{*2} \| < cq^2 \log^2\varepsilon; \quad \| \partial^2 w/\partial\varphi^*\partial I^*_j \| < c|q|^3 \log^3(1/\varepsilon);$$
$$\| \partial^2 w/\partial\varphi^*\partial\lambda \| < cq^2 \log^2\varepsilon.$$

This completes the proof of Proposition 2. A few remarks will perhaps clarify its importance. First, as we already pointed out, equation (24) shows that the nonlinear "pendulum" with damping and forcing constitutes a "universal" model for passage through resonance; this point will be further developed in the next section. Proposition 2 also underscores the great difficulty involved in constructing resonant normal forms (even to first order) outside the Hamiltonian formalism. With its explicit definition of the *inverse* of the desired diffeomorphism, the route we used above may seem tortuous, but it is not difficult to convince oneself that it is in fact the only one possible, assuming one wants to eliminate *all* the nonresonant terms while maintaining good control of the remainders. ▪

4.3 Passage trough resonance; the pendulum model

In this section we study passage through a fixed resonance using the resonant normal forms introduced above. For the sake of exposition, we redefine the resonant zone in terms of the variables $(I^*, \varphi^*, \lambda)$ as the set $V'(-q, p, \delta)$ of points in $K_1 - \eta$ satisfying $|d\gamma^*/dt| < 2c_3 q\delta(p, q, \varepsilon)$ (we assume that $q > 0$ for notational simplicity), where $\gamma^* = q\varphi + p\lambda$ is the resonant linear combination of phase variables. It is easy to see that this definition is valid and that V' contains V.

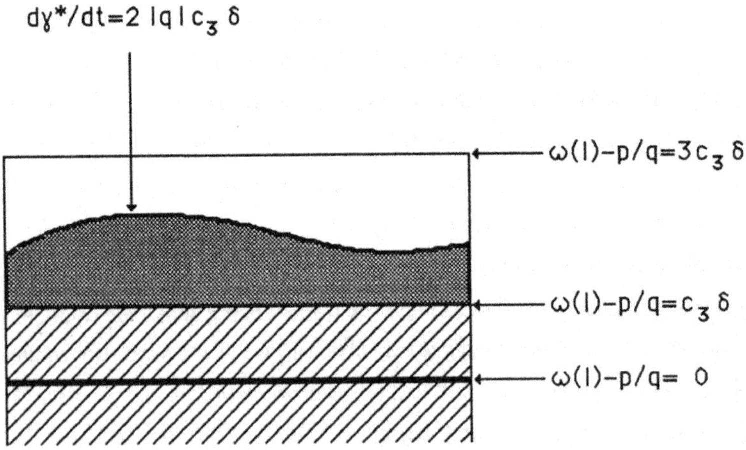

Figure 4.1: Nested resonant zones. The resonance (p, q) itself appears as a bold line. The narrow resonant zone $V(-q, p, \delta)$ (hatched) is enclosed in a larger resonant zone $V'(q, p, \delta)$ (shaded). The same conventions will be adopted in the following figures.

In fact, since the diffeomorphism is defined for $|\omega(I) - p/q| < 3c_3\delta$ and on this domain the inequality e) of Proposition 2 above is satisfied, $|dy^*/dt|$ may, for small enough ε, take any value less than $3c_3\delta cq^3\varepsilon\log(1/\varepsilon)$. But this latter quantity is greater than $2c_3q\delta$, so the new resonant zone is correctly defined. In addition, on the old resonant zone dy^*/dt is bounded in absolute value by $c_3q\delta + cq^2\varepsilon\log(1/\varepsilon)$ and thus by $2c_3q\delta$ for small enough ε; the new resonant zone thus contains the old one (see Figure 4.1). We will see the advantage of defining these two nested zones below.

Whereas condition A of Chapter 3 prohibited trajectories from visiting a given resonance more than once the exact trajectory of a system satisfying condition N may visit, a priori, a given resonance many times (Figure 4.2). Condition N simply ensures the existence of an *average* transverse force.

Introducing the two nested zones allows us to account for this difficulty, as we shall see. We first consider the very simple case of resonances with orders between N' and N ($N' < |p| + q \leq N$), where we recall that N' is an integer *independent* of ε such that:

$$(42) \qquad |s| = |p| + q > N' \quad \Rightarrow \quad |L(I) + \Phi_s(I, y)| > c_1/2,$$

for I on the resonance properly speaking.

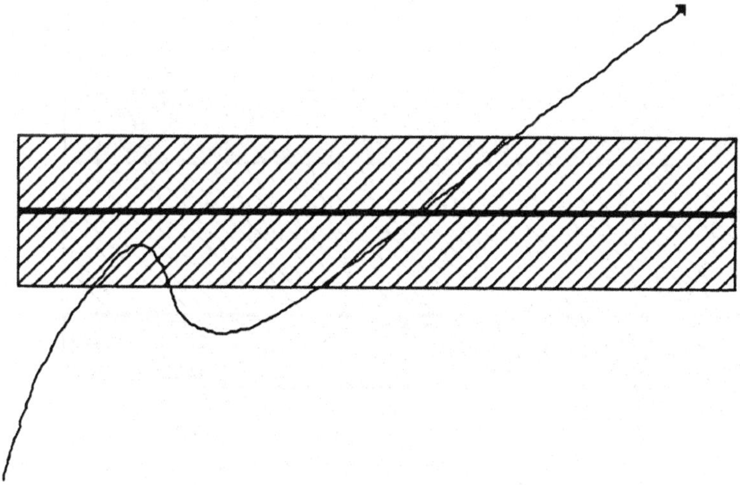

Figure 4.2: Crossing of a narrow resonant zone $V(-q, p, \delta)$ by a solution of the exact system (or rather its projection on the space of slow variables). Conventions are the same as in Figure 4.1.

It is easy to see that for $\varepsilon < \varepsilon_0$ independent of q:

(43) $\quad I \in V'(-q, p, \delta) \Rightarrow |L(I) + \Phi_s(I, \gamma)| > c_1/4.$

In other words, for I in the larger associated resonant zone, the width of which does not exceed $4qc_3\sqrt{\varepsilon}$, $|L(I) + \Phi_s(I, \gamma)| > c_1/4$.

Therefore, by Proposition 2, during its passage through this resonance the trajectory satisfies, for small enough ε, the inequality:

(44) $\quad |d^2\gamma^*/dt^2| > q\varepsilon[c_1/4 - cq\sqrt{\varepsilon}] > q\varepsilon c_1/8.$

Consequently, $d\gamma^*/dt$ varies monotonically during passage through the strip $V'(q, p, \delta)$, and this passage takes place, in a well defined sense, on a time interval bounded from above (by $32c_3\delta(p, q, \varepsilon)/(c_1\varepsilon)$). Also, the resonance may be traversed only once by the trajectory since $\omega(I) = p/q$ defines a surface which divides K into two pieces, and the direction of passage trough this surface is not arbitrary.

We next consider resonances of order less than N'. There are finitely many such resonances, independently of ε. We show that after passage through the larger zone V', during which it visits the smaller zone V, the trajectory may revisit V' but may no longer penetrate V; this will take care of troublesome "multiple passages" through resonance (Figure 4.3).

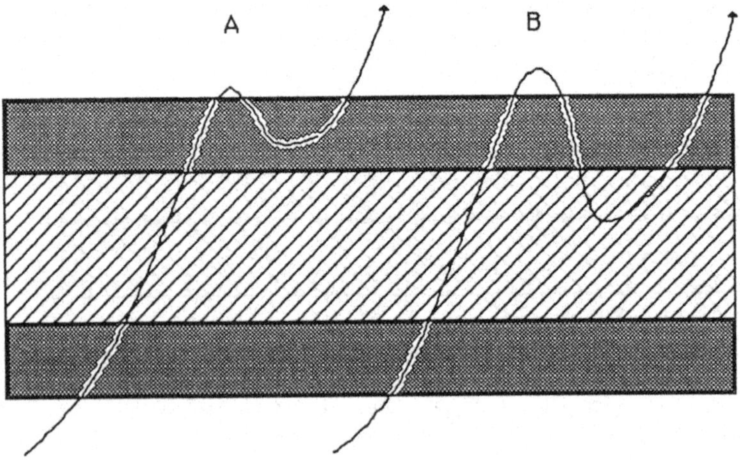

Figure 4.3: Schematic representation of the crossing of a resonant zone by a solution of the exact system. Only situation A may occur as a trajectory may not revisit the narrow resonant zone after exiting the larger one.

Now suppose that the trajectory leaves V at time T_0, leaves V' at time $T_1 > T_0$, reenters V' at time $T_2 > T_1$, and finally reenters V at time $T_3 > T_2$ (Figure 4.4). The trip outside of V takes place in a region of the nonresonant domain lying between the resonances (p, q) and (p', q').

On the time interval $[T_0, T_3]$, the trajectory is located in the nonresonant domain and we may use the change of variables $I \to P$. Let us show that by virtue of condition N, the function $|\, \omega(P(t)) - p/q \,|$ is increasing on $[T_0, T_3]$, provided that the constant c_3 is chosen sufficiently large. We have:

$$(45) \qquad d\omega/dt = \partial\omega(P)/\partial P . dP/dt$$
$$= \partial\omega(J)/\partial J . dJ/dt + \partial\omega(P)/\partial P . [dP/dt - dJ/dt] +$$
$$+ dJ/dt . [\partial\omega/\partial J - \partial\omega/\partial P]$$

where J is the averaged trajectory with the same initial condition (at T_0) as I and where $P(t) = I(t) + \varepsilon S(I(t), \varphi(t))$.

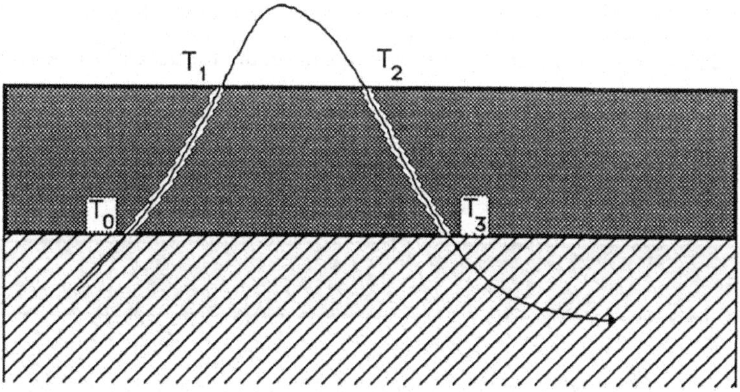

Figure 4.4: Proof that a trajectory cannot revisit the narrow resonant zone after exiting the larger one. Times of entrance or exit are indicated on this schematic figure.

By condition N, the first term is bounded below in norm by $c_1 \varepsilon$. The second and third terms are bounded respectively by $c\| dP/dt - dJ/dt \|$ and $c\varepsilon\| J - P \|$. We have:

$$(46) \qquad dP/dt - dJ/dt = dP/dt - f_0(P) + \varepsilon[f_0(P) - f_0(J)].$$

From this we deduce, using Proposition 2:

(47) $\| dP/dt - dJ/dt \| \leqslant c\varepsilon^2(1/\delta^2+1/\delta'^2)/c_3^2 + c\varepsilon\| P-J \|$

and on $[T_0, T_3]$, using Gronwall's lemma:

(48) $\| P - J \| \leqslant c\varepsilon(1/\delta^2 + 1/\delta'^2)/c_3^2.$

It follows that the second and third terms are both bounded in norm by $c\varepsilon^2(1/\delta^2 + 1/\delta'^2)/c_3^2$. Since δ and δ' are bounded below by $(\sqrt{\varepsilon}/\sqrt{N'}) \exp[-\sigma N'/2]$, for sufficiently large c_3 we have:

(49) $| d\omega(P(t))/dt | \geqslant c_1\varepsilon/2,$

which shows that $| \omega(P(t)) - p/q |$ is increasing on $[T_0, T_3]$. We note that the crucial parameter is the product $c_1c_3^2$ (which must be large enough), which is reasonable since c_3 characterizes the distance to the center of the resonance and c_1 the speed at which the trajectory leaves the resonance.

From this result and from Proposition 2, we deduce:

(50) $$\begin{aligned} | \omega(I(T_3)) - p/q | &\geqslant | \omega(P(T_3)) - p/q | - c\varepsilon(1/\delta+1/\delta')/c_3 \\ &\geqslant | \omega(P(T_2)) - p/q | - c\varepsilon(1/\delta+1/\delta')/c_3 \\ &\geqslant | \omega(I(T_2)) - p/q| - c\varepsilon(1/\delta+1/\delta')/c_3 . \end{aligned}$$

Since $| dy^*/dt | = 2c_3q\delta$ for $t = T_2$, by inequality e) of Proposition 2 we obtain:

(51) $| \omega(I(T_3)) - p/q | \geqslant 2c_3\delta - c\varepsilon(1/\delta+1/\delta')/c_3 - cq\varepsilon \log(1/\varepsilon) .$

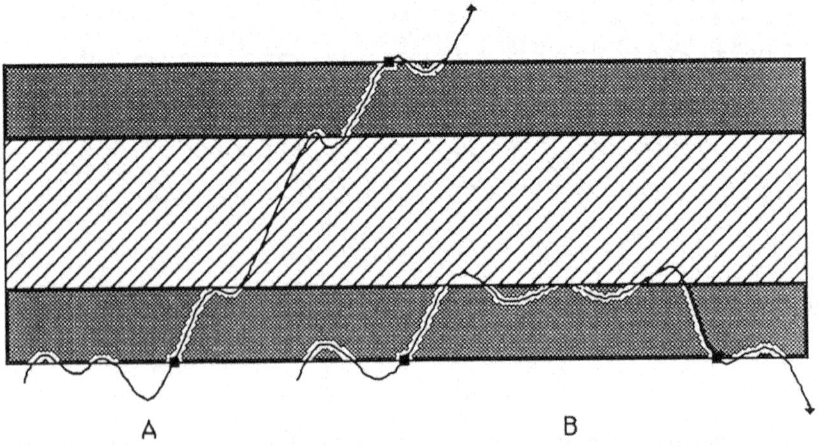

Figure 4.5: Schematic representation of the crossing of resonant zones. Actually only situation A may occur, situation B being forbidden by condition N.

Consequently, as $q \leqslant N'$ and δ and δ' are bounded below by $(\sqrt{\varepsilon}/\sqrt{N'})e^{-\sigma N'/2}$, for small enough ε and for c_3 chosen sufficiently large, we have:

$$(52) \qquad | \omega(I(T_3)) - p/q | > c_3\delta,$$

which contradicts the definition of T_3. Therefore the trajectory may not reenter V again after leaving. We thus see that, a priori, the only passages through resonance possible are of the kind represented in the figure above. We will see later that, as intuition suggests, condition N prevents trajectories from behaving like the one depicted on the right as well.

We now undertake a more detailed analysis of the behavior of trajectories during their passage through a resonant zone V' of order less than N'. Using equation (24) of Section 3, we will show that certain trajectories are in fact "trapped" in such a zone. We will adopt the scaled time $\tau = \sqrt{\varepsilon}t$. V' is then defined as the set of points satisfying $| dy^*/d\tau | \leqslant 2c_3 q\delta/\sqrt{\varepsilon}$, and in this zone trajectories satisfy the relations of Proposition 2, which we recall:

$$(53) \qquad d^2y^*/d\tau^2 = q[\Phi(I^*, y^*) + L(I^*) + \sqrt{\varepsilon}\alpha(I^*, \varphi^*, \lambda, \varepsilon)];$$

$$\| dI^*/d\tau \| < c\sqrt{\varepsilon}$$

where $\Phi = \Phi_{-q,p}$ depends on the resonance considered (as does α). Equation (53) is a perturbation of:

$$(54) \qquad d^2y^*/d\tau^2 = q[\Phi(I^*(0), y^*) + L(I^*(0))]; \qquad I^*(\tau) = I^*(0).$$

I^* is fixed and y^* satisfies the equation, depending on the parameter I^*, of a nonlinear pendulum subjected to a *constant* force $qL(I^*)$. The corresponding potential is:

$$(55) \qquad V(I^*, y^*) = - qL(I^*)y^* - q \int_0^{y^*} \Phi(I^*, u)\, du.$$

A priori, the phase portrait of the pendulum possesses elliptic and hyperbolic fixed points, homoclinic loops and open separatrix branches, with some possible nestings (Figure 4.6 below).

However, by virtue of condition N, all the open separatrix branches behave asymptotically like parabolic branches. In fact, condition N says that $| L(I^*) | > c_1$, and for large values of y^*, the periodic part of the potential is negligible compared to the linear part, so that the asymptotic behavior of the open branch in phase space is given by $(dy^*/d\tau)^2 \sim 2qL(I^*)y^*$. Condition N ensures the existence of an average driving force which dominates at high speeds; it is thus a *global* condition of minimal openness for branches of separatrices.

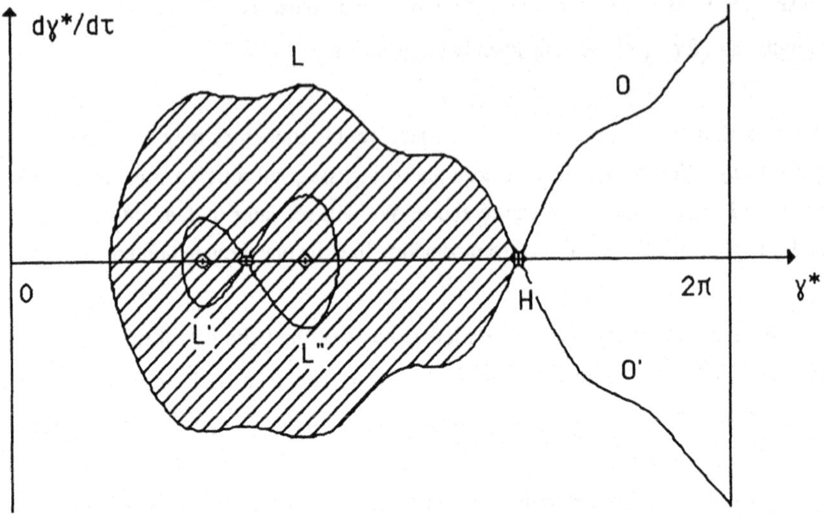

Figure 4.6: Typical phase portrait for a forced pendulum. To the hyperbolic fixed point H is associated a separatrix made of three pieces: the homoclinic loop L and the two open branches O and O'. Inside the maximal loop L are nested the homoclinic loops L' and L''.

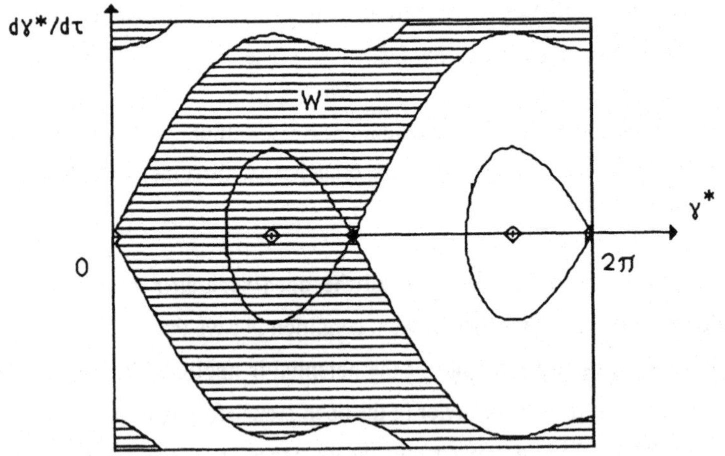

Figure 4.7: The pendulum phase space is divided into separate regions, such as the region W depicted here (hatched area) by open branches of separatrix.

For c_3 chosen large enough, all the homoclinic loops are entirely contained in the strip $|\,d\gamma^*/d\tau\,| < 2c_3 q\delta/\sqrt{\varepsilon}$, which implies that the smaller resonant zones are traversed rather than simply visited. These loops may be nested, and any loop is contained in a maximal loop. In the cylindrical phase space $T^1 \times \mathbb{R}$ of the unperturbed pendulum, the open separatrix branches partition the strip $|\,d\gamma^*/d\tau\,| < 2c_3 q\delta/\sqrt{\varepsilon}$ into different regions (Figure 4.7 above). We will now examine the motion in such a region outside the maximal loop it contains.

Consider an exact trajectory of the system which at time $t = 0$ penetrates inside the larger resonant zone $V'(-q, p, \delta)$ associated to a resonance $(-q, p)$ of low order $(|\,p\,| + q < N)$. We will determine the set of initial conditions which lead to passage through the zone in sufficiently short time. The trajectory penetrates V' at $(I^*(0), \gamma^*(0), \lambda(0))$ for which $|\,d\gamma^*/d\tau(0)\,| = 2c_3 q\delta$, and the subsequent motion satisfies equation (53) which may be rewritten:

$$(56) \qquad d^2\gamma^*/d\tau^2 = q[\phi(I^*(0),\gamma^*)+L(I^*(0))] \;+\; q[\phi(I^*(\tau),\gamma^*)-$$
$$- \phi(I^*(0),\gamma^*)] \;+\; q[L(I^*(\tau))-L(I^*(0))] \;+\; \sqrt{\varepsilon}q\alpha(I^*,\varphi^*,\lambda,\varepsilon).$$

Since $\|\,dI^*/d\tau\,\| \leqslant c\sqrt{\varepsilon}$, and since $\|\,\nabla_{I^*}\phi_{-q,p}\,\|$ admits a bound independent of p and q, this equation takes the form:

$$(57) \qquad d^2\gamma^*/d\tau^2 = q[\phi(I^*(0), \gamma^*) + L(I^*(0))] + q\sqrt{\varepsilon}\beta(I^*, \varphi^*, \lambda, \tau, \varepsilon)$$

with $|\,\beta(I^*, \varphi^*, \lambda, \tau, \varepsilon)\,| \leqslant cq + c\tau$.

This is the equation of a nonlinear pendulum perturbed by a weak time dependent force.

The initial condition of the trajectory is situated on the boundary of the larger resonant zone V' in one of the regions W (see Figure 4.8 below) defined by the open separatrix branches of the unperturbed pendulum. We may assume that $L(I^*(0))$ is positive and that c_3 has been chosen large enough to ensure that long trips through the resonant zone may only take place for trajectories entering the zone on the side where $d\gamma^*/d\tau = -2c_3 q\delta$.

The two open separatrix branches (associated to the hyperbolic points H and H') of the unperturbed pendulum, which bound the region W, intersect this side at two points, denoted here by γ^*_1 and γ^*_2 ($\gamma^*_2 > \gamma^*_1$).

We will now show

Proposition 1:

For $\gamma^*(0)$ in a segment $[\gamma^*_1 + c_4\sqrt{\varepsilon}, \gamma^*_2 - c_4\sqrt{\varepsilon}]$ or in other words, outside a neighborhood of order $\sqrt{\varepsilon}$ of the separatrix, the exact trajectory remains in the region W during its passage through the resonant zone, and traverses this zone in a time τ less than $c_5 \log(1/\varepsilon)$, the logarithmic behavior being characteristic of slowing in the neighborhood of a hyperbolic fixed point.

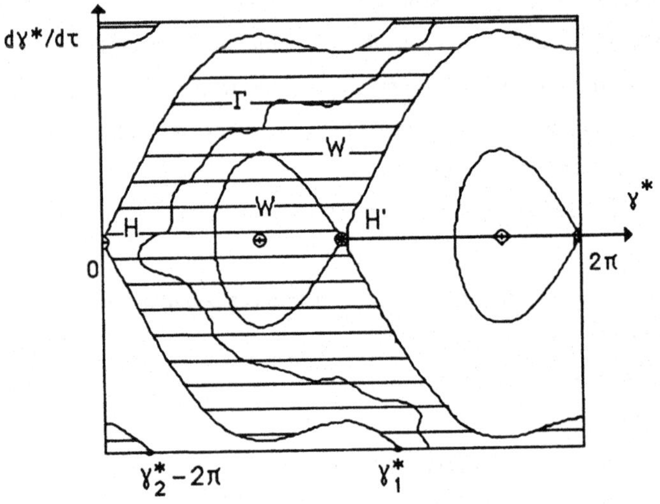

Figure 4.8: The trajectory Γ which enters the resonant zone outside an order $\sqrt{\varepsilon}$ neighborhood of the separatrices remains in the region W (and outside the homoclinic loop) during the whole passage.

The vectorfield of the unperturbed pendulum does not vanish in the part of W lying outside the maximal loop, which we denote by W', except at the hyperbolic points H and H'.

We introduce two small neighborhood V_ϱ and V'_ϱ of H and H'; V_ϱ is defined by the inequalities: $|\gamma^* - \gamma^*(H)| \leqslant \varrho$, $|d\gamma^*/d\tau| \leqslant \varrho$, and V'_ϱ is defined similarly. From the fact that $d\gamma^*/d\tau$ is extremal on the vertical lines passing through the fixed points, we see that an *unperturbed* trajectory which during its passage through W visits neither V_ϱ nor V'_ϱ

may be decomposed into three pieces (Figure 4.9): the first piece corresponding to passage through the region $dy^*/d\tau \leqslant -\rho$, the second corresponding to passage through that part of the intersection of W' and the strip $|dy^*/d\tau| \leqslant \rho$ which contains H and the third corresponding to the trajectory's passage through the region $dy^*/d\tau \geqslant \rho$.

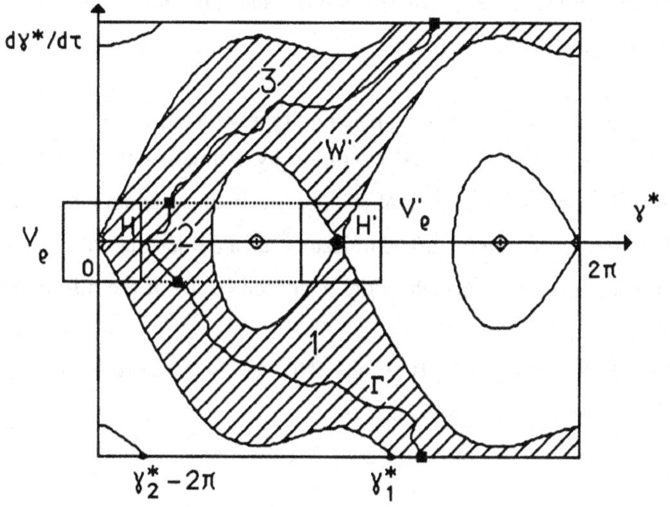

Figure 4.9: Decomposition of the unperturbed trajectory Γ into three pieces as explained in the text. W' is the part of region W lying outside the homoclinic loop, V_ρ and V'_ρ are neighborhoods of the hyperbolic fixed points H and H'.

In zones 1 and 3, the horizontal component of the vectorfield is greater than ρ and the trajectory spends a time less than c_6/ρ in these zones, where c_6 bounds the total variation of y^* along an arbitrary open trajectory of the unperturbed pendulum during passage through the domain. This bound is independent of the pendulum considered, in other words it is uniform in p, q, and I^*; it should be kept in mind that an open trajectory can make several turns around the cylinder before leaving the resonant zone, and that the bound is possible because of condition N.

In zone 2, for small enough ρ, the vertical component of the vectorfield is greater than $q\lambda(H)\rho/2$, where $\lambda(H)$ is the positive eigenvalue associated to the hyperbolic fixed point H. Since both the values of I^* for which the fixed points of the unperturbed pendulum may be degenerate (the set Θ), as well as a small neighborhood of this set (K_1

is contained in the interior of $K - \Theta$) have been excluded from the domain under consideration, $\lambda(H)$ is bounded below by a positive constant c_7 . This constant is independent of p, q ($|p| + q \leqslant N'$), I^* and of the fixed point considered. The time spent in region 2 is therefore less than the constant $4/c_7$.

The time spent in the resonant zone by an open unperturbed trajectory which remains at least ρ-distant from all hyperbolic fixed points is thus (uniformly) bounded by $4/c_7 + 2c_6/\rho \leqslant c_8/\rho$, for small enough ρ.

Let us now compare a trajectory of the preceding type with the trajectory of the *perturbed* pendulum having the same initial condition by estimating the separation between these trajectories on the time interval $[0, 2c_8/\rho]$. If (u, v) parametrizes the unperturbed trajectory, and (u', v') the perturbed trajectory, then the separation $(\Delta u, \Delta v)$ between the trajectories is the solution with initial condition $(0, 0)$ of the differential system:

(58) $$d(\Delta u)/d\tau = \Delta v \quad d(\Delta v)/d\tau = q[\Phi(I^*(0), u') - \Phi(I^*(0), u)] + \\ + q\sqrt{\varepsilon}\beta(I^*(0), \lambda(0), u', \tau).$$

Since β is bounded by $cq + c\tau$, the norm Δ of $(\Delta u, \Delta v)$ satisfies the differential inequality:

(59) $$d\Delta/d\tau \leqslant c\Delta + c\sqrt{\varepsilon}[q^2 + q\tau] \leqslant c_9\Delta + c\sqrt{\varepsilon}(1+\tau),$$

where c_9 bounds the norm $\sup(1, \partial\Phi/\partial y^*)$ of the Jacobian matrix of (58) at the hyperbolic fixed point H (this bound is uniform in p and q, and in the choice of H). Consequently, $\Delta(\tau)$ is bounded by $2c\sqrt{\varepsilon}\exp[c_9\tau]$ and thus, on the time interval considered, by $\rho/2$, for small enough ε.

Thus the perturbed trajectory enters the band $|dy^*/d\tau| \leqslant \rho/2$ only in the neighborhood of the hyperbolic fixed point H and does not visit the neighborhood $V_{\rho/2}$ of H on the time interval $[0, 2c_8/\rho]$. We may proceed as before by decomposing phase space into three regions (this time defined by the strip $|dy^*/d\tau| \leqslant \rho/2$), which allows us to establish the bound $2c_8/\rho$ for the time spent by the perturbed trajectory in the resonant zone (for ρ and ε small enough, ε depending on ρ).

It remains to account for trajectories which pass close to hyperbolic fixed points. The trajectories of the unperturbed pendulum which are not contained in the maximal separatrix loop may be classified as belonging to one of two categories, for ρ small enough: those which enter the box V_ρ without visiting V'_ρ , and those which visit the box V'_ρ once or twice (according to their distance from the separatrix) without entering V_ρ .

Condition N again restricts the number of passages through the neighborhood of a hyperbolic fixed point. In fact, consider for example a trajectory which has passed twice near H'. Its equation reads:

$$(60) \qquad (dy^*/d\tau)^2 = qL(I^*(0))y^* + \Theta(I^*(0),\ y^*)$$

where Θ has period 2π in y^*; at the second crossing of the vertical line which passes through H', $dy^*/d\tau$ is equal to $(dy^*/d\tau)_2$ and at the next crossing it is equal to $(dy^*/d\tau)_3$.

These values are connected via the relation:

$$(61) \qquad (dy^*/d\tau)_3{}^2 = (dy^*/d\tau)_2{}^2 + 4\pi qL(I^*(0))$$

and therefore $(dy^*/d\tau)_3$ is greater than $[4\pi qL(I^*(0))]^{1/2}$ which in turn is greater than ρ, for ρ small enough, since $|L| \geqslant c_1$ by condition N.

In order to study the behavior of perturbed trajectories in the neighborhood of fixed points, we linearize, again starting from the unperturbed pendulum expressed by:

$$(62) \qquad du/d\tau = v, \quad dv/d\tau = qL(I^*(\tau^*), \mu) + q\Phi(I^*(\tau^*), \mu, u),$$

where τ^* is an arbitrary point in the time interval $[0, 2c_{10} \log(1/\varepsilon)]$, and c_{10} is a sufficiently large constant. This system is formally obtained from the perturbed pendulum by dropping the term α and replacing the dependence on τ by dependence on the parameter μ. This pendulum admits a hyperbolic fixed point H_μ whose position varies with μ. For all values of μ, we may linearize the system (62) via a transformation $(x, y) = \psi_\mu(u, v)$, which is again in the spirit of normal forms techniques. In these new variables, the system is written:

$$(63) \qquad dx/d\tau = \lambda(I^*(\tau^*), \mu)\, x; \quad dy/d\tau = -\lambda(I^*(\tau^*), \mu)\, y$$

with $\lambda > 0$.

By restricting the values of the parameter μ to the interval $[\tau^*, \tau^* + c_{10} \log(1/\varepsilon)]$, $\tau^* < c_{10} \log(1/\varepsilon)$, and since the variation of the position of the hyperbolic fixed point H_μ is slight on this interval, we may define $\psi_\mu{}^{-1}$ on the box $|x| \leqslant 3c_{11}\rho$, $|y| \leqslant 3c_{11}\rho$, where c_{11} is a uniform bound of $\lambda(H_\mu)$ greater than 1. The following estimates, uniform in μ, then hold:

$$(64) \qquad \| d\psi_\mu{}^{-1}/d\mu \| \leqslant c\sqrt{\varepsilon}; \quad \| \psi_\mu{}^{-1}(x,y) - (u(H_\mu) + (x-y)/(2\lambda),\ (x+y)/(2\lambda)) \| \leqslant c\rho.$$

Then, for every value of μ in the interval considered, the image under $\psi_\mu{}^{-1}$ of the box just defined contains the box $V_{3\rho/2}(H_\mu)$.

This family of transformations allows us to define a τ-dependent transformation $(x, y) = \Psi(y^*, dy^*/d\tau), \tau)$. In the variables (x, y), the nonautonomous perturbed system (62) becomes:

$$(65)\qquad dx/d\tau = \lambda x + \sqrt{\varepsilon}X(I^*(\tau^*), y^*(\tau^*), \lambda(\tau^*), \varepsilon, \tau)$$
$$dy/d\tau = -\lambda y + \sqrt{\varepsilon}Y(I^*(\tau^*), y^*(\tau^*), \lambda(\tau^*), \varepsilon, \tau)$$

where X and Y are both uniformly bounded in norm by the constant c_{12}. These terms arise not only from the perturbation $q\sqrt{\varepsilon}\alpha$, but also from the derivative of Ψ with respect to τ.

In this new form, we may study the behavior of a perturbed trajectory near a hyperbolic point which at time τ^* enters the box $\{|x| \leqslant 3c_{11}\rho, |y| \leqslant 3c_{11}\rho\}$, with initial condition $(x(\tau^*), y(\tau^*))$, $x(\tau^*) < -\sqrt{\varepsilon}c_{12}/c_7$, $y(\tau^*) = -3c_{11}\rho$. Adopting the new time $\sigma = \int_{\tau^*}^{\tau} \lambda(I^*(\tau^*), s)ds$ (admissible since $d\sigma/d\tau$ is bounded below by the constant c_7), the system may be rewritten:

$$(66)\qquad dx/d\sigma = x + \sqrt{\varepsilon}X/\lambda; \quad dy/d\tau = y + \sqrt{\varepsilon}Y/\lambda.$$

In view of the initial condition, x decreases along the trajectory and we obtain the following estimates for $\sigma \geqslant 0$:

$$(67)\qquad x(\sigma) < [x(0) - \sqrt{\varepsilon}(c_{12}/c_7)]e^{\sigma}; \quad |y(\sigma) - y(0)e^{-\sigma}| < \sqrt{\varepsilon}c_{12}/c_7.$$

We deduce that the scaled (τ) time of passage through the box is bounded by $\log(1/\varepsilon)/c_7$ for $x(0) \leqslant (2c_{12}/c_7)\sqrt{\varepsilon}$.

We now proceed to estimate the time of passage through the resonant zone for a perturbed trajectory with an initial condition $y^*(0)$ on the boundary of the resonant zone near y^*_2 (which, we recall, is the intersection of an open separatrix branch with the resonant strip). We assume $y^*(0) < y^*_2 - c_4\sqrt{\varepsilon}$. Let Γ be the perturbed trajectory, and let Γ' be the trajectory of the unperturbed pendulum with the same initial condition.

Γ' enters the neighborhood V'_ρ of $H'(I^*(0))$ at time $\tau_1 \leqslant c_8/\rho$, and the separation at that instant between Γ' and Γ is less than $\rho/2$ for ε small enough. Consequently, at time τ_1 the perturbed trajectory Γ is located in the neighborhood $V'_{3\rho/2}$ of the hyperbolic fixed point $H'(I^*(\tau_1))$ of the pendulum equation:

$$(68)\qquad d^2y^*/d\tau^2 = qL(I^*(\tau_1)) + q\Phi(I^*(\tau_1), y^*)$$

and therefore also in the image of the domain of linearization.

In fact the trajectory Γ enters the domain at time $\tau_2 \leqslant \tau_1$, at the point $(x(\tau_2), -3c_{11}\rho)$. If c_4 is chosen large enough and ρ small enough, $x(\tau_2)$ will satisfy $x(\tau_2) < -2c_{12}\sqrt{\varepsilon}/c_7$; it is then possible to locally analyze the behavior of the perturbed trajectory, as we discussed above, and to deduce that the perturbed trajectory leaves the neighborhood $\{|x| \leqslant 3c_{11}\rho, |y| \leqslant 3c_{11}\rho\}$ at time $\tau_3 < \tau_2 + \log(1/\varepsilon)/c_7$. At that instant, the value of y is less than $x(\tau_2) + 2c_{12}\sqrt{\varepsilon}/c_7 < 0$.

Finally, we consider the trajectory Γ'' of the unperturbed pendulum equation:

(69) $$d^2y^*/d\tau^2 = qL(I^*(\tau_3)) + q\Phi(I^*(\tau_3), y^*)$$

which passes through the point $\psi^{-1}(x(\tau_3), y(\tau_3), \tau_3)$. From the possible values of $y(\tau_3)$, we see that this trajectory is located outside the maximal separatrix loop and at a distance of at least $1/2\,[x(\tau_2) + 2c_{12}\sqrt{\varepsilon}/c_7]$ from the separatrix.

On the other hand, since one of the variables x or y takes the value $-3c_{11}\rho$ on leaving the box, $(dy^*/d\tau)(\tau_3)$ is less than $-3c_{11}\rho/2$. Two cases are then possible:

- In the first case, the unperturbed trajectory Γ'' makes one trip around the separatrix loop without visiting the box $V'_\rho(H'(I^*(\tau_3))$, and so leaves the resonant zone in a time less than c_8/ρ. For ε small enough, the separation between Γ and Γ'' does not exceed $\rho/2$ on this time interval. Therefore, Γ does not visit the neighborhood $V'_{\rho/2}$ of $H'(I^*(\tau_3))$ and leaves the resonant zone in a time less than $2c_8/\rho$ (estimated starting from the exit from the domain of linearization).
- In the second case the unperturbed trajectory Γ'', after making a trip around the separatrix loop, enters the neighborhood $V'_\rho(H'(I^*(\tau_3))$ at time τ_4. It is easy to show that $\tau_4 - \tau_3$ is bounded by c/ρ (where $c = 2c_8/c_7$) and that the separation between Γ and Γ'' thus does not exceed $\rho/2$ on entering $V'_\rho(H'(I^*(\tau_3))$, for small enough ε. The perturbed trajectory thus enters the image of the domain of linearization at a time τ_5, before τ_4, and we may proceed as before with a local analysis showing that Γ leaves the domain of linearization at a time τ_6 less than $\tau_5 + \log(1/\varepsilon)/c_7$, and leaves the resonant zone after a time less than c_8/ρ.

These estimates complete the proof of Proposition 1 (trajectories with initial conditions y^* close to y^*_1 are treated similarly). ∎

Transforming from τ back to the original time t, we have shown that provided c_4 is chosen sufficiently large, trajectories entering a resonant zone of order less than N' at a distance greater than $c_4\sqrt{\varepsilon}$ from all open separatrix branches leave this zone in a time less than $c_5 \log(1/\varepsilon)/\sqrt{\varepsilon}$, where $c_5 = 3/c_7$, and c_7 measures the nondegeneracy of the hyperbolic fixed points.

4.4 Excluded initial conditions, maximal separation, average separation

Proof of Part b of Theorem 1:

The preceding analysis serves to determine the domain of initial conditions V'' arising in the statement of Theorem 1 (i.e., initial conditions that must be excluded) and to estimate its measure. Two kinds of initial conditions must be excluded:

- Those which are contained in a resonant zone of order less than N' (we recall that N' is an integer independent of ε) and are such that the associated exact trajectories remain trapped in the resonant zone or leave the zone after a time greater than $c_5 \log(1/\varepsilon)/\sqrt{\varepsilon}$.
- Those which belong to no resonant zone of order less than N', but which are such that the associated exact trajectories spend a fraction of the time interval $[0, 1/\varepsilon]$ greater than $c_5 \log(1/\varepsilon)/\sqrt{\varepsilon}$ in a resonant zone of order less than N'.

The measure of initial conditions of the first kind is bounded by $4\pi^2$ (surface area of the (φ, λ) torus) times the measure on the base of resonant zones of order less than N'. Since the width of a larger resonant zone is less than $6c_3\delta$, since δ is bounded by $\sqrt{\varepsilon}$, and since the number of resonances of order less than N' is less than N'^2, the total measure on the base of low order resonant zones is bounded by $c\sqrt{\varepsilon}$, and the measure of initial conditions of the first kind is less than $c\sqrt{\varepsilon}$.

Let us now consider initial conditions of the second kind. In the previous section, we saw that trajectories entering a low order resonant zone at a point $I_0{}^*$ situated on the boundary of the strip $| \, dy^*/dt \, | \leqslant 2c_3 q\delta$ traverse the zone in a time t less than $c_5 \log(1/\varepsilon)/\sqrt{\varepsilon}$, provided that the initial phase $y_0{}^*$ on entry is at least $c_4\sqrt{\varepsilon}$-distant from all open separatrix branches of the unperturbed pendulum equation:

$$(1) \qquad d^2y^*/d\tau^2 = q[\Phi_{-q,p}(I^*{}_0\,,\, y^*) + L(I^*{}_0)], \quad \tau = \sqrt{\varepsilon}t.$$

Since there are a finite number of low order resonances, and this number is independent of ε, we may consider a single resonance $(-q, p)$ and decompose the set of points of the base corresponding to the border of the resonant strip into a finite number of subsets $E_1(-q, p), \ldots, E_r(-q, p)$, such that for every I^* in E_j, the pendulum equation (1) with parameter I^* admits $n(j)$ open separatrix branches which depend differentiably on I^*, for I^* in E_j. We may consider only a single open separatrix branch of one of the subsets E_j which intersects the boundary of the resonant strip at the point $y^*(I^*)$.

The initial conditions to be excluded from V will then be those with trajectories that enter the resonant strip at a time prior to $1/\varepsilon$ with a value of I^* belonging to E_j and a value of y^* in the segment $D(I^*)$ of length $2c_4\sqrt{\varepsilon}$ surrounding $y^*(I^*)$ (see figures above). If we designate the flow of the exact system by T_t, these initial conditions form the set:

$$(2) \qquad V^*(1/\varepsilon) = V \cap \cup_{t\in[0,\, 1/\varepsilon]} T_{-t}(\cup_{I^*\in E_j}\{I^*\}\times D(I^*)).$$

We will show that the measure of this set is less than $c\sqrt{\varepsilon}$, which will in turn bound the measure of the set V'' of excluded initial conditions by $c\sqrt{\varepsilon}$.

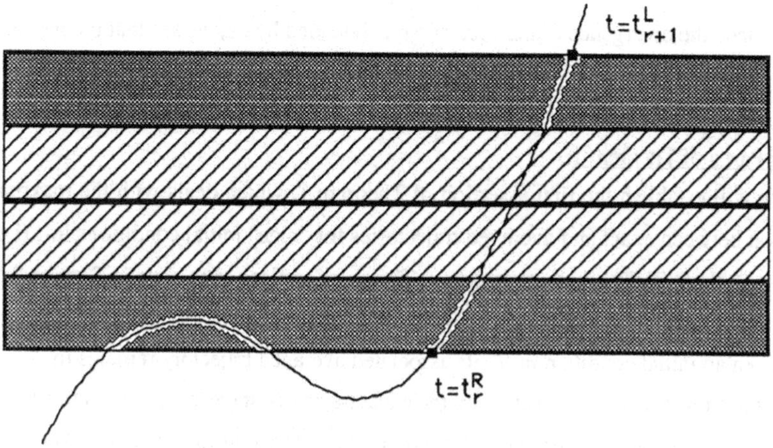

Figure 4.10: Passage of a trajectory trough a resonant zone during the time interval $[t_r^R, t_{r+1}^L]$. Conventions are the same as in Figure 4.1. The narrow resonant zone cannot be visited twice.

To do this, we first estimate the measure of the set $V^*(1)$ (defined like V with $1/\varepsilon$ replaced by 1) by using local coordinates I^*, y^*, and λ.

The measure of $D(I^*)$ on the torus is equal to $4\pi c_4\sqrt{\varepsilon}$. Since $\| dI^*/dt \| \leqslant c\varepsilon$, the maximal variation of I^* on the trajectory $I^*(-t)$ for $0 \leqslant t \leqslant 1$ is less than $c\varepsilon$ and the measure of the projection on the torus of the set $T_{-t}(\{I^*\} \times D(I^*))$ is less than $8\pi c_4\sqrt{\varepsilon}$ for $0 \leqslant t \leqslant 1$. In fact, by Proposition 2 of Section 4.2, for I^* fixed on the boundary of the resonant zone the evolution of y^* is given by the equation:

$$(3) \qquad dy^*/dt = q\omega(I^*(t)) - p + O(\varepsilon \log(1/\varepsilon)).$$

Using the estimate of the variation of I^*, we find that on the time interval $[-1, 0]$:

$$(4) \qquad y^*(t) = y^*(0) + [q\omega(I^*(0)) - p]t + O(\varepsilon \log(1/\varepsilon)).$$

The evolution of y^* differs from translation only by a term of order $\varepsilon \log(1/\varepsilon)$, and λ only differs from t by a constant (which still makes a difference, as this constant enters into the initial conditions). Since the measure of $D(I^*(0))$ is equal to $4\pi c_4\sqrt{\varepsilon}$, for I^* fixed in E_j, t in $[0, 1]$, and ε small enough, the measure of the projection on the torus of $T_{-t}(\{I^*\} \times D(I^*))$ is less than $8\pi c_4\sqrt{\varepsilon}$.

From this result on the variation of I^*, we deduce that the measure of $V^*(1)$ is less than $c\varepsilon^{3/2}$ for the measure $dI^*dy^*d\lambda$, and also less than $c\varepsilon^{3/2}$ for the original measure $dId\varphi d\lambda$. Since the divergence of the vectorfield is bounded by $c\varepsilon$, we see that the measure of $V^*(1/\varepsilon)$ is less than $c\sqrt{\varepsilon}$. ■

Proof of Part a of Theorem 1:

We will establish the first assertion of Theorem 1, which deals with the maximal separation between exact and averaged trajectories for "nice" initial conditions in I (i.e., those in V'); the initial phases are arbitrary. The proof is close to the one in Section 3.5; we will therefore stress only the points where the estimates differ.

Given an initial condition in V', its associated averaged trajectory remains in $K_1 - \eta$ at least until time $1/\varepsilon$; on the other hand, the exact trajectory stays in K_1 only until time T, which a priori may be less than $1/\varepsilon$. We will assume for the present that $T = 1/\varepsilon$. One may easily prove by a "bootstrap argument" that this assumption is valid.

As in Chapter 3, the time interval $[0, 1/\varepsilon]$ may be decomposed into nonresonant segments $[t_r^L, t_r^R]$ and resonant segments $[t_r^R, t_{r+1}^L]$; here the latter segments correspond to the trajectory's passage through the larger resonant zones $V'(s, \delta)$ *during which the smaller resonant zone* $V'(s, \delta)$ *is traversed*. Therefore two such segments cannot correspond to passage through the *same* resonance.

As in Chapter 3, we set $I(t_r^L) = I_r^L$ and $I(t_r^R) = I_r^R$ and we introduce the averaged solutions $J_r(t)$ with initial conditions $J_r(t_r^L) = I_r^L$ (Figure 4.11). On $[t_s^L, t_{s+1}^L]$, we may then write:

$$(5) \qquad \| I(t) - J(t) \| \leq \| I(t) - J_s(t) \| + \| J_s(t) - J_{s-1}(t) \| + \dots + \| J_2(t) - J_1(t) \|.$$

On $[t_{r+1}^{L}, 1/\varepsilon]$, we have:

(6) $$\| J_{r+1}(t) - J_r(t) \| \leqslant c \| J_{r+1}(t_{r+1}^{L}) - J_r(t_{r+1}^{L}) \|$$
$$\leqslant c \| I_{r+1}^{L} - I_r^{L} \| + \| J_r(t_r^{R}) - J_r(t_{r+1}^{L}) \| +$$
$$+ \| I_r^{R} - J_r(t_r^{R}) \|.$$

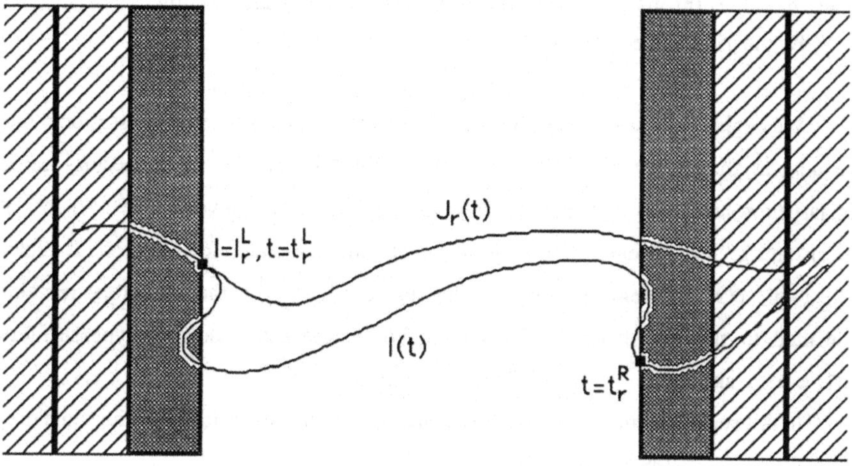

Figure 4.11: Passage trough the nonresonant domain. The trajectory $I(t)$ leaves the extended resonant zone $V'(p', q', \delta')$ at time t_r^{L} (first exit from this zone) to enter the resonant zone $V'(p'', q'', \delta'')$ at time t_r^{R} (last entrance in that zone).

The first term represents the variation of I on the resonant segment $[t_r^{R}, t_{r+1}^{L}]$. Therefore by Proposition 1, if the order of the resonance is between N' and N, it is bounded by $c\delta$, the product of the rate of variation of I with the duration of passage through the resonant zone, and by $c\sqrt{\varepsilon}\log(1/\varepsilon)$ if the order of the resonance is less than or equal to N'.

The second term represents the variation of the averaged trajectory J_r on the same resonant segment. By condition N, this trajectory crosses the resonant zone transversely and with finite speed. The crossing time is less than $c\delta/(c_1\varepsilon)$, and the variation of J_r during this time is bounded by $c\delta$ (independently of the order of the resonance considered).

Finally, the last term represents the separation, at the extremity t_r^R of the nonresonant segment $[t_r^L, t_r^R]$, between $I(t)$ and the averaged solution J_r which coincides with $I(t)$ at t_r^L. It is considerably more difficult to establish a sufficiently refined estimate of this term; we will show that it is bounded by $c(\delta + \delta'')$, where δ' and δ'' are the half widths of the resonant zones visited by the trajectory just before and just after passage through the nonresonant domain divided by the constant c_3.

By the triangle inequality:

$$(7) \qquad \| I(t) - J_r(t) \| \leq \| I(t) - P(t) \| + \| P(t) - J_r(t) \|.$$

We begin by estimating the first term. As the system considered possesses only two frequencies, the resonant surfaces do not intersect. At the instant t_r^L, the trajectory leaves the large resonant zone associated to the resonance with resonant vector $k' = (-q', p')$; at the instant t_r^R it then enters the large resonant zone associated to the resonant vector $k'' = (-q'', p'')$, different from k'. On the segment $[t_r^L, t_r^R]$, the trajectory remains confined to the region of the base bounded by the two resonant hypersurfaces $\omega(I) = p'/q'$ and $\omega(I) = p''/q''$, far from any other resonance.

In this region we may estimate the norm of S, which is the solution of the equation:

$$(8) \qquad \omega.\partial S/\partial\varphi + \partial S/\partial\lambda + f_N = 0$$

and is therefore defined by the Fourier series:

$$(9) \qquad S = i \sum_{0<|k|\leq N} f_k(I)(k_1\omega(I) + k_2)^{-1} \exp\{i(k_1\varphi + k_2\lambda)\}.$$

The only small divisors arise from the resonances associated to k' and k'', so that S may be reexpressed as:

$$(10) \qquad S = if_{k'}(I)e^{ip'\lambda - iq'\varphi}/(-q'\omega(I)+p') + if_{k''}(I)\, e^{ip''\lambda - iq''\varphi}/(-q''\omega(I)+p'') + \\ + S'(I, \varphi, \lambda)$$

where the Fourier series of S' contains no small divisors.

Using the exponential decay of the Fourier coefficients f_k (cf. Appendix 1), we obtain the estimate:

$$(11) \qquad \| S \| \leq c + ce^{-\sigma|k'|}/(c_3|q'|\delta') + ce^{-\sigma|k''|}/(c_3|q''|\delta'').$$

We will now justify our earlier choice of the function $h(p, q)$, which we recall was used in defining δ:

$$(12) \qquad \delta = h(p, q)\sqrt{\varepsilon} = (\varepsilon/|q|)^{1/2} e^{-\sigma(|p|+|q|)/2}$$

We see that this choice allows $\| \varepsilon S \|$ to be bounded by the expression $c\varepsilon + c(\delta' + \delta'')$. For the resonances taken into account, δ remains greater than $\sqrt{\varepsilon}\, \varepsilon^{\sigma c_3/2}/(c_3 \log(1/\varepsilon))$ and thus greater than $\varepsilon^{3/2}$ for σ small enough, and the norms of εS and $I(t) - P(t)$ are therefore bounded by $c(\delta' + \delta'')$, for sufficiently small ε.

It remains to examine the term $\| P(t) - J(t) \|$, which may be estimated by means of a differential inequality. In fact, since $P = I + \varepsilon S$:

(13) $$dP/dt - dJ_r/dt = \varepsilon[f_0(P) - f_0(J_r)] + b(t)$$

where

(14) $$b(t) = \varepsilon[f(I, \varphi, \lambda, \varepsilon) - f(I, \varphi, \lambda, 0)] + \varepsilon R_N + \varepsilon^2 \partial S/\partial\varphi . g + \varepsilon^2 \partial S/\partial I . f.$$

Consequently:

(15) $$| d/dt \| P - J_r \| | \leqslant c\varepsilon \| P - J_r \| + | b(t) |.$$

On the domain under consideration, $\| \partial S/\partial\varphi \| \leqslant c(1/\delta' + 1/\delta'')$ and $\| \partial S/\partial I \|$ is bounded by:

$$c(\delta' + \delta'')/\varepsilon + c \| f_{k'}(I) \| q'/(p' - q'\omega(I))^2 + c \| f_{k''}(I) \| q''/(p'' - q''\omega(I))^2$$

or, setting $\xi' = p'/q'$ and $\xi'' = p''/q''$:

(16) $$\varepsilon^2 \| \partial S/\partial I \| \leqslant c\varepsilon(\delta' + \delta'') + c\varepsilon^2 e^{-\sigma | k' |}/[q'(\omega(I(t)) - \xi')^2] + $$
$$+ c\varepsilon^2 e^{-\sigma | k'' |}/[q''(\omega(I(t)) - \xi'')^2]$$
$$\leqslant c\varepsilon(\delta' + \delta'') + c\varepsilon\delta'^2/(\omega(I(t)) - \xi')^2 +$$
$$+ c\varepsilon\delta''^2/(\omega(I(t)) - \xi'')^2.$$

$b(t)$ is therefore bounded in norm by that same quantity, and on the time interval $[t_r^L, t_r^R]$:

(17) $$\| (P - J_r)(t) \| \leqslant \| (P - J_r)(t_t^L) \| e^{c\varepsilon t}$$
$$+ c e^{c\varepsilon t} \varepsilon \int_{t_r^L}^{t} [(\delta' + \delta'') + \delta'^2/(\omega(I(u)) - \xi')^2 +$$
$$+ \delta''^2/(\omega(I(u)) - \xi'')^2] e^{-c\varepsilon u} \, du.$$

Since $I(t_r^L) = J_r(t_r^L)$, $\| (P - J_r)(t_r^L) \|$ is bounded by $\varepsilon \| S \|$ and thus:

(18) $$\| (P - J_r)(t) \| \leqslant c(\delta' + \delta'') + c\varepsilon \int_{trL}^{t} du \, \delta'^2/(\omega(I(u)) - \xi')^2 +$$
$$+ c\varepsilon \int_{trL}^{t} du \, \delta''^2/(\omega(I(u)) - \xi'')^2$$

since $t_r^R - t_r^L \leqslant 1/\varepsilon$:

The second and third terms of the right hand side is of the same order as the first term. This fact is far from obvious at first sight: Indeed $\omega(I(u)) - \xi'$ (respectively $\omega(I(u)) - \xi''$) may a priori remain in the neighborhood of $c_3\delta'$ (respectively $c_3\delta''$) during an appreciable part of the interval $[0, 1/\varepsilon]$ (the trajectory is caught in the neighborhood of a resonant strip), in which case one is faced with a small divisor problem which did not occur in Chapter 3, where condition A ensured that the exact trajectory passed through resonant zones with order ε speed.

In the present case, we rely once again on the fact that $\omega(P(t)) - \xi$ is monotone in the neighborhood of a resonance $\omega(I) = \xi$; this follows from condition N, as we saw in Section 4.3. We may rewrite the first integral on the right-hand side of (18) in the form:

$$c\varepsilon\delta'^2 \int_{t_r^L}^{t} [1/(\omega(P(u))-\xi')^2]du + c\varepsilon\delta'^2 \int_{t_r^L}^{t} [1/(\omega(P(u))-\xi')^2 - 1/(\omega(I(u))-\xi')^2]\, du$$

or:

$$c\varepsilon\delta'^2 \int_{t_r^L}^{t} [1/(\omega(P)-\xi')^2]du + c\varepsilon\delta'^2 \int_{t_r^L}^{t} [\omega(I)+\omega(P)-2\xi']^2 \times [\omega(I)-\omega(P)]^2/[\omega(I)-\xi']^2[\omega(P)-\xi']^2\, du.$$

Since $|\omega(P(u)) - \xi'| > c\varepsilon(u - t_r^L)$, the first term is less than $c\delta'$. The second term is of higher order. Indeed we may write it as the sum of three integrals:

$$c\varepsilon\delta'^2 \int_{t_r^L}^{t} du\, [\omega(I)-\omega(P)]^2/[\omega(P)-\xi']^2 + c\varepsilon\delta'^2 \int_{t_r^L}^{t} du\, [\omega(I)-\omega(P)]^2/[\omega(I)-\xi']^2 + 2c\varepsilon\delta'^2 \int_{t_r^L}^{t} du\, [\omega(I)-\omega(P)]^2/[\omega(I)-\xi'][\omega(P)-\xi'].$$

Using the monotonicity of $\omega(P(u))$, the definition of $P = I + \varepsilon S$, and the fact that $t - t_r^L \leqslant 1/\varepsilon$, one easily obtains the bounds $c\delta' \| \varepsilon S \|^2$, $c\|\varepsilon S \|^2$ and $\varepsilon\delta'\|\varepsilon S \|^2 | \log(\varepsilon\delta') |$ for these three integrals. As $\| \varepsilon S \| \leqslant c\varepsilon + c(\delta'+\delta'')$, the integral we started with is less in norm than $c(\delta' + \delta'')$ and a similar estimate is easily derived for the second integral on the right-hand side of (18). These bounds in turn give the inequality

$$\| (I - J_r)(t) \| < c(\delta' + \delta'')$$

on the segment $[t_r^G, t_r^D]$.

From this estimate and from inequality (6), we deduce that for $t > t_{r+1}^L$:

$$\| J_{r+1}(t) - J_r(t) \| \leqslant c\,[\delta' + c(\delta'+\delta'')] \tag{19}$$

if the resonant zone (of half-width $c_3\delta'$) traversed during the time interval $[t_r^R, t_{r+1}^L]$ is associated to a resonance of order between N' and N, and that:

$$\| J_{r+1}(t) - J_r(t)\| \leqslant c[\sqrt{\varepsilon}\, \log(1/\varepsilon) + c(\delta'+\delta'')] \tag{20}$$

if the order of this resonance is less than N', in which case $c_3\delta''$ is the half-width of the resonant zone crossed during the time interval $[t_{r+1}^R, t_{r+2}^L]$.

We now have only to put these estimates end to end to obtain the bound a) of Theorem 1. Since a given resonant zone may only be visited once, using inequality (5) we find that for $t \in [0, 1/\varepsilon]$:

$$\| I(t) - J(t)\| \leqslant c\textstyle\sum_{0<|k|\leqslant N'} \sqrt{\varepsilon}\, \log(1/\varepsilon) + c\sum_{N'<|k|\leqslant N} \delta_k + c\sum_{0<|k|\leqslant N} \delta_k\, . \tag{21}$$

The first term corresponds to passage through resonances of order less than N', the second term to passage through resonances of order between N' and N, and the last term to passage through the nonresonant domain.

Consequently:

(22) $$\| I(t) - J(t) \| \leqslant c\sqrt{\varepsilon}\,\log(1/\varepsilon) + c\textstyle\sum_{0<|k|\leqslant N} \delta_k .$$

Since the series $\sum \delta_k$ converges we thus obtain for $0 \leqslant t \leqslant 1/\varepsilon$:

(23) $$\| I(t) - J(t) \| \leqslant c\sqrt{\varepsilon}\,\log(1/\varepsilon) + c\sqrt{\varepsilon} \leqslant c\sqrt{\varepsilon}\log(1/\varepsilon).$$

To complete the proof of part a) of Theorem 1, it suffices to use a bootstrap argument similar to the one in Chapter 3 to show that the trajectory actually remains in K_1 until time $1/\varepsilon$. ∎

We note that if we had chosen the size of the resonant zones independently of the order of the associated resonance, the series $\sum \delta_k$ would not converge and the estimate would be on the order of $\sqrt{\varepsilon}\,(\log\varepsilon)^2$, the product of the order $\sqrt{\varepsilon}$ size of a resonant zone with the number of resonances of order less than the ultraviolet cutoff N, which is of order N^2, like the estimate obtained in Chapter 3 for systems satisfying condition A.

We further note that the order $\sqrt{\varepsilon}\,\log(1/\varepsilon)$ estimate originates in the passage through low order resonances (of order less than N'), in other words from the slowing of the exact trajectory in the neighborhood of hyperbolic fixed points of certain "pendulums" described in the preceding section. On the other hand, the contribution to the separation from passage through higher order resonances and from passage through the nonresonant domain is of order $\sqrt{\varepsilon}$. For this reason, whenever condition A is satisfied the maximal separation is of order $\sqrt{\varepsilon}$, which we show below in proving Theorem 3.

Proof of Theorem 3:

We assume that condition A from Chapter 3 is satisfied. Since condition A implies condition N, the partial results established above still hold. However, an important difference arises in that, as we saw in Section 4.1, systems satisfying condition A have no degeneracies.

This means that no domain restrictions are necessary and that all resonant zones are traversed in a time of order δ/ε. No initial conditions need be excluded (V'' is empty), and the separation between exact and averaged trajectories, estimated by the same sum as before, in other words in accordance with inequality (21), no longer contains terms corresponding to slow passages through low order resonances (the first term in the right hand side of (21)), and is thus order $\sqrt{\varepsilon}$.

Finally, to show the optimality of this result, it will suffice to consider the system mentioned in the previous chapter (equations (2) of Section 3.5) to which we add another phase λ which evolves via the equation $d\lambda/dt = 1$.

Proof of Part c of Theorem 1:

Here we establish the result on the *average* separation between exact and averaged trajectories with the same initial condition; we claimed this separation was:

$$(24) \qquad M(\varepsilon) =_{def} \operatorname{Sup}_{t\in[0,\,1/\varepsilon]} 1/\mu(V).\int_V \| I(t) - J(t) \| \, dI_0 d\varphi_0 d\lambda_0 < c\sqrt{\varepsilon}.$$

For initial conditions in V'' and for times less than $1/\varepsilon$, the separation between exact and averaged trajectories is bounded by a constant. Since the measure of V'' is less than $c\sqrt{\varepsilon}$, the part of the integral in (24) which is taken over V'' is at most order $\sqrt{\varepsilon}$.

For initial conditions in V', we saw in the preceding section that the separation at time t is less than $c\sqrt{\varepsilon} + c\varepsilon(T_1 + \ldots + T_p)$, where p is the number of resonant zones of order less than N' visited by the exact trajectory on the time interval $[0, 1/\varepsilon]$, and where T_i is the time spent in the ith zone. Recalling that the number of low order resonances is less than N'^2, we see that the part of the integral in (24) taken over V' is bounded by the expression:

$$(25) \qquad c\sqrt{\varepsilon} + c\varepsilon\mu(V')N'^2\langle T\rangle$$

where $\langle T\rangle$ designates a uniform bound on the average time for crossing a low order resonant zone, which we will estimate.

Let $(-q, p)$ be a resonance of order less than N' and consider a trajectory entering the resonant strip $| dy^*/dt | \leq 2c_3q\delta$ at the point $I^*(0)$ at $t = 0$. As in section 4.3, we construct the "pendulum" equation:

$$(26) \qquad d^2y^*/d\tau^2 = q[L(I^*(0)) + \Phi_{-q,p}(I^*(0), y^*)]$$

and we surround each hyperbolic fixed point by a square neighborhood V_ρ of size ρ. Proceeding exactly as in Section 4.3, we easily establish that:

- For $y^*(0)$ farther than $c_{13}\rho$ away from every intersection of an open branch of an invariant manifold with the boundary of the resonant strip, the exact trajectory visits none of the neighborhoods of hyperbolic fixed points and traverses the strip in a time τ less than c/ρ, the constants being independent of the resonance considered.
- If $y^*(0)$ is a distance x from the intersection of an open branch of an invariant manifold

with the boundary of the resonant strip, and if x satisfies $c_{14}\sqrt{\varepsilon} < x < c_{13}\rho$, the trajectory visits the neighborhood of a hyperbolic fixed point at most twice, by condition N, and traverses the resonant strip in a scaled time less than $c|\log[c\rho/(x-c_{14}\sqrt{\varepsilon})]|$, a result which is again obtained by a local *linear* analysis in the neighborhood of the hyperbolic fixed point.

We also show by proceeding as at the beginning of this section that, because the system is nearly conservative, the measure of the set of initial conditions in V' such that $y^*(0)$ is situated at a distance between x and $x + dx$ from the intersection of an open branch of a stable manifold with the boundary of the resonant strip is on the order of dx.

We then deduce from these results that the average scaled time for crossing the resonance is less than:

$$(27) \qquad c/\rho + c\int_{c_{14}\sqrt{\varepsilon}}^{c_{13}\rho} |\log[c\rho/(x - c_{13}\sqrt{\varepsilon})]|\, dx$$

where the integral converges. Returning to the original time, we obtain the bound $\langle T\rangle \leqslant c\sqrt{\varepsilon}$ for the average crossing time. Since the measure of V is finite, the average separation $M(\varepsilon)$ on the time interval $[0, 1/\varepsilon]$ is still less than $c\sqrt{\varepsilon}$. ∎

We remark that this result derives essentially from the convergence of the integral $\int_0^1 |\log(x)|\,dx$, where the logarithmic behavior is due to the time of passage in the proximity of a hyperbolic fixed point, which itself can be read off from the linearized equation.

4.5 Optimality of the results

We first show the optimality of Parts b and c of Theorem 1 concerning the measure of excluded initial conditions and the average separation. More precisely, we show that in general, for any estimate of the maximal separation $\mathrm{Sup}_{t\in[0,\,1/\varepsilon]}\| I(t) - J(t)\|$ tending to 0 with ε, the result $\mathrm{mes}(V'') < c\sqrt{\varepsilon}$ may not be improved. To do this, it suffices to study an example satisfying condition N and to prove that, for a set A of initial conditions with measure greater than $c\sqrt{\varepsilon}$, the separation between exact and averaged trajectories is greater than a constant c. This will ensure the optimality of the estimate of the average separation, since the integral of the separation over all initial conditions is bounded below by the integral over the initial conditions in A alone; it will thus be at least order $\sqrt{\varepsilon}$.

The simplest and most physical example is that of a forced nonlinear pendulum subjected to fluid damping:

$$(1)\qquad d^2\varphi/d\tau^2 = 1/2 + \sin\varphi - \sqrt{\varepsilon}\, d\varphi/d\tau,$$

which is equivalent to the system:

$$(2)\qquad dI/dt = \varepsilon(1/2 + \sin\varphi - I);\quad d\varphi/dt = I;\quad d\lambda/dt = 1\quad (\tau = \sqrt{\varepsilon}t).$$

The damping causes the separatrix loop to open (see Figure 4.12), thus trapping the trajectories in the single "resonance" $I = 0$ of the system.

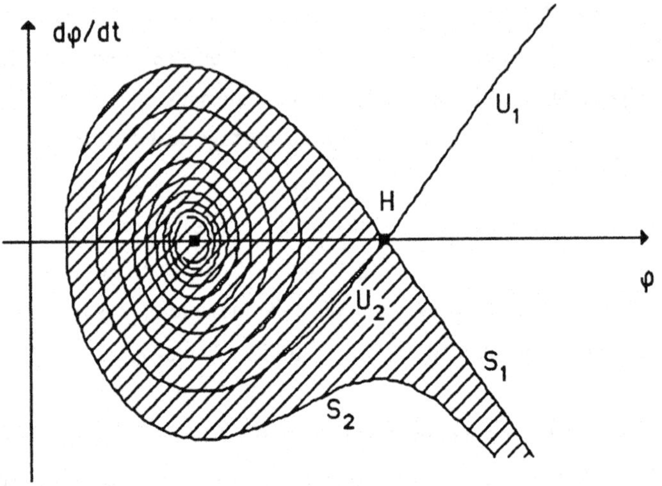

Figure 4.12: Phase portait of a damped pendulum picturing the trapping of trajectories (hatched area) due to separatrix splitting. H is a hyperbolic fixed point, S_1 and S_2 are the branches of its stable manifold, U_1 and U_2 the branches of its unstable manifold.

The fluid damping is in some sense the most effective possible, since a damping of size ε provokes a contraction of volume in phase space of the same order ε (the trace of the Jacobian matrix is equal to $-\varepsilon$). For more details concerning this elementary example, the reader may consult the book by Andronov and Khaïkine [And], the first edition of which appeared in 1937 and which, like the book by Bogoliubov and Mitropolski [Bog], has long been a classic in the analysis of nonlinear phenomena. A more recent and well-documented reference is the book by Guckenheimer and Holmes [Gu], the stated intention of which is in part to provide a sequel to [And]. Some of the essential results on classical nonlinear equations, such as Van der Pol's and Duffing's equations, are contained in [Gu], as well as a detailed development of Melnikov's method ([Mel1-3], [Hol1–5]), which we introduce below in its simplest form to estimate the measure of

initial conditions corresponding to trapped trajectories.
We write system (2) in the form:

$$(3) \qquad du/d\tau = f_1(u, v) + \sqrt{\varepsilon} g_1(u, v); \quad dv/d\tau = f_2(u, v) + \sqrt{\varepsilon} g_2(u, v)$$

where f is the unperturbed vectorfield ($f_1 = v$, $f_2 = \sin(u) + 1/2$) and g is the perturbation ($g_1 = 0$, $g_2 = -v$). We will use this overly general form at the outset to give an overview of the method.

Around the hyperbolic fixed point H, we define a box V_ρ of size ρ. We are concerned with estimating the separation between stable and unstable invariant manifolds arising from the opening of the separatrix loop; we estimate this separation near a point $q_0(0)$ lying outside V_ρ on the unperturbed separatrix $q_0(\tau)$.

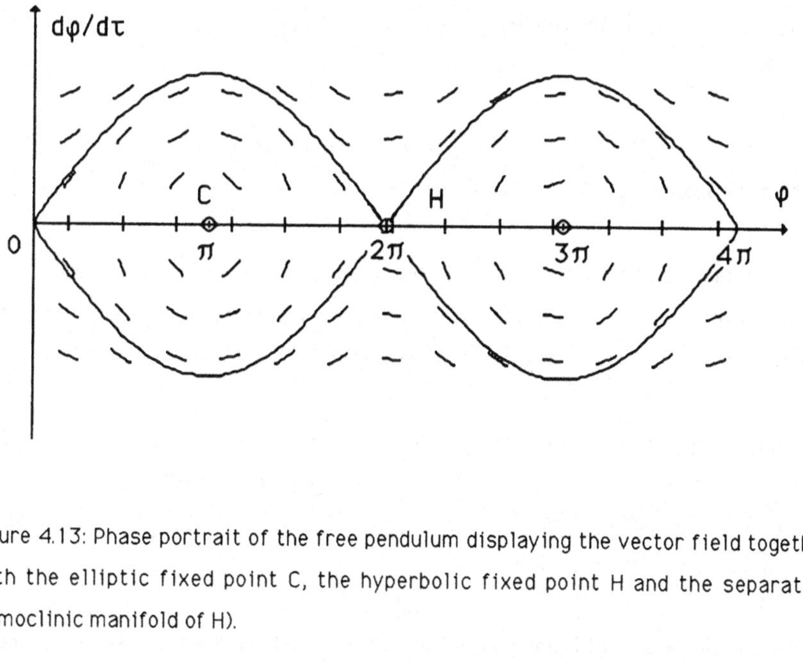

Figure 4.13: Phase portrait of the free pendulum displaying the vector field together with the elliptic fixed point C, the hyperbolic fixed point H and the separatrix (homoclinic manifold of H).

By $q_\varepsilon^s(0)$ we denote the point of the stable manifold lying on the normal to the unperturbed separatrix at the point $q_0(0)$; the analogous point on the unstable manifold is denoted $q_\varepsilon^i(0)$. We may then define the first order separations $q_1^s(\tau)$ and $q_1^i(\tau)$ such that:

$$(4) \qquad | q_\varepsilon^s(\tau) - q_0(\tau) - \sqrt{\varepsilon} q_1^s(\tau) | = O(\varepsilon), \text{ uniformly for } \tau \geq 0,$$

and

$$| q_\varepsilon^i(\tau) - q_0(\tau) - \sqrt{\varepsilon} q_1^i(\tau) | = O(\varepsilon), \text{ uniformly for } \tau \leq 0.$$

$q_1^s(\tau)$ and $q_1^i(\tau)$ are solutions of the variational system:

$$(5) \qquad dq_1(\tau)/d\tau = Df(q_0(\tau))q_1(\tau) + \sqrt{\varepsilon}g(q_0(\tau)),$$

where Df is the matrix of first derivatives of the unperturbed vectorfield. To lowest order in ε, the algebraic separation at the point $q_0(0)$ between the stable manifold $q_\varepsilon^s(\tau)$ and the unstable manifold $q_\varepsilon^i(\tau)$ is:

$$(6) \qquad d(q_0(0)) = q_\varepsilon^i(0) - q_\varepsilon^s(0) = \sqrt{\varepsilon}[f(q_0(0))/\| f(q_0(0)) \|] \wedge$$
$$\wedge\ (q_1^i(0) - q_1^s(0)),$$

where the exterior product is defined by $a \wedge b = a_1b_2 - a_2b_1$. This may be written:

$$(7) \qquad d(q_0(0)) = \sqrt{\varepsilon}[\Delta^i(0) - \Delta^s(0)]/\| f(q_0(0)) \|,$$

where $\Delta^s(\tau) =_{def} f(q_0(\tau)) \wedge q_1^s(\tau)$ and $\Delta^i(\tau) =_{def} f(q_0(\tau)) \wedge q_1^i(\tau)$.

$\Delta^s(\tau)$ is the solution to the equation:

$$(8) \qquad d\Delta^s/d\tau = [Df(q_0(\tau))dq_0/d\tau]\wedge q_1^s(\tau) + f(q_0(\tau))\wedge dq_1^s/d\tau$$

or, since $dq_0/d\tau = f(q_0(\tau))$:

$$(9) \qquad d\Delta^s/d\tau = [Df(q_0).f(q_0)]\wedge q_1^s + f(q_0)\wedge[Df(q_0).q_1^s] + f(q_0)\wedge g(q_0)$$

or, rewriting again:

$$(10) \qquad d\Delta^s/d\tau = [\text{Trace } Df(q_0)]\wedge\Delta^s + f(q_0)\wedge g(q_0).$$

The trace of Df vanishes here since the unperturbed system is Hamiltonian (with Hamiltonian $H(u, v) = v^2/2 - u/2 + \cos u$ in the conjugate variables u and v) and the evolution of $\Delta^s(\tau)$ is therefore given by the equation:

$$(11) \qquad d\Delta^s/d\tau = f(q_0(\tau)) \wedge g(q_0(\tau)).$$

Integrating this equation from $\tau = 0$ to $\tau = +\infty$ and taking into account the fact that $f(q_0(\tau)) \to 0$ as $\tau \to +\infty$, since q_0 tends to the fixed point H, we find:

$$(12) \qquad \Delta^s(0) = - \int_0^\infty f(q_0) \wedge g(q_0)\, d\tau.$$

In the same way, we find that:

$$(13) \qquad \Delta^i(0) = \int_{-\infty}^0 f(q_0) \wedge g(q_0)\, d\tau.$$

These formulas are valid in the case of an unperturbed Hamiltonian system, but more complicated formulas can also be obtained when the unperturbed field is not divergence free.

From these formulas, and from (7), we deduce that to lowest order in ε, $d(q_0(0))$ is given by:

$$(14) \qquad d(q_0(0)) = \sqrt{\varepsilon} \int_{-\infty}^{+\infty} f(q_0) \wedge g(q_0)\, d\tau \,/\, \| f(q_0(0)) \|$$
$$= \sqrt{\varepsilon} \int_{-\infty}^{+\infty} (f_1g_2 - f_2g_1)(q_0)\, d\tau \,/\, \| f(q_0(0)) \|$$
$$= \sqrt{\varepsilon} \int_\Gamma (g_2du - g_1dv) \,/\, \| f(q_0(0)) \|,$$

where the integral is calculated along the unperturbed separatrix Γ.
This yields:

(15) $$d(q_0(0)) = \sqrt{\varepsilon} \iint \mathrm{Trace}(Dg)\, du\, dv \,/\, \| f(q_0(0)) \|,$$

where the domain of integration is the interior of the unperturbed homoclinic loop.

In the present case, $\mathrm{Trace}(Dg) = -1$ and so, to lowest order in ε:

(16) $$d(q_0(0)) = -\sqrt{\varepsilon} A(\Gamma)/\| f(q_0(0)) \|,$$

where $A(\Gamma)$ designates the area of the unperturbed separatrix loop. Since $q_0(0)$ is situated outside the neighborhood V_ρ of the hyperbolic fixed point, we obtain:

(17) $$-2cA(\Gamma)\sqrt{\varepsilon}/\rho < d(q_0(0)) < -cA(\Gamma)\sqrt{\varepsilon}/\rho.$$

This result shows that the algebraic separation between the stable and unstable manifolds, calculated in the neighborhood of a point on the unperturbed separatrix outside of V_ρ, is of order $\sqrt{\varepsilon}$, and that the stable manifold is further displaced from the center than the unstable manifold, which was evident to begin with in view of the dissipative term.

The branch of the stable manifold is thus at a distance of order $\sqrt{\varepsilon}$ from the branch of the unstable manifold outside the neighborhood V_ρ, in the region $y \leqslant 0$. A simple local analysis in the neighborhood of the hyperbolic point assures that this separation remains order $\sqrt{\varepsilon}$ inside V_ρ. A straightforward argument (cf. Section 4.4) then serves to show that the measure of initial conditions giving rise to trapped trajectories is also of order $\sqrt{\varepsilon}$.

We next consider for further illustration another one-frequency case, that of a pendulum subjected to a force diminishing linearly with time, represented by:

(18) $$d^2\varphi/d\tau^2 = (1 - I_2(0) - \sqrt{\varepsilon}\tau) + \sin\varphi,$$

where $I_2(0)$ is a constant. This equation is equivalent to the system:

(19) $$dI_1/dt = \varepsilon(1 - I_2 + \sin\varphi); \quad dI_2/dt = \varepsilon; \quad d\varphi/dt = I_1 \quad (\tau = \sqrt{\varepsilon}t),$$

with associated averaged system:

(20) $$dJ_1/dt = \varepsilon(1 - I_2); \quad dJ_2/dt = \varepsilon.$$

It is easy to verify that condition N is satisfied away from the straight line $I_2 = 1$; condition A is satisfied for $I_2 < 0$. It is useful to keep in mind the phase portraits of the family of forced pendulums represented by:

(21) $$d^2\varphi/d\tau^2 = \mu + \sin\varphi, \quad \mu \geqslant 0,$$

beginning with the free pendulum, $\mu = 0$ (see Fig. 4.13 to 4.17).

For $0 < \mu < 1$, the hyperbolic fixed point admits an unstable branch and an open stable branch, as well as a homoclinic loop (Figures 4.14 and 4.15) and the size of the separatrix loop decreases with increasing μ.

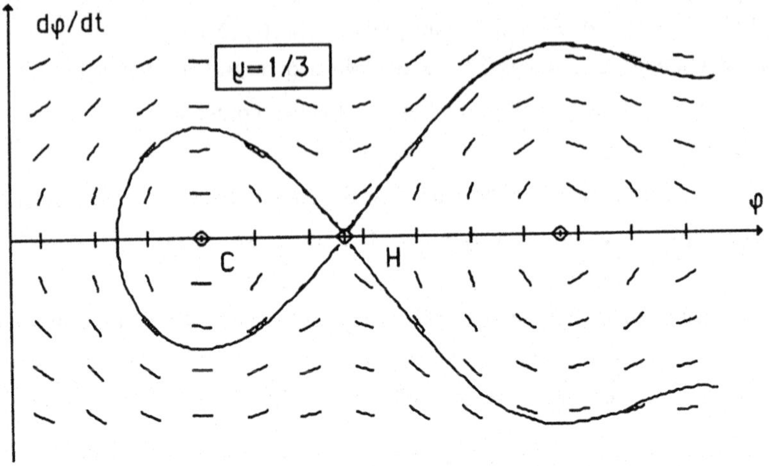

Figure 4.14: Phase portrait of the forced pendulum for constant driving force $\mu = 1/3$. The separatrix now consists of a homoclinic loop and two open branches.

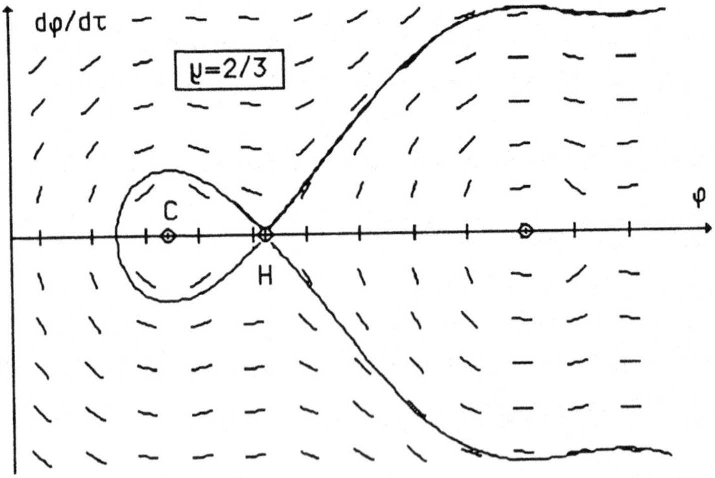

Figure 4.15: Phase portrait of the forced pendulum for higher driving force.

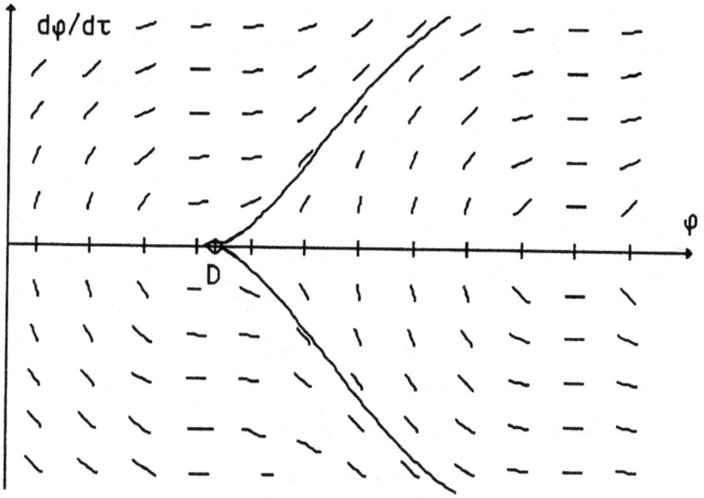

Figure 4.16: Phase portrait of the forced pendulum for the bifurcation value of the constant driving force ($\mu = 1$). The fixed points C and H of Figure 4.15 have coalesced into a single degenerate fixed point D.

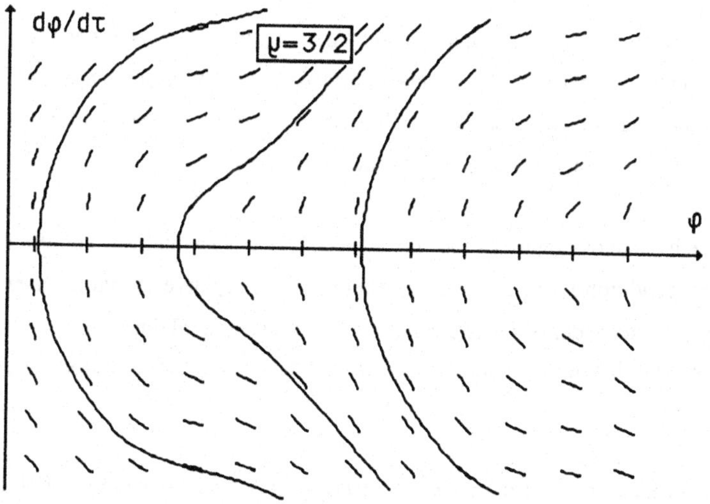

Figure 4.17: Phase portrait of the forced pendulum for high values of the constant driving force.

For $\mu = 1$, the bifurcation value, the fixed point is degenerate and the separatrix loop disappears (Figure 4.16) and for $\mu > 1$, there is no more fixed point, the only possible behavior being libration (Figure 4.17).

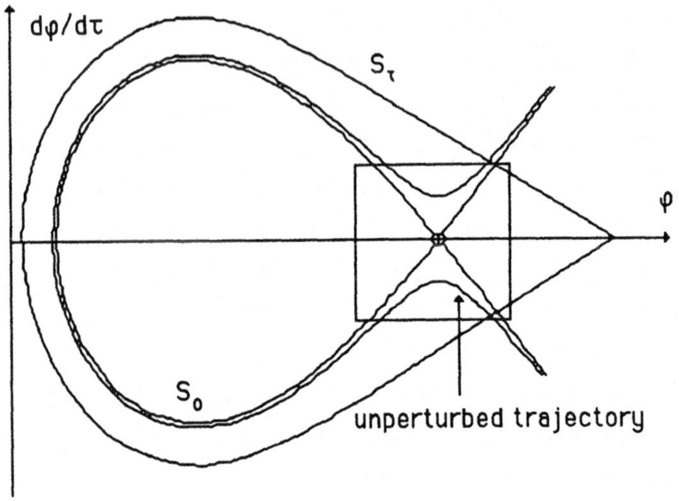

Figure 4.18: Explanation of the trapping mechanism. S_0 is the "frozen" separatrix at time $\tau = 0$ and S_τ the "frozen" separatrix at a later time. The exact trajectory lies close to the unperturbed trajectory also included in this figure. It starts outside S_0 but is ultimately trapped inside the loop which grows with time.

The mechanism of capture by the "resonance" $I_1 = 0$ is an intuitive one. We begin with an initial condition $I_1(0)$, $I_2(0)$, $\varphi(0)$, where $I_2(0)$ is negative so that the averaged trajectory remains in the domain where condition N is satisfied until time $1/\varepsilon$.

Suppose that this initial condition is such that the corresponding trajectory of the pendulum equation:

$$(22) \qquad d^2\varphi/d\tau^2 = 1 - I_2(0) + \sin\varphi$$

executes one turn around the separatrix loop, passing very close to the hyperbolic fixed point $H(I_2(0))$ (passage through the neighborhood of this point will thus take a long time). Since the force applied to the pendulum diminishes with time, the "instantaneous" separatrix, obtained by considering time as a parameter in the right hand side of (18), grows with time, so that the exact trajectory finds itself trapped inside the loop as it passes near $H(I_2(0))$.

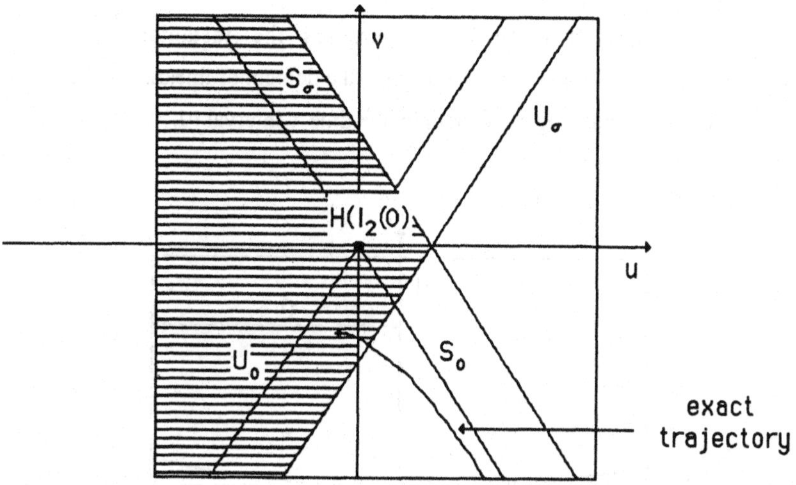

Figure 4.19: The trapping mechanism. One may linearize the system in the vicinity of the hyperbolic fixed point: U_0 and S_0 are respectively the "frozen" unstable and stable manifolds at time σ, U_σ and S_σ the same manifolds at some later time. The trapping area is hatched.

To estimate the measure of the set of initial conditions trapped in this way, we undertake a local analysis near the fixed point $H(I_2(0))$ by using a near-identity change of variables to linearize equation (22) in a neighborhood of $H(I_2(0))$ of size ρ. In the new variables x and y, this yields the system:

(23) $dx/d\tau = y; \quad dy/d\tau = \chi^2 x, \quad \text{where} \quad \chi = (1 - I_2(0)^2)^{1/4}.$

The perturbed trajectory enters the domain of linearization at time τ^* at a point defined by $y + \chi x = -\gamma$, $y = -\rho$ ($y + \chi x = 0$ is the equation of the stable manifold of $H(I_2(0))$). We translate the origin of time to τ^* by defining the new time $\sigma = \tau - \tau^*$. The subsequent evolution of the trajectory, for as long as it remains in the domain of linearization, is then given by:

(24) $dx/d\sigma = y; \quad dy/d\sigma = \chi^2 x - \sqrt{\varepsilon}\sigma,$

from which it follows that:

$$(25)\qquad \begin{aligned} x(\sigma) &= (\rho-y)/\chi\ \mathrm{ch}(\chi\sigma) - (\rho+\sqrt{\varepsilon}/\chi^2)/\chi\ \mathrm{sh}(\chi\sigma) + \sqrt{\varepsilon}\sigma/\chi^2 \\ y(\sigma) &= -\ (\rho+\sqrt{\varepsilon}/\chi^2)\ \mathrm{ch}(\chi\sigma) + (\rho-y)\ \mathrm{sh}(\chi\sigma) + \sqrt{\varepsilon}/\chi^2. \end{aligned}$$

The trajectory is trapped if it crosses the instantaneous trajectory $y - \chi x + \sqrt{\varepsilon}\sigma/\chi = 0$ at a time τ prior to leaving the domain of linearization, and prior to $\tau = 1/\sqrt{\varepsilon}$, as we are studying the system on the time τ interval $[0, 1/\sqrt{\varepsilon}]$.

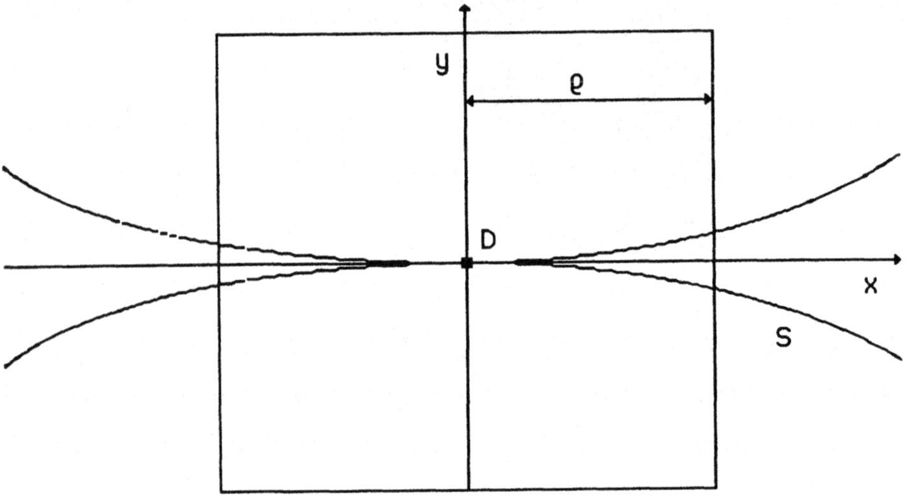

Figure 4.20: Shape of the separatrix S near the degenerate fixed point D.

It is not hard to see that this is the case for $y \in [0, \sqrt{\varepsilon}/\chi^2]$. Proceeding again as in Section 4.4, one shows that the measure of the set of initial conditions with exact trajectories which satisfy the inequality:

$$(26)\qquad 0 \geq y + \chi x \geq -\sqrt{\varepsilon}/\chi^2$$

on entering the domain of linearization around the hyperbolic point $H(I_2(0))$ is of order $\sqrt{\varepsilon}$. ■

It remains to establish the optimality of the estimate of maximal separation. In other words, we must show that upon excluding a set of initial conditions with measure no greater than order $\sqrt{\varepsilon}$, the estimate of $\sqrt{\varepsilon}\log(1/\varepsilon)$ for the maximal separation of the trajectories remains the best possible.

To do this it will suffice to reconsider the example of the pendulum with constant forcing:

(27) $d^2\varphi/d\tau^2 = 1/2 + \sin\varphi,$

which is equivalent to:

(28) $dI/dt = \varepsilon(1/2 + \sin\varphi\ ); \quad d\varphi/dt = I \quad (\tau = \sqrt{\varepsilon}t).$

In a neighborhood V_ρ of the hyperbolic fixed point, we reduce this equation to the form:

(29) $du/d\tau = v; \quad dv/d\tau = \sigma^2 u,$

where σ is the positive eigenvalue associated to the fixed point.

Consider a trajectory which enters V_ρ at time $\tau = 0$ at the point $(u = \rho,\ v + \sigma u = c_{15}\sqrt{\varepsilon})$ (for simplicity we assume that $\sigma < 1$). Its equation is:

(30) $v = -[\sigma^2 u^2 + C]^{1/2},$ where $C = (\sigma\rho + c_{15}\sqrt{\varepsilon})^2 - \sigma^2\rho^2,$

and it crosses V_ρ after time:

(31) $\tau^* = \int_{-\rho}^{+\rho} du/|\,v\,| = (2/\sigma)\ \mathrm{Arcsinh}(\rho\sigma/\sqrt{C}).$

Since C is less than $4c_{15}\sqrt{\varepsilon}\sigma\rho$, τ^* is greater than $(1/4\sigma)\log(1/\varepsilon)$ for ε small enough, and the original time t corresponding to τ^* is greater than $\log(1/\varepsilon)/(4\sigma\sqrt{\varepsilon})$.

The variation of I on the time interval $[0,\ \tau^*/\sqrt{\varepsilon}]$ is less than $2\sqrt{\varepsilon}\rho$, whereas, since $dJ/dt = \varepsilon/2$, the variation of J along the averaged trajectory with identical initial condition surpasses the value $\sqrt{\varepsilon}\ \log(1/\varepsilon)/(8\sigma)$ at time $\tau^*/\sqrt{\varepsilon}$. This proves the optimality of the estimate $\sqrt{\varepsilon}\ \log(1/\varepsilon)$ of the maximal separation. ▪

4.6 The case of a one-dimensional base

Proof of Theorem 2:

In the particular case of a single scalar variable I, it is possible to establish Theorem 2 of Section 4.1 without restricting the domain in the nongeneric case where degeneracies exist. The result thus obtained is optimal for the class of all systems (even degenerate ones) satisfying the usual conditions of regularity.

We begin by pointing out certain characteristics peculiar to dimension one:

- By condition N, ω is a strictly monotone function of I, and along an averaged trajectory $J(t)$ is also monotonic. Consequently, if degeneracies exist and are not too widely separated, the points corresponding to them will necessarily be visited by certain trajectories on the time interval $[0, 1/\varepsilon]$.

- The resonant zones are segments, and the resonances are reduced to points, a finite number of which may correspond to degeneracies.

- As before it is possible to analyze the motion in a resonant zone, *whether it is associated to a degenerate resonance or not.* We were previously restricted to nondegenerate resonances, but this restriction was only useful in analyzing passage through resonance, and was not necessary for reducing the system to a resonant normal form. The motion is again described by the equations (cf. Section 4.2, Proposition 2):

$$(1) \qquad d^2\gamma^*/dt^2 = q\varepsilon[\Phi_s(I^*(t),\gamma^*(t)) + L(I^*(t)) + \sqrt{\varepsilon}\alpha(I^*(t), \varphi^*(t), \lambda(t), \varepsilon)]$$

with $|\alpha| < cq$ and $|dI^*/dt| < c\varepsilon$.

Using I_0^* to designate the point where $\omega(I^*(0))$ takes the value p/q, we have, since a resonant zone is a segment of length less than $2c_3(\varepsilon/q)^{1/2}e^{-\sigma/2|k|}$:

$$(2) \qquad d^2\gamma^*/dt^2 = q\varepsilon[\Phi_s(I^*_0, \gamma^*(t))+L(I^*_0)] + \varepsilon^{3/2}\beta(I^*(t), \varphi^*(t), \lambda(t), \varepsilon)$$

where $|\beta| < cq$.

It should be noted that in the case where the dimension of the base is strictly greater than one, in general the best possible estimate is $|\beta| < cq(1 + \sqrt{\varepsilon}t)$. The variation of the action variables perpendicular to the resonance is of order $\sqrt{\varepsilon}$, but for the drift parallel to the resonance we have only the estimate $|dI^*/dt| < c\varepsilon$.

The previous results on passage through nondegenerate resonances and through the nonresonant domain remain valid. To prove Theorem 2 it will therefore suffice to examine passage through degenerate resonances and, since there are finitely many of them, we may restrict our attention to the resonance with strongest degeneracy.

In terms of the scaled time $\tau = \sqrt{\varepsilon}t$, this resonance is described by the equation:

$$(3) \qquad d^2\gamma^*/d\tau^2 = q[\Phi_s(I^*_0, \gamma^*(t))+L(I^*_0)] + \sqrt{\varepsilon}\beta.$$

The corresponding unperturbed pendulum admits a degenerate fixed point D. Far from this point we may proceed as before (cf. Section 4.3). To study the effect of the degeneracy we may limit ourselves to a local analysis near D, making use of a near-identity change of variables on the preimage of V_ρ ($|x| \leqslant \rho$, $|y| \leqslant \rho$). In the new coordinates (x, y), the unperturbed pendulum is described by the equations:

$$(4) \qquad dx/d\tau = y; \quad dy/d\tau = ax^{2p+1}.$$

The origin is taken to be the fixed point; the equation of the separatrix is $y^2 = ax^{2(p+1)}/(p+1)$, which contacts the axis $y = 0$ to order p (in general, the equation of the unperturbed trajectories in the box is $y^2 = ax^{2(p+1)}/(p+1) + E$).

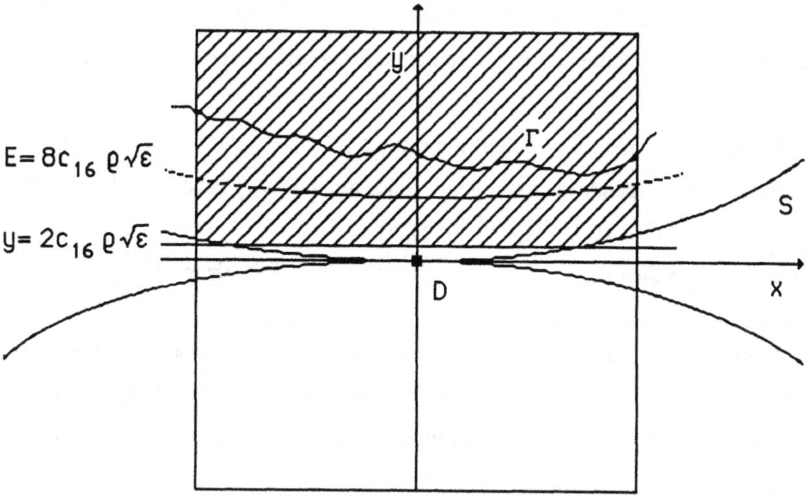

Figure 4.21: Passage of the perturbed trajectory Γ through a neighborhood of the degenerate fixed point D. Γ stays above the separatrix S and the curve $E = 8c_{16}\rho\sqrt{\varepsilon}$.

Consider a perturbed trajectory which enters V_ρ in the fourth quadrant at a point situated above the unperturbed separatrix (we will treat only this case, as the others are similar). In V_ρ, the perturbed pendulum is described by the equations:

(5) $\quad dx/d\tau = y + \sqrt{\varepsilon}X; \quad dy/d\tau = ax^{2p+1} + \sqrt{\varepsilon}Y$

where $|X| \leqslant c_{16}\rho$ and $|Y| \leqslant c_{16}$.

We introduce the variable $z = y^2 - ax^{2(p+1)}/(p+1)$ which measures the distance to the separatrix. We assume that ρ is very small and we consider only perturbed trajectories with initial conditions satisfying:

(6) $\quad x(0) = -\rho; \quad 16c_{16}\rho\sqrt{\varepsilon} < z(0) < \rho.$

In V_ρ, the evolution of z is given by the equation:

(7) $\quad dz/d\tau = 2\sqrt{\varepsilon}[yY - ax^{2p+1}X].$

As long as the perturbed trajectory remains in the part of V_ρ situated above the separatrix and the straight line $y = 2c_{16}\rho\sqrt{\varepsilon}$, we have:

(8) $dx/d\tau > y/2$ (x is therefore monotonic) and

$| ax^{2p+1}X/y | < c_{16}\rho^{p+1}[a(p+1)]^{1/2}$.

Consequently:

(9) $| dz/d\tau | \leqslant 2\sqrt{\varepsilon}[| Y | + a| x |^{2p+1}| X |/y] \leqslant 4c_{16}\sqrt{\varepsilon}$.

The maximal variation of z is thus less than $8c_{16}\rho\sqrt{\varepsilon}$. Since $z(0) > 16c_{16}\rho\sqrt{\varepsilon}$, as the perturbed trajectory traverses V_ρ, it remains *above* the unperturbed trajectory (with equation $y^2 - ax^{2(p+1)}/(p+1) = 8c_{16}\rho\sqrt{\varepsilon}$).

This result allows us to bound the time $\tau^*[z(0)] = \int_{-\rho}^{\rho} dx/[y + \sqrt{\varepsilon}X]$ which the perturbed trajectory spends in crossing V_ρ. Since the trajectory is contained in the region:

(10) $y^2 - ax^{2(p+1)}/(p+1) \geqslant 8c_{16}\rho\sqrt{\varepsilon}$,

the time τ^* is less than $(2\rho/c_{16})^{1/2}\varepsilon^{-1/4}$ (for any value of the degeneracy p).

To summarize, a perturbed trajectory which enters V_ρ at a distance (as measured by z) greater than $16c_{16}\rho\sqrt{\varepsilon}$ from the unperturbed separatrix leaves this domain in a time τ^* less than $(2\rho/c_{16})^{1/2}\varepsilon^{-1/4}$ and remains a distance of at least $8c_{16}\rho\sqrt{\varepsilon}$ from the unperturbed separatrix throughout its passage through V_ρ.

Since, by condition N, an unperturbed trajectory outside the maximal separatrix loop may not pass through the neighborhood of the degenerate fixed point more than twice, it is easy to show that every perturbed trajectory which enters the strip $| d\gamma^*/d\tau | \leqslant 2c_3 q\delta$ in the phase plane of the unperturbed pendulum at a distance greater than $c\sqrt{\varepsilon}$ from the open separatrix branches of the unperturbed pendulum crosses this strip in a time t^* less than $c\varepsilon^{-3/4}$.

For such initial conditions, the separation between an exact and averaged trajectory does not exceed $c\varepsilon^{1/4}$ at the instant when the perturbed trajectory leaves the resonant strip. Proceeding once again as in Section 4.4, this allows us to establish the first part of Theorem 2 (maximal separation of order $\varepsilon^{1/4}$).

We estimate the measure of the initial conditions to be excluded (the set V'') as in Section 4.4 and we obtain the same bound $mes(V'') < c\sqrt{\varepsilon}$, which proves the second part of the theorem.

Estimating the average separation also proceeds as in Section 4.4. The part of the separation integral over the domain V'' is no trouble, since the measure of V'' is bounded by $c\sqrt{\varepsilon}$.

As for the integral $\int_{V'} | I(t) - J(t) | dI_0 \, d\varphi_0 d\lambda_0$ over V', it is bounded by $c\sqrt{\varepsilon} + \varepsilon N'^2 \langle T \rangle \, mes(V')$, where $\langle T \rangle$ is a uniform bound of the crossing time for a low-order resonance, starting from an initial condition in V'.

Let us estimate $\langle T \rangle$. We consider a resonance of order less than N' and a trajectory with initial condition in V' which enters the resonant strip at $t = 0$ at the point I_0^*. Its subsequent evolution is given by equation (3), and using the normal form (4), we show that:

- If the initial phase $y^*(0)$ of the trajectory is located a distance x from the intersection of the separatrix with the boundary of the resonant zone, and x is between $c_{17}\sqrt{\varepsilon}$ and $c_{18}\rho$, then the time τ which the trajectory takes to cross the resonant strip is less than $c(x - c_{17}\sqrt{\varepsilon})^{-p/(2p+2)}$.
- If $y^*(0)$ is located a distance greater than $c_{18}\rho$ from all open branches of stable manifolds, then the scaled crossing time is less than c/ρ.

From these facts we deduce that τ, the scaled average crossing time for a low-order resonance, is bounded by $c/\rho + c\int_{c_{17}\sqrt{\varepsilon}}^{c_{18}\rho} dx/(x-c_{17}\sqrt{\varepsilon})^{p/(2p+2)}$.

Since the integral $\int_0^1 z^{-p/(2p+2)} \, dz$ converges, we see that in terms of the unscaled time t, we may choose $\langle T \rangle \leqslant c\sqrt{\varepsilon}$ provided that c is large enough, which shows that the average separation $M(\varepsilon)$ does not exceed $c\sqrt{\varepsilon}$.

Remark:

We note that it is possible to slightly improve the above estimate $\tau \sim \varepsilon^{-1/4}$ of the time of passage through the neighborhood of a degenerate fixed point, and to obtain a more refined bound on the maximal separation which depends on the order of the degeneracy. In fact, the scaled time of passage τ^* satisfies the inequality:

$$(11) \qquad \tau^* \leqslant 2\int_{-\rho}^{\rho} dx/y \leqslant \int_{-\rho}^{\rho} dx/[ax^{2(p+1)}/(p+1) + 8c_{16}\rho\sqrt{\varepsilon}]^{1/2} \leqslant c\varepsilon^{-p/4(p+1)}.$$

This allows us to show that for initial conditions in V':

$$(12) \qquad \operatorname{Sup}_{t\in[0,\,1/\varepsilon]} | I(t) - J(t) | < c\varepsilon^{(p+2)/4(p+1)},$$

where p is the maximal order of degeneracy in the problem.

We turn finally to the proof of the optimality of the results obtained. It will suffice to treat the estimate of maximal separation, as optimality for the two other results is shown as before, using nondegenerate systems. For this purpose, we consider the equation:

$$(13)\qquad d^2\varphi/d\tau^2 = \varepsilon(1/2 + \sin\varphi)(\sin\varphi)^{2p+1}$$

which is equivalent to the system:

$$(14)\qquad dI/dt = \varepsilon(1/2 + \sin\varphi)(\sin\varphi)^{2p+1};\quad d\varphi/dt = I \quad (\tau = \sqrt{\varepsilon}t).$$

In a neighborhood V_ϱ of the fixed point $(0, 0)$ with index -1 and degeneracy of order p, the equations take the normal form:

$$(15)\qquad dx/d\tau = y;\quad dy/d\tau = x^{2p+1}/2.$$

A trajectory entering V_ϱ at the point $(x = \rho,\ y = -\rho^{p+1}/(p+1)^{1/2} - c_{19}\sqrt{\varepsilon})$, with initial condition situated under the open branch of a stable manifold at a distance of order $\sqrt{\varepsilon}$, is represented in V_ϱ by the equation:

$$(16)\qquad y = -[x^{2p+2}/(p+1) + C]^{1/2},$$

where $C = \{\rho^{p+1}/(p+1)^{1/2} + c_{19}\sqrt{\varepsilon}\}^2 - \rho^{2p+2}/(p+1)$.

The trajectory crosses V_ϱ in the scaled time:

$$(17)\qquad \tau^* = 2\int_0^\varrho dx\ /\ [x^{2p+2}/(p+1) + C]^{1/2} > c\varepsilon^{-p/(4p+4)}.$$

We have $|\, I(\tau^*) - I(0)\,| \leqslant 2\rho\sqrt{\varepsilon}$, whereas:

$$(18)\qquad |\, J(\tau^*) - I(0)\,| \geqslant c\varepsilon^{p/(4p+4)}$$

since the averaged system is $dJ/dt = \varepsilon/2$.

Taking p arbitrarily large, this result shows that the estimate in Theorem 2 (and the estimate (12)) are optimal. ∎

CHAPTER 5: N FREQUENCY SYSTEMS; NEISTADT'S RESULT BASED ON ANOSOV'S METHOD

5.1 Introduction and results

This chapter and the next are devoted to n frequency systems in which the fast variables φ belong to T^n, the n dimensional torus, and the unperturbed motion is quasiperiodic. The various resonant surfaces are no longer disjoint (cf. Appendix 3) and a detailed study of passage through resonance leading to global estimates is no longer possible. At least two approaches are possible: The first, which is the subject of this chapter, is based on Neistadt's application [Nei3] of Anosov's work to n frequency systems (see Chapter 2). The second approach, discussed in the following chapter and also due to Neistadt [Nei4], relies on Kasuga's idea of an average (L^1) solution to the linearized equation; the original version of this idea which arised in the context of ergodic systems may be found in Chapter 9. We should point out that the second approach furnishes the optimal result, and for this reason the present chapter may be viewed as an illustrative application of Anosov's general theorem presented in Chapter 2.

In this and the next chapter we will still be concerned with systems of the form:

(1) $\qquad dI/dt = \varepsilon f(I, \varphi, \varepsilon); \quad d\varphi/dt = \omega(I) + \varepsilon g(I, \varphi, \varepsilon)$

where $I \in \mathbb{R}^m$ and here $\varphi \in T^n$ $(n > 1)$.

The associated averaged system is of course:

(2) $\qquad dJ/dt = \varepsilon f_0(J); \quad f_0(J) = 1/(2\pi)^n . \oint f(J, \varphi, 0)\, d\varphi.$

Anosov's theorem (cf. Chapter 2) applies in this particular case provided that a nondegeneracy condition is satisfied. We recall that this theorem states that for arbitrary fixed ρ, the measure of the set of initial conditions leading to a separation greater than ρ between exact and averaged trajectories tends to 0 with ε. A (nonoptimal) estimate of $O([\rho^2 \log(1/\varepsilon)]^{-1})$ for the measure of this set in the case of quasiperiodic systems of the form (1) was first obtained by Kasuga, using methods from functional analysis. Employing techniques inspired by Anosov's method and taking into account the quasiperiodic motion on the torus, Neistadt improved this estimate to $O([\varepsilon/\rho^3]^{1/2})$. As we mentioned, this result is not optimal either, but was further refined to the optimal bound $O(\sqrt{\varepsilon}/\rho)$ by Neistadt (cf. Chapter 6), who made use of methods from Kasuga's work on the adiabatic ergodic theorem.

In what follows, I belongs to K, a smooth compact subset of $\mathbb{R}^m$ such that $K - \eta$ is non empty for small enough η and we again work on $K' \times T^n$, where K' is the set of initial conditions such that the corresponding solutions of the averaged system remain in $K - \eta$ on the time interval $[0, \tau^*/\varepsilon]$; η is a some small positive constant and we set $\tau^* = 1$ again for simplicity (which amounts to a scaling of $\varepsilon \in [0, \varepsilon_0]$).

All functions are assumed C^1 in I, φ and ε, and ω does not vanish on K. Moreover, the function $f(I, \varphi, 0)$ (and so also $\partial f/\partial I_k$) will first be assumed analytic in φ for $|\mathrm{Im}\varphi| < \sigma$, but in Section 5.3, we will show that this may be relaxed to finite differentiability. We also assume that *the frequencies satisfy a nondegeneracy condition* (cf. Appendix 3); in this chapter we adopt Kolmogorov's condition, which in particular imposes $m \geq n$, but it is also possible to work with Arnold's nondegeneracy condition.

For fixed $\varepsilon > 0$ and $\rho(\varepsilon) > 0$ (here ρ may depend on ε), we partition $D = K' \times T^n$ into two disjoint subsets $V'(\varepsilon, \rho)$ and $V''(\varepsilon, \rho)$. $V''(\varepsilon, \rho)$ is the set of ρ-deviant initial conditions, that is, the set of values $(I_0, \varphi_0) = (I(0), \varphi(0))$ for which the maximal separation between the exact trajectory with initial condition (I_0, φ_0) and the averaged trajectory with initial condition I_0 is greater than ρ:

$$(I_0, \varphi_0) \in V''(\varepsilon, \rho) \quad \Leftrightarrow \quad \sup_{[0, 1/\varepsilon]} \| I(t) - J(t) \| > \rho.$$

The result obtained by Neistadt is the following:

<u>Theorem</u> :

For $c_1 \varepsilon^{1/3} < \rho(\varepsilon) < c_2$, the measure of $V''(\varepsilon, \rho)$ is less than $c(\varepsilon/\rho^3)^{1/2}$.

Before giving the proof, it is perhaps useful to point out the link between the resonances and what we called in Chapter 2 the "ergodization" of a function, or the convergence of its temporal average to its spatial average.

For a nondegenerate n frequency system, satisfying, say, Kolmogorov's condition $(\mathrm{rank}\ (\partial\omega_i/\partial I_j)_{i,j} = n)$, the unperturbed motion is quasiperiodic and ergodic on almost every fiber. Ergodicity is lacking only on the resonances ($k.\omega = 0$, $k \in \mathbb{Z}^n$ and $k \neq 0$); in this case, additional integrals of the motion exist which foliate the torus, and the motion is ergodic on lower dimensional tori T^{n-r}; r is the dimension of the resonant module (cf. Appendix 3).

Since the original fiber is the n torus T^n, Fourier analysis may be used and we restrict ourselves to examining the rate of ergodization of functions belonging to the complete basis $\{e^{ik.\varphi}\}_{k\in\mathbb{Z}^n}$. The time average of such a function for the unperturbed motion over a nonresonant torus ($k.\omega \neq 0$) is:

$$(3) \qquad 1/T \int_0^T e^{ik.\varphi(t)}\, dt = e^{ik.\varphi(0)}\, 1/T.\int_0^T e^{i(k.\omega(I))t}\, dt$$
$$= e^{ik.\varphi(0)} [e^{i(k.\omega)T} - 1]/[(k.\omega)T], \quad k \neq 0.$$

As T tends to ∞ this converges to 0, which is the spatial average over the torus, and the rate of convergence may be estimated by means of the trivial bound:

$$(4) \qquad | 1/T. \int_0^T e^{ik.\varphi(t)}\, dt | \leq 2\, T^{-1} | k.\omega |^{-1}.$$

The time required to "ergodize" the function $e^{ik.\varphi}$ to within χ is therefore:

$$T_0 = (2/\chi) | k.\omega |^{-1}.$$

It is inversely proportional to the distance $| k.\omega |$ to the resonance with resonant vector k.

If we surround the resonance by a resonant zone of size δ defined by $| \omega(I).k | \leq \delta$, then outside this zone the function $e^{ik.\varphi}$ will be ergodized to within χ after time $T_0 = 2/(\delta\chi)$. We thus see that it is equivalent to consider nonresonant domains, as Arnold and Neistadt do, or zones where the basis functions $e^{ik.\varphi}$ are rapidly ergodized, as Anosov does in the more general case. The size of the resonant zones and the time of ergodization play inverse roles in the two points of view, and this gives rise to two different ways of expressing the elementary lemmas on the decrease of oscillating integrals (essentially the Riemann-Lebesgue lemma).

We note that the rate of ergodization of basis elements increases with their index ($|1/T.\int_0^T e^{ink.\varphi(t)}\, dt | \leq 2/(nT).| k.\omega |^{-1}$), which is a reflection of the lesser importance of high-order resonances.

In view of these remarks, the principal steps in the proof of the theorem are clear:

- One takes only a finite number of resonances into account (the ultraviolet cutoff N). As we shall see shortly, N is on the order of $\log(1/\rho)$ because of the exponential decrease of the Fourier coefficients of f. One then defines a resonant domain R around these resonances.
- One introduces a priori, as in Chapter 2, a partition of D into two subsets:

$$D = LT(\zeta, R, \varepsilon) \cup ST(\zeta, R, \varepsilon);$$

$LT(\zeta, R, \varepsilon)$ will be the set of initial conditions with associated trajectories that spend a *long time* in the resonant domain; trajectories starting in $ST(\zeta, R, \varepsilon)$ spend a *short time* in R.
- One shows (Lemma 1) that for every initial condition in $ST(\zeta, R, \varepsilon)$, the separation on

the base between corresponding exact and averaged trajectories is small. This is the analogue of Anosov's Lemma 1 from Chapter 2. To prove it, one uses arguments analogous to Arnold's reasoning for two frequency systems (see Chapter 3). In the nonresonant zone, the usual change of variables is used to show that the separation remains small. In the resonant domain, one shows that the trajectories spend too little time to drastically increase the separation. Lemma 1 thus indicates that exact trajectories with ρ-deviant initial conditions spend a long time in the resonant domain; in other words:

$$V''(\rho, \varepsilon) \subset LT(\zeta, R, \varepsilon).$$

- To complete the proof of the theorem it thus suffices to estimate the measure of the set $LT(\zeta, R, \varepsilon)$. This is the purpose of Lemma 2 which is the analogue of Anosov's Lemma 2 in Chapter 2, and whose proof is almost identical.

5.2 Proof of the theorem

We first define the ultraviolet cutoff frequency. $f(I, \varphi, 0)$ may be expressed in terms of its Fourier series, with the Nth order remainder denoted by $R_N(I, \varphi)$:

(1) $\qquad f(I, \varphi, 0) = \sum_k f_k(I)e^{ik.\varphi}, \quad R_N(I,\varphi) = \sum_{|k|>N} f_k(I)e^{ik.\varphi}.$

Here we adopt the norm $|k| = \sup_i |k_i|$ to measure the order of a resonance.

For fixed χ, $0 < \chi < 1/2$, the norm of R_N is less than χ for $N \geqslant N(\chi) = [c \log(1/\chi)]$, by reason of the analyticity of $f(I, \varphi, 0)$ (cf. Appendix 1). Henceforth we shall take into account only resonances of order less than N, χ being determined below.

Around each of these resonances, we define a resonant zone $V(\delta, k)$ of size δ as the set of points I in K such that $|k.\omega(I)| \leqslant \delta/|k|^n$. The resonant domain is the finite union of these resonant zones: $R(\delta, N) = \{ \cup V(\delta, k),\ 1 \leqslant |k| \leqslant N \}$, and its measure is given by:

Proposition 1:

For sufficiently small δ_0, $\delta \leqslant \delta_0$ implies $\text{mes}[R(\delta, N)] < c\delta\, \text{mes}(K)$ (ordinary Lebesgue measure on $\mathbb{R}^m$).

For the (elementary) proof of this proposition, we refer to Appendix 4; we point out however that the nondegeneracy condition is of course indispensable at this stage. ▪

As in Chapter 2, to every initial condition $(I_0, \varphi_0) = (I(0), \varphi(0))$ in D we associate the subset $S(\varepsilon, I_0, \varphi_0, R)$ of $[0, 1/\varepsilon]$ consisting of times which the exact trajectory spends in the resonant zone $R(\delta, N)$ prior to leaving $K - \eta/4$ for the first time, at a time we denote by T and which, a priori, may be less than $1/\varepsilon$:

$$S(\varepsilon, I_0, \varphi_0, R) = \{\tau, 0 \leqslant \tau \leqslant 1/\varepsilon, \tau \leqslant T(I_0, \varphi_0, \varepsilon, \eta) \text{ and } I(\tau) \in R(\delta, N)\}.$$

We also introduce, for $0 < \zeta < 1$, the decomposition:

$$D = LT(\varepsilon, \zeta, R) \cup ST(\varepsilon, \zeta, R)$$

mentioned above: $LT(\varepsilon, \zeta, R) = \{(I_0, \varphi_0) \in D$ such that $\mathrm{mes}[S(\varepsilon, I_0, \varphi_0, R)] \geqslant \zeta/\varepsilon\}$.

Lemma 1 estimates the separation between exact and averaged trajectories with the same initial condition in $ST(\varepsilon, \zeta, R)$:

Lemma 1:

For every initial condition in $ST(\varepsilon, \zeta, R)$, $I(t)$ remains in $K - \eta/4$ (and thus in K) until at least time $t = 1/\varepsilon$, and

$$(2) \qquad \sup_{t\in[0, 1/\varepsilon]} \| I(t) - J(t) \| \leqslant \lambda$$

provided that $\lambda = c_3(\zeta + \chi + \varepsilon/\delta^2 + \varepsilon\zeta |\log \chi|^{n+1}/\delta^2)$ is strictly less than $\eta/2$ and that the parameters ε, δ, χ and ζ are correctly chosen (see below).

The proof will be broken into several subsidiary propositions more or less parallel to those in Chapter 2. We first consider an initial condition (I_0, φ_0) in $ST(\varepsilon, \zeta, R)$. The corresponding averaged trajectory remains in $K - \eta$ until at least time $t = 1/\varepsilon$, but the exact trajectory leaves $K - \eta/4$ at a time T which, a priori, may be less than $1/\varepsilon$. As usual, we carry out the first estimates of separation on the interval $[0, T]$, then complete the proof of Lemma 1 by using a bootstrap argument to show that in fact $T \geqslant 1/\varepsilon$.

Associated to the initial condition under consideration is the set $S = S(\varepsilon, I_0, \varphi_0, R)$ of time intervals which the trajectory spends in R. For $0 < \xi < 1$, we introduce the subset $LP(\xi, S)$ of S (LP for long passages) comprising the union of all segments of length greater than ξ/ε. Because $\mu(S) < \zeta/\varepsilon$, since the initial condition is in $ST(\varepsilon, \zeta, R)$, $LP(\xi, S)$ contains at most $[\zeta/\xi]$ segments. $SP(\xi, S)$ (SP for short passages) the closure of the complement of $LP(\xi, S)$ in $[0, 1]$, contains at most $[\zeta/\xi] + 1$ segments, which we denote from now on by $[a_i, b_i]$. Each of these segments corresponds to a time interval which the trajectory spends in the nonresonant domain, with the possible exception of brief excursions of duration less than ξ/ε (see Figure 5.1).

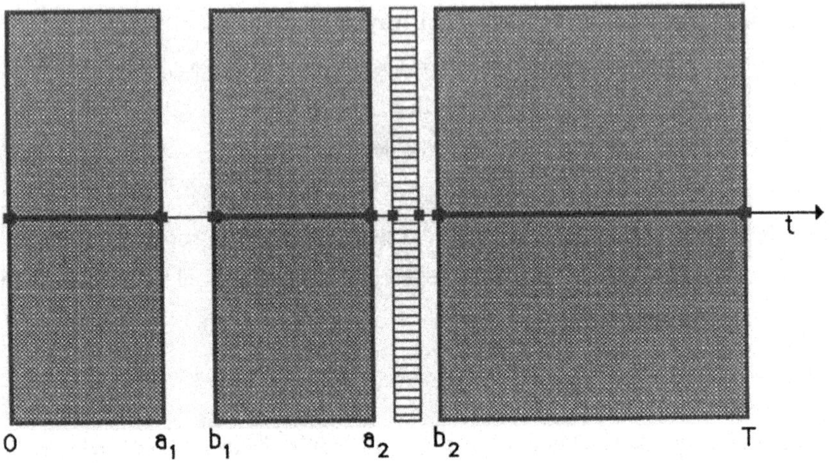

Figure 5.1: Partition of the segment [0, T]. Resonance crossings (subset $S(\varepsilon, I_0, \varphi_0, R)$ of [0, T]) appear as blocks in this schematic representation: Shaded blocks are used for long passages (subset $LP(\xi, S)$ of S) and hatched blocks for short passages (subset $SP(\xi, S)$ of S).

For technical reasons, we define a new nonresonant domain, larger than the previous one, by considering half-size resonant zones. This new domain is defined as:

$$NR = NR(\delta, N) = K - R(\delta/2, N).$$

We fix the value of ξ at $\xi = \delta/(c_4 N^{n+1})$, where c_4 is an upper bound of $\sum_k |f(I, \varphi, \varepsilon).\partial\omega_k/\partial I|$ for $(I, \varphi, \varepsilon)$ in $K \times T^n \times [0, \varepsilon_0]$. We may now state

<u>Proposition 2</u>:

For the above choice of ξ, the exact trajectory is located in the nonresonant domain $NR(\delta, N)$ at all points in time belonging to $SP(\xi, S)$.

This may be seen as follows: $SP(\xi, S)$ is the union of $[0, T] - S$ and the closure in S of $S - LP(\xi, S)$. If t does not belong to S, then by definition:

$$(3) \qquad |k.\omega(I)| > \delta/|k|^n > \delta/(2|k|^n),$$

and $I(t)$ belongs to $NR(\delta, N)$. If t belongs to S and is in the closure of the complement of $LP(\xi, S)$, so that t corresponds to a brief excursion of the trajectory into $R(\delta, N)$, then there exists a time t' which does not belong to S such that $|t' - t| < \xi\varepsilon/2$. We then have:

$$(4) \qquad |k.\omega(I(t))| > |k.\omega(I(t'))| - c_4\xi|k|/2$$
$$> \delta/|k|^n - \delta|k|/(2N^{n+1}) > \delta/(2|k|^n),$$

and $I(t)$ is again in $NR(\delta, N)$. ∎

We now proceed to the usual change of variables $P = I + \varepsilon S(I, \varphi)$ on the nonresonant domain $NR(\delta, N)$, which will eliminate the oscillating terms of f to first order (to within χ here). The truncated Fourier series of $f(I, \varphi, 0)$ is again defined by:

$$(5) \qquad f_N(I, \varphi) = f(I, \varphi, 0) - f_0(I) - R_N(I, \varphi).$$

S will be the trigonometric polynomial solution of the equation:

$$(6) \qquad \omega(I).\partial S(I, \varphi)/\partial\varphi = - f_N(I, \varphi).$$

The Fourier coefficients of S are written $S_k(I) = if_k(I)(k.\omega(I))^{-1}$, and we easily obtain, using the φ analyticity of $f(I, \varphi, 0)$, the following bounds on $K \times T^n$:

$$(7) \qquad \mathrm{Sup}_i |S_i| < c_5/\delta, \quad \mathrm{Sup}_{i,k} |\partial S_i/\partial I_k| < c/\delta^2, \quad \mathrm{Sup}_{i,k} |\partial S_i/\partial\varphi_k| < c/\delta.$$

In addition, for $I(t)$, $\varphi(t)$ such that $P(I(t), \varphi(t))$ is in K, we have the following bound:

<u>Proposition 3</u>:

$$\| dP/dt - f_0(P) \| < c(\varepsilon\chi + \varepsilon^2/\delta^2).$$

We write:

$$(8) \qquad dP_i/dt = \textstyle\sum_k(\delta_{ik}+\varepsilon\partial S_i/\partial I_k)dI_k/dt + \varepsilon\sum_k d\varphi_k/dt\, \partial S_i/\partial\varphi_k$$
$$= \varepsilon \textstyle\sum_k(\delta_{ik}+\varepsilon\partial S_i/\partial I_k) f_k(I, \varphi, \varepsilon) + \varepsilon\sum_k \partial S_i/\partial\varphi_k (\omega_k(I)+$$
$$+\varepsilon g_k(I, \varphi, \varepsilon)).$$

The desired bound is now obtained by using the estimate of $\| R_N \|$ and the fact that S solves (6) and satisfies the inequalities (7). ∎

Suppose T is the time (less than $1/\varepsilon$) at which the exact trajectory leaves $K - \eta/4$. We assume that $\varepsilon c_5/\delta$ is less than $\eta/4$ (we will see later how this is brought about), which ensures that on the time interval $[0, T]$, $I(t)$ is in $K - \eta/4$, $J(t)$ is in $K - \eta$ and $P(I(t), \varphi(t))$ is in K. We may thus work in the domain K. Proceeding recursively, we first estimate the separation between $I(t)$ and $J(t)$ on this time interval (to avoid repetition, we give an alternate to the proof in Chapter 3).

For a segment $[a_i , b_i]$ corresponding to passage through the nonresonant domain $NR(\delta, N)$, we have:

Proposition 4:

On the segment $[a_i , b_i]$:

(9) $$\| I(t) - J(t) \| \leqslant e^{c\varepsilon(b_i - a_i)} \| I(a_i) - J(a_i) \| + \varepsilon c(\chi + \varepsilon/\delta^2)(b_i - a_i) + \\ + c\varepsilon/\delta.$$

According to Proposition 3, on such a segment:

(10) $$\| dJ/dt - dP/dt \| < c\varepsilon \| J - P \| + c\varepsilon(\chi + \varepsilon/\delta^2).$$

Using Gronwall's lemma, we obtain:

(11) $$\| J(t) - P(t) \| < e^{c\varepsilon(b_i - a_i)} [\| J(a_i) - P(a_i) \| + c\varepsilon(b_i - a_i)(\chi + \varepsilon/\delta^2)]$$

and, since $\|I - J\| \leqslant \varepsilon \| S \| \leqslant c_5 \varepsilon/\delta$, (9) is in fact satisfied. ■

The above proposition will allow us to estimate the separation between $I(t)$ and $J(t)$ on $[0, T]$. For simplicity, we assume that $a_1 = 0$ and $b_p = T$. By Proposition 4, on the nonresonant segment $[0, b_1]$ we have the bound:

(12) $$\| I(t) - J(t) \| < c[b_1(\varepsilon\chi + \varepsilon^2/\delta^2) + \varepsilon/\delta].$$

The succeeding interval $[b_1 , a_2]$ corresponds to a passage through resonance. On this interval we have:

(13) $$\begin{aligned} \|I(t) - J(t)\| &\leqslant \|I(b_1) - J(b_1)\| + \| I(t) - I(b_1)\| + \| J(t) - J(b_1) \| \\ &\leqslant \|I(b_1) - J(b_1) \| + c\varepsilon(a_2 - b_1), \end{aligned}$$

and therefore, if $M(x, y)$ is the maximum of $\| I(t) - J(t) \|$ on $[x, y]$:

(14) $$M(0, a_2) \leqslant M(0, b_1) + c\varepsilon(a_2 - b_1).$$

The succeeding interval $[a_2 , b_2]$ corresponds to passage through the nonresonant domain and, proceeding as before:

(15) $$M(a_2 , b_2) \leqslant e^{c\varepsilon(b_2 - a_2)} M(0, a_2) + \varepsilon c[(\chi + \varepsilon/\delta^2)(b_2 - a_2) + 1/\delta],$$

from which it follows that:

(16) $$M(0, b_2) \leqslant e^{c\varepsilon(b_2 - a_2)} M(0, b_1) + \varepsilon c(a_2 - b_1) e^{c\varepsilon(b_2 - a_2)} + \\ + \varepsilon c[(\chi + \varepsilon/\delta^2)(b_2 - a_2) + 1/\delta].$$

Generally, one obtains the recurrence relation:

(17) $$M(0, b_{r+1}) \leqslant \alpha_{r+1} M(0, b_r) + \beta_{r+1}$$

where $\alpha_{r+1} = e^{c\varepsilon(b_{r+1}-a_{r+1})}$ and

$$\beta_{r+1} = c\varepsilon(b_{r+1}-a_{r+1})e^{c\varepsilon(b_{r+1}-a_{r+1})} + c\varepsilon/\delta + c\varepsilon(b_{r+1}-a_{r+1})(\chi+\varepsilon/\delta^2).$$

From this we deduce that:

$$(18) \qquad M(0, T) \leq e^{c\varepsilon T}[cp\varepsilon/\delta + c\varepsilon\textstyle\sum_{2\leq k\leq p}(a_k - b_{k-1}) + \\ + \varepsilon c(\chi+\varepsilon/\delta^2)\textstyle\sum_{1\leq k\leq p}(b_k-a_k)].$$

Since the initial condition is in $ST(\varepsilon, \zeta, R)$, $\sum_{2\leq k\leq p}(a_k-b_{k-1})$ is bounded by ζ/ε. Since also $\sum_{1\leq k\leq p}(b_k-a_k) \leq T$, we obtain the bound:

$$(19) \qquad M(0, T) \leq e^{c\varepsilon T}[c\varepsilon T(\chi+\varepsilon/\delta^2) + cp\varepsilon/\delta + c\zeta].$$

Since p is less than $1 + \zeta/\xi$ and since $\xi = \delta/(c_4N^{n+1})$, assuming $\zeta > \xi$, for ε_0 small enough we have:

$$(20) \qquad M(0, T) \leq ce^{c\varepsilon T}[\varepsilon T(\chi+\varepsilon/\delta^2) + \varepsilon\zeta\,|\log\chi|^{n+1}/\delta^2 + \zeta].$$

Since $T \leq 1/\varepsilon$, this leads to inequality (2) of Lemma 1.

Lemma 1 is now established on the interval $[0, T]$, and it is easy to extend the estimate to time $1/\varepsilon$. For this purpose, assume that the various parameters are such that $\lambda < \eta/2$ and suppose T is strictly less than $1/\varepsilon$. We then have $M(0, T) < \eta/2$ and, since $J(t)$ remains in $K - \eta$ for $t \leq T$, $I(t)$ belongs to $K - \eta/2$ for $t \leq T$. $I(t)$ therefore remains in $K - \eta/4$ until time T' strictly greater than T, which contradicts the definition of T as the instant at which the exact trajectory leaves $K - \eta/4$. Therefore T is greater than or equal to $1/\varepsilon$ and $M(0, 1/\varepsilon) \leq \lambda \leq \eta/2$. ∎

Lemma 1 is thus proved with the following restrictions on the parameters:

$$(21) \qquad \varepsilon < \varepsilon_0 \ (\varepsilon_0 \text{ small enough}),\ \chi < 1/2,\ \delta < \delta_0 \ (\delta_0 \text{ small enough}), \\ \xi = \delta/(c_4N^{n+1}),\ \xi < \zeta < 1, \text{ and } \varepsilon c_5/\delta < \eta/4.$$

The last restriction may be rewritten $4c_4c_5\varepsilon N^{n+1} < \eta$. It is satisfied for η and χ fixed and for ε_0 small enough.

It remains to bound the measure of $LT(\varepsilon, \zeta, R)$ by means of

<u>Lemma 2</u>:

The measure of $LT(\varepsilon, \zeta, R)$ is bounded by $c\ \mathrm{mes}(R(\delta, N))/\zeta$ (Lebesgue measure on $\mathbb{R}^m$).

The proof is very close to the proof of Lemma 2 in Chapter 2, and we will not repeat all the details. Let $\mu = dI \times d\varphi$ be the product measure on $\mathbb{R}^m \times T^n$, and let μ' be the measure $\mu \times dt$ on $\mathbb{R}^m \times T^n \times [0, 1/\varepsilon]$. T_ε^τ will be the flow of the exact system, and T'_ε will be the mapping of $D \times [0, 1/\varepsilon]$ into $\mathbb{R}^m \times T^n \times [0, 1/\varepsilon]$ defined by $T'_\varepsilon(I, \varphi, \tau) = (T_\varepsilon^\tau(I, \varphi), \tau)$.

As in Chapter 2, we have:

$$(22) \qquad \mu' [T'^{-1}_\varepsilon(R \times T^n \times [0, 1/\varepsilon])] = \int_0^{1/\varepsilon} \mu [(T_\varepsilon^\tau)^{-1}((R \times T^n) \cap T_\varepsilon^\tau(D))]\, d\tau \leqslant \int_0^{1/\varepsilon} c\, \mu(R \times T^n)\, d\tau \leqslant c\, \mathrm{mes}(R)/\varepsilon.$$

and, by definition of $S = S(\varepsilon, I_0, \varphi_0, R)$:

$$(23) \qquad \int_D \mathrm{mes}[S(\varepsilon, I_0, \varphi_0, R)]\, d\mu \leqslant \mu'[T'^{-1}_\varepsilon(R \times T^n \times [0, 1/\varepsilon])] \leqslant c\, \mathrm{mes}(R)/\varepsilon.$$

Integrating over the smaller domain $LT(\varepsilon, \zeta, R)$ and using its definition, we obtain the desired result:

$$(24) \qquad \mu[LT(\varepsilon, \zeta, R)]\, \zeta/\varepsilon \leqslant c.\mathrm{mes}(R(\delta, N)/\varepsilon. \quad \blacksquare$$

Lemmas 1 and 2 provide an almost immediate proof of Neistadt's theorem.

The parameters ρ, δ, ζ, ξ, and χ will be positive functions, continuous in ε and defined as follows:

$$(25) \qquad \zeta = \rho(\varepsilon)/(4\, c_3), \quad \chi = \zeta, \quad \delta = (4 c_3 \varepsilon/\rho(\varepsilon))^{1/2}$$
$$\xi = \delta/(c_4 N^{n+1}), \quad \text{where} \quad N = [c\,|\log\chi\,|].$$

Thus, for a judicious choice of c_1, c_2, and ε_0, $\lambda < \rho(\varepsilon) < \eta/2$, and all the conditions (21) required of the parameters for the validity of Lemma 1 are satisfied. We deduce from this lemma that all the initial conditions of ρ-deviant trajectories are contained in the set $LT(\varepsilon, \zeta, R(\delta, N))$.

According to Lemma 2 the measure of this set is bounded by $c\, \mathrm{mes}[R(\delta, N)]/\zeta$. By Proposition 1, $\mathrm{mes}[R(\delta, N)] \leqslant c\delta\mu(K)$, and it follows that:

$$(26) \qquad \mu[LT(\varepsilon, \zeta, R(\delta, N))] \leqslant c\, \mathrm{mes}(K)(\delta/\zeta) \leqslant c(\varepsilon/\rho(\varepsilon)^3)^{1/2},$$

which proves the theorem. $\blacksquare$

5.3 Proof for the differentiable case

Neistadt's theorem was proved assuming that the function $f(I, \varphi, 0)$ is φ-analytic, which implied the exponential decrease of their Fourier coefficients. This led in turn to simple estimates of the remainders R_N of the Fourier series and to the bounds (7) on the function S. We show in this section that the result remains valid if $f(I, \varphi, 0)$ is of class

C^{2n+1} in φ and $\partial f(I, \varphi, 0)/\partial I$ of class C^q with $q > 3n/2$. To establish these relaxed differentiability conditions, we will make use of the results in Appendix 1 on the convergence of Fourier series for functions with finite differentiability; this section is in part included to illustrate (a contrario) how analyticity helps to greatly simplify the proofs.

We adopt here a definition of the resonant zones which differs slightly from the definition in the analytic case. For $\chi > 0$ we choose $N = [c/\chi^{1/2n}]$ in such a way that, by Proposition 8 of Appendix 1, $\| R_N \| \leq \chi$ (χ is determined by (25) in Section 5.2). To each resonance of order less than $2nN$ we associate the resonant zone $V(k, \delta) = \{ I \in K, | \omega(I).k | \leq \delta/| k |^{n-1+\kappa} \}$, $(0 < \kappa < 1$ fixed) and we define the resonant domain $R(\delta, 2nN)$ as the union of these resonant zones. Proposition 1 of last section remains valid:

$$\mathrm{mes}(R(\delta, 2nN)) \leq c\delta \ \mathrm{mes}(K).$$

The nonresonant domain $NR(\delta, 2nN)$ is then defined as the complement of $R(\delta/2, 2nN)$. Henceforth, we fix the value of ξ at $\xi = \delta/(c_4[2nN]^{n+\kappa})$ so that Proposition 2 of Section 5.2 remains true.

We next proceed to bound the norm of the function S, defined a priori on $NR(\delta, N) \times T^n$, where $NR(\delta, N)$ is the complement of $R(\delta/2, N)$, along with the norms of its derivatives on the domain $NR(\delta, 2nN)$, and we will obtain the same estimates as before (cf. (7) of last section). We recall that the Fourier coefficients $S_k(I)$ of S are zero for $| k | > N$ and are equal to $if_k(I)/k.\omega(I)$ for $1 \leq | k | \leq N$. We restrict ourselves to $n \geq 3$ since better results are obtained, as we saw, for n equal to 1 or 2.

We first show that $\| S \| \leq c/\delta$. By the definition of the nonresonant domain, we have the bound:

$$(1) \qquad \| S \| \leq c\textstyle\sum_{k\neq 0} \| f_k \|_e \| k \|_e^{n-1+\kappa}/\delta,$$

where $\| \ \|_e$ denotes the Euclidean norm, which in this section we use for the frequency vector k as well. Let r be such that $n < 2r \leq 2(n+1)$ and such that $n + r$ is even. Then:

$$(2) \qquad \| S \| \leq c\textstyle\sum_{k\neq 0} [(\| f_k \|_e \| k \|_e^{n}]/\delta \leq c\sum_{k\neq 0} [(\|f_k \|_e \| k \|_e^{2p})/\| k \|_e^{r}]/\delta$$

where we have set $n + r = 2p$.

By the choice of r it follows that the sequence $1/\| k \|_e^r$ is in l^2. Since f is of class C^{2n+1}, $\Delta^p f$ is a function of class C^{n-r+1} (Δ is the Laplacian) and hence belongs to $L^2(T^n)$. The series $\| f_k \|_e \| k \|_e^{2p}$ of the norms of its Fourier coefficients belongs to $l^2(\mathbb{Z}^n)$. It then follows from Schwartz's inequality that the sequence $\| f_k \|_e\| k \|_e^n$ is in l^1. We thus obtain the desired estimate for $\| S \|$, and this estimate remains valid if f is of class C^q with

$q > 3n/2$.

An analogous result is valid for $\partial S_i/\partial\varphi_l$. We have the bound:

(3) $\qquad \| \partial S_i/\partial\varphi_l \| \leqslant c\sum_{k\neq 0} \| f_k \|_e \| k \|_e^{n+\kappa}/\delta.$

Consequently:

(4) $\qquad \| \partial S_i/\partial\varphi_l \| \leqslant c\sum_{k\neq 0}[(\| f_k \|_e \| k \|_e^{n+r+1})/\| k \|_e^{r}]/\delta$

where r is such that $n < 2r \leqslant 2n$ and $n + r$ is even. Proceeding as before, we deduce that $\| \partial S_i/\partial\varphi_l \| \leqslant c/\delta$. To obtain this bound it suffices to require that f be of class C^q, where $q \geqslant 3n/2 + 1$.

It is more difficult to estimate the norm of $\partial S/\partial I$. We have:

(5) $\qquad \partial S/\partial I_j = \sum_{0<|k|\leqslant N} \partial f_k(I)/\partial I_j e^{ik.\varphi}/\omega.k - (k.\partial\omega/\partial I_j)f_k e^{ik.\varphi}/(\omega.k)^2.$

Consequently, since $\partial f/\partial I_j$ is of class C^q in φ with $q > 3n/2$, proceeding as before it is possible to prove that:

(6) $\qquad \mathrm{Sup}_{i,j} \| \partial S_i/\partial I_j \| \leqslant c/\delta + c\sum_{0<|k|\leqslant N} \| k \|_e \| f_k \|_e /(\omega.k)^2.$

We must estimate the sum on the right hand side, and the simple techniques we have used up to now are unequal to the task. We turn instead to the procedure used in Rüssmann's paper [Rü].

We set $a(k) = 1/(\omega.k)^2$ and $b(k) = \| k \|_e \| F_k \|_e$. We then have

<u>Proposition 5</u>:

For $m \leqslant N$, $S_m = \sum_{0\leqslant |k| \leqslant m} a(k)$ is less than $cm^{2(n-1+\kappa)}$.

This is the most technical proposition, whose proof the reader may well skip if he is not interested in its technicalities. Let I belong to $NR(\delta, 2nN)$. One of the components of $\omega(I)$ has absolute value greater than the average $\sum_i |\omega_i|/n$; we assume this component is ω_n.

On the other hand, for $|k| \leqslant 2nN$, $|\omega'.q + \omega_n p| \geqslant \delta/(2|k|^r)$, where $r = n - 1 + \kappa$, p is the nth component of k, q is the vector in $\mathbb{Z}^{n-1}$ consisting of the first $n - 1$ components of k, and ω' is the vector in $\mathbb{R}^{n-1}$ consisting of the first $n - 1$ components of ω. We then deduce that:

(7) $\qquad |\omega'.q + \omega_n p| \geqslant \delta/(2n^r|q|^r)$

for $|k| \leqslant 2nN$, in other words $|q| \leqslant 2nN$ and $|p| \leqslant 2nN$.

The sum S_m may be decomposed as:

(8) $$S_m = \sum_{q=0,\ 0<|p|\leq m} a(k) + \sum_{0<|q|\leq m,\ |p|\leq m} a(k).$$

The first term $\sum_{0<|p|\leq m} 1/(\omega_n p)^2$ is less than the constant $n^2\pi^2/(3\,\mathrm{Inf}_K(\sum|\omega_i|))$.

To evaluate the second term, we introduce the function $D(q) = \mathrm{Inf}_{p\in\mathbb{Z}} |\omega'.q + \omega_n p|$ on $\mathbb{Z}^{n-1}$. For $0 < |q| \leq 2N$, we have:

(9) $$D(q) \geq \delta/(2n^r |q|^r).$$

In fact, $D(q)$ attains its minimum for a value of p which we denote by $p_0(q)$ and which satisfies $|p_0(q) - [\omega'.q/\omega_n]| \leq 1$ (here the brackets [ ] denote the integer part). Consequently, $|p_0(q)| \leq 2nN$. Since the point I lies in the nonresonant zone $NR(\delta, 2nN)$, it follows that $D(q) \geq \delta/(2n^r|q|^r)$. In other words, a vector k in $\mathbb{Z}^n$ may not belong to the resonant module of ω (cf. Appendix 3) if its projection onto $\mathbb{Z}^{n-1}$ is too small.

Also, since the real scalar $\omega'.q/\omega_n$ differs from the integer $p_0(q)$ by less than 1/2, $D(q)$ is less than $\omega_n/2$. We also note that, for every value of p different from $p_0(q)$, $|\omega'.q + \omega_n p| > \omega_n |p - p_0(q)|/2$.

The sum under consideration is thus bounded by $\sum_{0<|q|\leq m} \pi^2/(4D(q)^2)$. To see this, write:

(10) $$\sum_{0<|q|\leq m,|p|\leq m} 1/(\omega'.q + \omega_n p)^2 \leq$$
$$\leq \sum_{0<|q|\leq m}\{(1/\omega_n^2).\sum_{p=-\infty}^{+\infty} 1/(p + D(q)/\omega_n)^2\}.$$

Since:

(11) $$\sum_{p=-\infty}^{+\infty} 1/(a+p)^2 = \pi^2/(\sin\pi a)^2 \leq \pi^2/(4a^2),$$

for $|a| \leq 1/2$ and $|D(q)/\omega_n| \leq 1/2$, the desired bound is established.

Let us estimate the sum $\sum_{0<|q|\leq m} 1/D(q)^2$ by setting:

(12) $$c_k = |\omega_n|(2/3)^k/2, \quad Q(k) = \{q \text{ such that } 0 < |q| \leq m$$
$$\text{and } c_{k+1} < D(q) \leq c_k\}, \quad d_k = n\,(4c_k/\delta)^{1/r}.$$

Let s be such that $c_{s+1} < \delta/(2(nm)^r) \leq c_s$. Since $D(q) \leq c_0$ for q in $Q(m)$, we have:

(13) $$\sum_{0<|q|\leq m} 1/D(q)^2 \leq \sum_{i=0}^{s} \mathrm{card}(Q(i))/c_{i+1}^2.$$

In addition, d_i is less than m for $i \leq s$. To estimate the cardinality of $Q(i)$, we consider the open cubes in $\mathbb{Z}^{m-1}$, with edges d_i, centered on the elements q in $Q(i)$ (see Figure 5.2):

$$W_q = \{k \in \mathbb{Z}^{n-1} \text{ such that } |k - q| < d_i/2\}.$$

Each cube has volume d_i^{n-1} in $\mathbb{R}^{n-1}$, and each is contained in the cube $W = \{k$ such that

$|k| \leqslant m + d_i/2\,\}$ with volume $(2m + d_i)^{n-1}$ in $\mathbb{R}^{n-1}$.

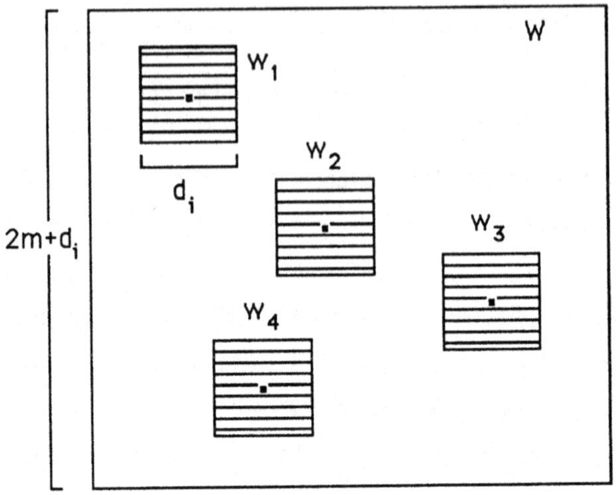

Figure 5.2: Each point of $Q(i)$ (black dots) is the center of an open cube w_i with edges d_i. Those cubes are contained in W and are pairwise disjoint.

The cubes are pairwise disjoint. Indeed, consider two distinct cubes w_j and $w_{j'}$ in $\mathbb{R}^{n-1}$, centered respectively on the points q and q' of $Q(i)$, and with edges d_i. Suppose these cubes intersect. We then have $|q' - q| < d_i$, but since $|q' - q| \leqslant 2m \leqslant 2N$, $D(q' - q) \geqslant \delta/(2n^r|q' - q|^r)$. Since also $D(q' - q) \leqslant D(q') + D(q)$ and since q and q' are both in $Q(i)$, the bound

$$\delta/(2n^r|q' - q|^r) \leqslant 2c_i$$

holds, from which we deduce that $|q' - q| \geqslant d_i$, in contradiction to the assumption that the cubes intersect.

The number of cubes (equal to the cardinality of $Q(i)$) is thus less than $(3m/d_i)^{n-1}$ and:

$$(14) \qquad \sum_{0<|q|\leqslant m} 1/D(q)^2 \leqslant (3m)^{n-1} \sum_{i=0}^{s} 1/(c_{i+1}^{2}\, d_i^{n-1}).$$

In view of the definitions (12), $c_{i+1} d_i$ is greater than $\delta m^{n-1}/(3(nm)^r)$. Consequently, the sum on the right hand side of (14) is bounded by:

$$(3\,nm)^r/(\delta m^{n-1}))\sum_{i=0}^{s} 1/c_{i+1}\,,$$

and:

$$(15) \qquad \sum_{0<|q|\leqslant m} 1/D(q)^2 \leqslant (3^n (nm)^r/\delta)\sum_{i=0}^{s} 1/c_{i+1}\,.$$

It follows that:

$$(16) \qquad \sum_{0<|q|\leqslant m} 1/D(q)^2 \leqslant 6(3^n(nm)^r/(\omega_n\delta))(3/2)^{s+1} \leqslant 3^{n+2}(nm)^{2r}/\delta^2,$$

the second inequality resulting from the definition of s. Finally:

$$(17) \qquad \sum_{0<|q|\leqslant m,\ |p|\leqslant m} 1/(\omega'.q + \omega_n p)^2 \leqslant \pi^2 3^{n+2}(nm)^{2r}/(4\delta^2),$$

which proves Proposition 5. ∎

The bound on $\| \partial S_i/\partial I_j \|$ now follows easily. First of all, we have:

$$(18) \qquad \sum_{0<|k|\leqslant N} \| k \|_e \| F_k \|_e/(\omega.k)^2 \leqslant c\sum_{p=1}^{N} \sum_{|k|=p} a(k)b(k).$$

Since f is of class C^{2n+1}, we have $b(k) \leqslant c/| k |^{2n}$. Consequently:

$$(19) \qquad \sum_{0<|k|\leqslant N} \| k \|_e \| F_k \|_e/(\omega.k)^2 \leqslant c\sum_{p=1}^{N} \sum_{|k|=p} a(k)/p^{2n}$$
$$\leqslant c\sum_{p=1}^{N} S_p/p^{2n}.$$

By Proposition 5 it follows that:

$$\sum_{0<|k|\leqslant N} \| k \|_e \| F_k \|_e/(\omega.k)^2 \leqslant c(1/\delta^2)\sum_{p=1}^{\infty} 1/p^{2(1-\kappa)}.$$

For $\kappa < 1/2$, the series on the right hand side converges, and we obtain the desired bound:

$$(20) \qquad \mathrm{Sup}_{i,j} | \partial S_i/\partial I_j | \leqslant c/\delta + c/\delta^2 \leqslant c/\delta^2. \quad ∎$$

Neistadt's theorem in the differentiable case follows easily from these estimates.

Propositions 1 and 2 from Section 5.2 remain valid and the function S and its derivatives satisfy the same bounds ((7) in section 5.2) as before. This permits us to prove the following Lemmas 3 and 4 for the differentiable case, analogues of Lemmas 1 and 2 in the analytic case.

Lemma 3:

For every initial condition in $ST(\varepsilon, \zeta, R(\delta, 2nN))$, $I(t)$ remains in $K - \eta/4$ (and thus in K) at least until time $1/\varepsilon$ and:

$$(21) \qquad \sup_{t \in [0, 1/\varepsilon]} \| I(t) - J(t) \| \leqslant \lambda,$$

where $\lambda \leqslant c_6(\zeta + \chi + \varepsilon/\delta^2 + \varepsilon\zeta N^{n+1}/\delta^2) \leqslant c(\zeta + \chi + \varepsilon[1+\zeta\chi^{-(n+1)/2n}]/\delta^2)$, provided λ is strictly less than $\eta/2$.

Lemma 4:

The measure of $LT(\varepsilon, \zeta, R(\delta, 2nN))$ is bounded by $c\ \mathrm{mes}(R(\delta, 2nN))/\zeta$ and thus by $c\delta/\zeta$.

We may choose $\zeta = \chi = c_9\rho$ and $\delta = (4c_9\varepsilon/\rho)^{1/2}$ so that λ is less than ρ; we then use Lemmas 3 and 4 to deduce Neistadt's result in the differentiable case in the same way that Lemmas 1 and 2 were used to deduce it in the analytic case. ∎

CHAPTER 6: N FREQUENCY SYSTEMS; NEISTADT'S RESULTS BASED ON KASUGA'S METHOD

6.1 Statement of the theorems

In the preceding chapter, we showed that for a nondegenerate system with n frequencies, on an appropriate time interval the separation between an exact solution and the corresponding solution of the averaged system does not exceed $\rho(\varepsilon)$, except for a set of initial conditions of measure $O([\varepsilon/\rho^3]^{1/2})$. Following Neistadt once again (cf. [Nei4]), we will now show that the measure of this set is in fact $O(\sqrt{\varepsilon}/\rho)$ and that this result is optimal; we will also improve upon the regularity conditions. We make use of a function which approximates the solution to the linearized equation in L^1 norm on all of the domain $K \times T^n$, which constitutes the principle innovation of the chapter. We encounter this technique again (in its original form) during the proof of the adiabatic theorem for Hamiltonians with flows ergodic on the energy surface (cf. Chapter 9).

As before we are concerned with systems in the standard form (see (1) of Section 5.1). We use the same notation as in Chapter 5, with the following qualifications:

a) We assume that $f(I, \varphi, 0)$ and $\partial f(I, \varphi, 0)/\partial I$ are of class C^{n+1} in the variables φ, rather than analytic or C^{2n+1} as in Section 5.3; f is still only C^1 in the slow variable(s) I.

b) We adopt Arnold's rather than Kolmogorov's nondegeneracy condition for the frequencies (see Appendix 3), our only reason being to give an example of its use. In what follows, to simplify the proofs we will often assume that one of the frequencies is nonvanishing on all of K. This is not restrictive in view of the fact that the original domain may be subdivided into finitely many parts on which this assumption holds.

We denote by $E(I_0, \varphi_0, \varepsilon) = \mathrm{Sup}_{[0, 1/\varepsilon]} \| I(t) - J(t) \|$ the maximal separation on the base between the exact solution with initial condition $(I_0, \varphi_0) = (I(0), \varphi(0))$ and the averaged solution with initial condition $I(0)$. For every positive continuous function $\rho(\varepsilon)$, $V'(\varepsilon, \rho)$ is again the set of initial conditions such that the separation is less than ρ and $V''(\varepsilon, \rho)$ is its complement.

As in the preceding chapters, we work on $D =_{def} K' \times T^n$ where K' is defined in the usual way; namely it is the subset of K of initial conditions such that the trajectory of the *averaged* system remains in $K - \eta$ ($\eta > 0$ small enough) for the time interval τ^*/ε

($\tau^* > 0$ small enough). Again for notational simplicity, we set $\tau^* = 1$ which amounts to a possible rescaling of ε. The first result affirms that the *average* separation (that is the average over the initial conditions for the ordinary measure $dI d\varphi$) for initial conditions in D is small and that the measure of $V''(\varepsilon, \eta/2)$ is also small:

Theorem 1:

a) $\int_D E(I_0, \varphi_0, \varepsilon)\, dI_0 d\varphi_0 < c\sqrt{\varepsilon}$.

b) $\mathrm{mes}(V''(\varepsilon, \eta/2)) < c\sqrt{\varepsilon}$.

c) These estimates are optimal.

The second result is an improvement over the analogous result of the preceding chapter.

Theorem 2:

a) $\mathrm{mes}(V''(\varepsilon, \rho)) < c\sqrt{\varepsilon}/\rho$ for $c_1\sqrt{\varepsilon} \leqslant \rho(\varepsilon) \leqslant c_2$.

b) This result is optimal.

It should be noted that in the first theorem, ρ is a constant ($\rho = \eta/2$), whereas in the second, it is a function subject only to the restrictions in Part a. The first assertion of this latter theorem in fact follows immediately from Theorem 1; to see this, choose c_2 so that ρ is less than $\eta/2$ to avoid domain problems. $V'(\varepsilon, \eta/2) \cap V''(\varepsilon, \rho)$ is a priori nonempty, and according to Theorem 1:

$$(1) \qquad \int_{V'(\varepsilon,\eta/2)\cap V''(\varepsilon,\rho)} E(I_0, \varphi_0, \varepsilon)\, dI_0 d\varphi_0 \leqslant c\sqrt{\varepsilon}.$$

Since $E(I_0, \varphi_0, \varepsilon)$ is greater than ρ for initial conditions in $V''(\varepsilon, \rho)$, it follows that:

$$(2) \qquad \rho\, \mathrm{mes}(V'(\varepsilon, \eta/2) \cap V''(\varepsilon, \rho)) \leqslant c\sqrt{\varepsilon}.$$

On the other hand, again by Theorem 1:

$$(3) \qquad \mathrm{mes}[V''(\varepsilon, \eta/2) \cap V''(\varepsilon, \rho)] \leqslant \mathrm{mes}(V''(\varepsilon, \eta/2) \leqslant c\sqrt{\varepsilon}.$$

Estimate a) of Theorem 2 then follows from these two inequalities. ∎

The optimality is proved by analysis of an example to be presented in Section 6.4.

6.2 Proof of Theorem 1

Proof of Parts a and b:

An example proving optimality (Part c) will be given in the next section.

In the preceding chapter, we used the exact solution of the equation $\omega.\partial S/\partial\varphi + f_N = 0$ as an approximate solution to the linearized equation $\omega.\partial S/\partial\varphi + \tilde{f} = 0$; as usual $\tilde{f}$ is the oscillating part of $f(I, \varphi, 0)$, f_N is the trigonometric polynomial obtained by truncating the Fourier series of f, and N denotes the order of the ultraviolet cutoff. This solution was defined only on the complement of resonant zones of order less than N, and on this domain it satisfied the inequality: $\| \omega.\partial S/\partial\varphi + \tilde{f} \| \leqslant \chi$ (χ small).

For S, we will use here a C^1 function *globally* defined on all of $K \times T^n$ which is an average (L^1) approximate solution to the linearized equation. In other words S will minimize the integral: $\int \| \omega.\partial S/\partial\varphi + \tilde{f} \| dI d\varphi$.

It is constructed as an Nth order trigonometric polynomial (the precise cutoff order N is defined later) with Fourier coefficients:

(1) $\qquad S_k = i\Psi(k.\omega) f_k\, e^{ik.\varphi}$,

where Ψ is a C^1 function which is close to $(\omega.k)^{-1}$ away from resonances and which vanishes with $\omega.k$. We have:

(2) $$f + \omega.\partial S/\partial\varphi = \sum_{0<|k|\leqslant N(\delta)} f_k [1-(\omega.k)\Phi(\omega.k)]\, e^{ik.\varphi} + \sum_{|k| > N(\delta)} f_k\, e^{ik.\varphi}$$
$$= R_1(I, \varphi, \delta) + R_2(I, \varphi, \delta).$$

We seek to estimate the L^1 norm of this expression as $O(\delta)$ $(0 < \delta < 1/2)$ where δ caracterizes the width of the resonances (see below). For this purpose, we first choose the ultraviolet cutoff $N(\delta)$ in such a way that:

(3) $\qquad \sum_{|k|>N(\delta)} \| f_k \| < \delta$,

which will imply $\| R_2(I, \varphi, \delta) \|_{L^1} < \delta \,\mathrm{mes}(K \times T^n)$. By the assumed regularity of f and Proposition 7 of Appendix 1, the ultraviolet cutoff $N(\delta) = [c/\delta]$ will suffice.

We next choose Ψ so that R_1 has small L^1 norm. It is natural to take Ψ to be odd; we choose $\Psi(x) = 1/x$ for $|x| \geqslant \delta$, thus avoiding any regularization procedures on the nonresonant domain $|\omega.k| \geqslant \delta$ where small divisor problems do not arise. It remains only to define Ψ as a cutoff function on the segment $[-\delta, \delta]$, tending to 0 with x (i.e., on approaching the resonance) in such a way that the L^1 norm of R_1 is small on the *resonant* domain, which here is defined as the set of I such that $|\omega(I).k| \leqslant \delta$ for a k with norm less than $N(\delta)$. R_1 vanishes elsewhere, by the choice of Ψ. Here, we make use of Neistadt's interpolation $\Psi(x) = 2x/\delta^2 - x^3/\delta^4$, which is the simplest C^1 choice (see Figure 6.1).

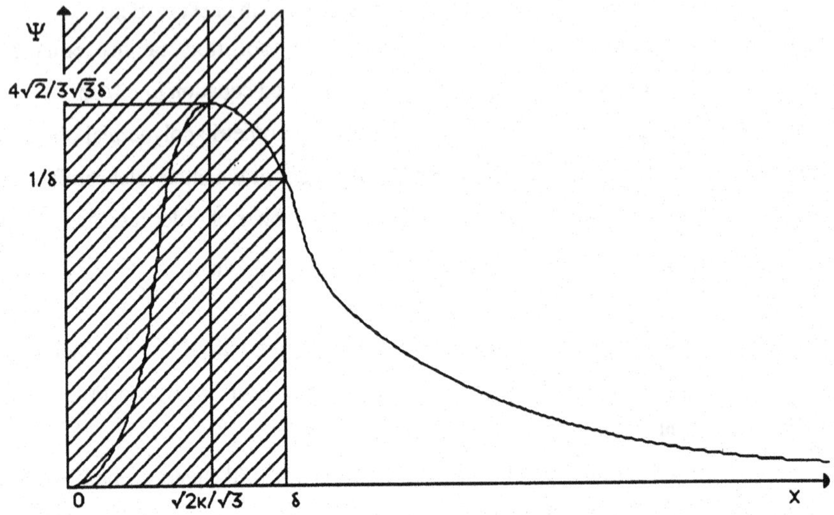

Figure 6.1: The function Ψ is equal to $1/x$ far from the resonance. It is defined as a cut-off function inside the resonant zone $|x| \leq \delta$ (hatched portion).

The L^1 norm of R_1 is then bounded by:

$$(4) \qquad \| R_1 \|_{L^1} \leq \sum_{0<|k|\leq N(\delta)} \| f_k \| \int (1 - [(\omega(I).k)/\delta]^2)^2 \, dI d\varphi.$$

In order to estimate the integrals appearing in this expression, we first recall the almost obvious observation (cf. Appendix 3) that the measure $M(k, \delta)$ of a resonant zone $V(k, \delta) = \{I \in K \text{ such that } |\omega(I).k| \leq \delta\}$ is $O(\delta)$ for all $k \neq 0$ (here we use the nondegeneracy condition).

Since the integration is restricted to the resonant zones, the right hand side of (4), and thus also the L^1 norm of R_1, are bounded by $\sum_k \| f_k \| (2\pi)^n M(k, \delta)$, which is less than $c\delta$ since the Fourier coefficients of f decrease at least like $|k|^{-(n+1)}$. This proves the inequality:

$$(5) \qquad \| \tilde{f} + \omega.\partial S/\partial\varphi \|_{L^1} \leq c\delta.$$

We next estimate the pointwise (C^0) norm of the function $S = S_\delta$ along with the L^1 norm of its derivatives for all δ in $(0, 1/2)$.

Lemma 1:

There exists a constant c such that:

a) $\| S \| \leqslant c/\delta$.

b) $\| \partial S/\partial\varphi \|_{L^1} \leqslant c| \log \delta |$.

c) $\| \partial S/\partial I \|_{L^1} \leqslant c/\delta$.

Since $S_k = i\Psi(k.\omega)f_k$ and $| \Psi(k.\omega) | \leqslant c/\delta$, we have $\| S_k \| \leqslant c\| f_k \|/\delta$ and inequality a) follows from the convergence in norm of the Fourier series of f.

To prove inequality b) of the lemma, we begin with the simplest case where the derivative is taken with respect to a variable φ_i different from φ_n (we are still assuming ω_n does not vanish on K). Then:

(6) $$\| \partial S/\partial\varphi_i \|_{L^1} \leqslant (2\pi)^n \sum_{0<|k|\leqslant N(\delta)} | k_i |.\| f_k \| \int_K | \Psi(k.\omega) | \, dI.$$

As before, we may decompose this expression into a sum of integrals over the resonant zones (denoted $\int^*$) and a sum of integrals over the complement of these zones (denoted $\int^{**}$). In the first sum, $\int^* | \Psi(k.\omega) | \, dI \leqslant c$ and $\| f_k \| \leqslant c(| k_i |.| k |^n)^{-1}$ (see Appendix 1). Consequently:

(7) $$(2\pi)^n \sum_{0<|k|\leqslant N(\delta)} | k_i |.\| f_k \| \int^* | \Phi(k.\omega) | \, dI \leqslant c \log(N(\delta)) \leqslant c \log(1/\delta),$$

where we have used the fact that the number of vectors in $\mathbb{Z}^n$ with fixed norm $| k |$ is less than $c| k |^{n-1}$.

To estimate the second sum, we replace the ith coordinate by $(\omega.k)$ in the neighborhood of resonances (but ouside the resonant zones of width δ) and, taking into account the factor $1/| k_i |$ arising from the Jacobian of this transformation, we find:

(8) $$(2\pi)^n \sum_{0<|k|\leqslant N(\delta)} | k_i |.\| f_k \|.\int^{**} |k.\omega|^{-1} \, dI \leqslant \leqslant c\sum_{0<|k|\leqslant N(\delta)} \| f_k \| \log(c| k |/\delta) \leqslant c \log(1/\delta),$$

which establishes estimate b) of Lemma 1 for variables φ_i with $i \neq n$.

It is more difficult to estimate $\| \partial S/\partial\varphi_n \|_{L^1}$, because, at least in its practical use, Arnold's condition singles out one of the angle variables, here φ_n. The decomposition into sums of resonant and nonresonant integrals is still possible, and the resonant sum is bounded as before by $c \log(1/\delta)$.

The nonresonant sum may be further reduced to $n + 1$ partial sums. The first partial sum $\sum_1$ is limited to vectors k such that $|k_1| \geq |k_n|$; the pth partial sum $\sum_p$ $(p \leq n-1)$ is limited to k satisfying $|k_i| < |k_n|$ $(i=1,\dots,p-1)$, $|k_p| \geq |k_n|$; the nth sum $\sum_n$, to k such that $|k_i| < |k_n|$ $(i=1,\dots,n-1)$, where at least one of the components $k_1,\dots,k_{n-1}$ is nonzero; and the last sum $\sum_{n+1}$ is taken over vectors k having k_n as their only nonzero component.

In order to bound $\sum_1,\dots,\sum_{n-1}$, it suffices to proceed as outlined above to obtain $\sum_p \leq c\log(1/\delta)$. For $\sum_{n+1}$, we have the bound $\sum_{n+1} \leq c\sum_k \| f_k \| \leq c$, since $\int^{**} |\Psi(k.\omega)|\, dI < c/|k_n|$.

It remains to bound $\sum_n$, which we (again) decompose into $n-1$ partial sums $\sum^1 + \dots + \sum^{n-1}$, where $\sum^p$ is the sum over vectors having k_p as the component (other than k_n) with largest absolute value.

To estimate $\sum^p$, we proceed as in the estimation of $\| \partial S/\partial\varphi_p \|_{L1}$:

$$(9) \qquad \textstyle\sum^p \leq c\sum |k_n|.\| f_k \| \log(c|k_n|/\delta)/|k_p| \leq c\sum \log(c|k_n|/\delta)/(|k_p|.|k_n|^n).$$

We fix $|k_n|$ and let $Q(j)$ denote the set of k such that $|k_n|/2^{j+1} < |k_p| \leq |k_n|/2^j$. For $j > c_3\log(|k_n|)$, $Q(j)$ is empty. We therefore have:

$$(10) \qquad c\textstyle\sum^{|k_n|} \log(c|k_n|/\delta)/(|k_p|.|k_n|^n) \leq$$
$$\leq 1/|k_n|^n \textstyle\sum_{j=0}^{[c_3\log|k_n|]} \operatorname{card} Q(j)\ \log(c|k_n|/\delta)/|k_p|$$
$$\leq 1/|k_n|^{n+1} \textstyle\sum_{j=0}^{[c_3\log|k_n|]} 2^{j+1}\operatorname{card} Q(j)\ \log(c|k_n|/\delta),$$

where $\sum^{|k_n|}$ designates a sum with $|k_n|$ fixed. Since $\operatorname{card} Q(j)$ is less than $c|k_n|^{n-1}/2^j$, this expression is bounded by $c\log|k_n|\log(c|k_n|/\delta)/|k_n|^2$, and by summing over the values of $|k_n|$, we find that $\sum^p \leq c\log(1/\delta)$.

Summing lastly over p we obtain $\sum_n \leq c\log(1/\delta)$, and from this bound and those already established for $\sum_1,\dots,\sum_{n-1}$ and $\sum_{n+1}$, we deduce that:

$$\| \partial S/\partial\varphi_n \|_{L1} \leq c\log(1/\delta),$$

which completes the proof of Part b of Lemma 1.

Part c is proved similarly. We consider the derivative with respect to the variable I_j:

$$(11) \qquad \| \partial S/\partial I_j \|_{L1} \leq \textstyle\sum_{0<|k|\leq N(\delta)} \| \partial f_k/\partial I_j \| \int |\Phi(k.\omega)|\, dI d\varphi + c\sum_{0<|k|\leq N(\delta)} \| f_k \| (\sum_{i=1}^m |k_i|) \int |\Phi'(k.\omega)|\, dI.$$

Since $\partial f_k/\partial I_j$ is of class C^{n+1}, the first sum may be estimated by the following procedure. We express each of the integrals $\int |\Phi(k.\omega)|\, dI d\varphi$ as a sum of an integral over the resonant zone $V(k,\delta)$ (denoted $\int^*$) and an integral over the complement of this zone in K (denoted $\int^{**}$). Since $|\Phi(k.\omega)|$ is bounded by c/δ and the measure in K of the resonant zone is $O(\delta)$, the first integral is bounded by a constant and, since the Fourier series of $\partial f/\partial I_j$ is convergent in norm:

$$(12) \qquad \| \partial S/\partial I_j \|_{L^1} \leq c + (2\pi)^n \textstyle\sum_{0<|k|\leq N(\delta)} \| \partial f_k/\partial I_j \| \int^{**} | \Phi(k.\omega) | \, dI d\varphi$$
$$+ c\textstyle\sum_{0<|k|\leq N(\delta)} \| f_k \| (\sum_{i=1}^m |k_i|) \int | \Phi'(k.\omega) | \, dI.$$

Outside the resonant zone, $\Phi(k.\omega) = 1/(k.\omega)$. By introducing the variable $\omega.k$ (the algebraic distance to the resonant surface) and still assuming ω_n is nonvanishing on K, we obtain the inequality:

$$(13) \qquad \int^{**} | \Phi(k.\omega) | \, dI d\varphi \leq c\int_\delta^{c|k|} d(k.\omega)/k.\omega,$$

where the upper limit of integration is greater than $\mathrm{Sup}_K | \omega(I).k |$. In this way we obtain:

$$(14) \qquad \| \partial S/\partial I_j \|_{L^1} \leq c + c\textstyle\sum_{0<|k|\leq N(\delta)} \| \partial f_k/\partial I_j \| \log(c|k|/\delta)$$
$$+ c\textstyle\sum_{0<|k|\leq N(\delta)} \| f_k \| (\sum_{i=1}^m |k_i|) \int | \Phi'(k.\omega) | \, dI$$
$$\leq c \log(1/\delta) + c\textstyle\sum_{0<|k|\leq N(\delta)} \| f_k \| (\sum_{i=1}^m |k_i|) \times$$
$$\times \int | \Phi'(k.\omega) | \, dI.$$

To estimate the last term with $i \neq n$ in the last sum, we again use $\omega.k$ as one of the variables, and we find:

$$(15) \qquad \textstyle\sum_{0<|k|\leq N(\delta)} \| f_k \| . | k_i | \int | \Phi'(k.\omega) | \, dI \leq$$
$$\leq c \textstyle\sum_{|k|\leq N(\delta)} \| f_k \| \int_{-\infty}^{+\infty} | \Phi'(x) | \, dx \leq c \int | \Phi'(x) | \, dx \leq c/\delta.$$

It remains to treat the similar expression containing $| k_n |$. As in Part b we define $n + 1$ partial sums $\Sigma_1, \ldots, \Sigma_{n+1}$ and, again proceeding as in b), we first bound Σ_i, $1 \leq i \leq n - 1$ by c/δ. Σ_n is also evaluated as above (by decomposing as $\Sigma^1, \ldots, \Sigma^{n-1}$), and we arrive again to the bound c/δ.

The remaining expression is Σ_{n+1}, the partial sum over vectors having k_n as their only nonzero component. Since $k.\omega = k_n\omega_n$ and since ω_n does not vanish on K:

$$| \Phi'(k.\omega) | \leq c$$

and:

$$(16) \qquad \textstyle\sum \| f_k \| . | k_n | \int | \Phi'(k.\omega) | \, dI \leq c \,\mathrm{mes}(K) \sum \| f_k \| . | k_n |.$$

Since $\| f_k \| \leq c/| k_n |^{n+1}$, the right hand sum, which is restricted to vectors having k_n as their only nonzero component, converges, and Σ_{n+1} is therefore less than a constant. This completes the proof of Part c) of Lemma 1. ∎

We will now prove the first part of Theorem 1. We introduce the increasing function:

$$z(t) =_{\mathrm{def}} \int_D \mathrm{Sup}_{[0,\,t]} \| I(u) - J(u) \| \, dI_0 d\varphi_0$$

defined on $[0, 1/\varepsilon]$.

From Lemma 1 we deduce:

<u>Lemma 2</u>:

$z(t)$ satisfies the inequality:

(17) $$z(t) \leq c\varepsilon \int_0^t z(\tau)\, d\tau + c\sqrt{\varepsilon}.$$

To prove this we write:

(18) $$\begin{aligned} I(t) - J(t) &= \varepsilon \int_0^t [\, f(I(u), \varphi(u), \varepsilon) - f_0(J(u)) \,]du \\ &= \varepsilon \int_0^t [\, f_0(I(u)) - f_0(J(u))]du - \varepsilon\, [S(I(t), \varphi(t)) - \\ &\quad - S(I(0), \varphi(0))] + \varepsilon \int_0^t R\, (I(u), \varphi(u), \varepsilon)\, du\,, \end{aligned}$$

where R is the remainder we met before in Chapter 3:

$$R(I, \varphi, \varepsilon) = f(I, \varphi, \varepsilon) - f_0(I) + \omega(I).\partial S/\partial\varphi + \varepsilon[f.\partial S/\partial I + g.\partial S/\partial\varphi],$$

which is equivalent to introducing the intermediate variable $P = I + \varepsilon\, S(I, \varphi)$.

From (18) we deduce:

(19) $$\begin{aligned} \operatorname{Sup}_{u\in[0,\, t]} \| I(u) - J(u) \| &\leq c\varepsilon \int_0^t \operatorname{Sup}_{v\in[0,\, u]} \| I(v) - J(v) \|\, du + \\ &+ \varepsilon \| S(I(t), \varphi(t)) \| + \varepsilon \| S(I(0), \varphi(0)) \| + \\ &+ \varepsilon \int_0^t \| R(I(u), \varphi(u), \varepsilon) \|\, du. \end{aligned}$$

Integrating over the initial conditions and reversing the order of integration, we obtain:

(20) $$\begin{aligned} z(t) &\leq c\varepsilon \int_0^t z(u)\, du + \varepsilon \| S \|_{L^1} + \varepsilon \int_D \| S(I(t), \varphi(t)) \|\, dI_0 d\varphi_0 + \\ &+ \varepsilon \int_0^t du \int_D \| R(I(u), \varphi(u), \varepsilon) \|\, dI_0 d\varphi_0\,. \end{aligned}$$

The second and third terms are both bounded by $\varepsilon\, \mathrm{mes}(D) \| S \|$. As for the last term, we use the change of variables $(I(t), \varphi(t)) \to (I(0), \varphi(0))$ to reduce it to $\| R \|_{L^1}$. The Jacobian of this transformation is:

(21) $$\mathrm{Jac}(t) = \exp[\varepsilon \int_0^t \{\mathrm{Tr}\ \partial f/\partial I + \mathrm{Tr}\ \partial g/\partial\varphi\}\, du],$$

(where Tr is the trace), so that $\operatorname{Inf}_{t\in[0,\, 1/\varepsilon]} \mathrm{Jac}(t)$, is bounded below by a positive constant. We therefore obtain:

(22) $$z(t) \leq c\varepsilon \int_0^t z(u)\, du + c\varepsilon \| S \| + c\varepsilon t \| R \|_{L^1}.$$

From the definition of R and from Lemma 1, we deduce:

(23) $$\| R \|_{L^1} \leq c[\varepsilon + \delta + \varepsilon/\delta],$$

and:

(24) $$z(t) \leq c\varepsilon \int_0^t z(u)\, du + c\varepsilon/\delta + c\varepsilon t[\varepsilon + \delta + \varepsilon/\delta].$$

Choosing $\delta = \sqrt{\varepsilon}$ (δ being a free parameter up to now), we arrive to inequality (17) for $t \leq 1/\varepsilon$. ∎

The first assertion of Theorem 1 follows immediately from this lemma. In fact, using the lemma we deduce that $z(t) \leqslant c\sqrt{\varepsilon}e^{c\varepsilon t}$ and, since $\int_{V'(\varepsilon,\eta/2)} E(I_0, \varphi_0, \varepsilon)\, dI_0 d\varphi_0$ is equal to $\mathrm{Sup}_{t\in[0,\, 1/\varepsilon]}\, z(t)$, assertion a) of Theorem 1 holds. ∎

We move to the proof of Part b) of Theorem 1 which says that the measure of the set of initial conditions such that the separation between exact and averaged trajectories exceeds $\eta/2$ is small (of order $\sqrt{\varepsilon}$).

For all T in $[0, 1/\varepsilon]$, we introduce the set $V'(\varepsilon, \eta/2, T)$, containing $V'(\varepsilon, \eta/2)$, of initial conditions in D such that $\mathrm{Sup}_{t\in[0,\, T]} \| I(t) - J(t) \| \leqslant \eta/2$. For each initial condition in this set $I(t)$ remains in $K - \eta/2$ on the time interval $[0, T]$ and in K on the time interval $[0, T + \beta]$, where $\beta = \eta/(2c_4\varepsilon)$ and c_4 bounds $\| f \|$. We will denote the complement of $V'(\varepsilon, \eta/2, T)$ in D by $V''(\varepsilon, \eta/2, T)$.

For an initial condition in $V'(\varepsilon, \eta/2, T)$, $\mathrm{Sup}_{t\in[0,\, T+\beta]} \| I(t) - J(t) \|$ is well defined and:

(25) $\displaystyle\int_{V'(\varepsilon,\eta/2,T)} \mathrm{Sup}_{t\in[0,\, T+\beta]} \| I(t) - J(t) \|\, dI_0 d\varphi_0 \leqslant c\sqrt{\varepsilon},$

from which it follows that:

(26) $\displaystyle\int_{V'(\varepsilon,\eta/2,T)\,\cap\, V''(\varepsilon,\eta/2,T+\beta)} \mathrm{Sup}_{t\in[0,\, T+\beta]} \| I(t) - J(t) \|\, dI_0 d\varphi_0 \leqslant c\sqrt{\varepsilon}.$

For an initial condition in $V''(\varepsilon, \eta/2, T+\beta) \cap V'(\varepsilon, \eta/2, T)$, for which $\| I(t) - J(t) \|$ becomes "bad" between $t = T$ and $t = T + \beta$, $I(t)$ and $J(t)$ remain in K until time $T + \beta$ and $\mathrm{Sup}_{t\in[0,\, T+\beta]} \| I(t) - J(t) \| > \eta/2$. Therefore by (26):

(27) $\mathrm{mes}[V'(\varepsilon, \eta/2, T) \cap V''(\varepsilon, \eta/2, T+\beta)] \leqslant c\sqrt{\varepsilon}/\eta.$

We may cover $V''(\varepsilon, \eta/2)$, which is simply $V''(\varepsilon, \eta/2, 1/\varepsilon)$, by the sets $V''(\varepsilon, \eta/2, j\beta) \cap V'(\varepsilon, \eta/2, (j-1)\beta)$, where j runs from 1 to $[1/(\beta\varepsilon)] + 1$. Each of these sets corresponds to initial conditions such that the separation between trajectories remains small until time $(j - 1)\beta$, but exceeds $\eta/2$ at a certain point between $(j - 1)\beta$ and $j\beta$ (see Figure 6.2 below).

Consequently, by inequality (27) above:

(28) $\mathrm{mes}(V''(\varepsilon, \eta/2)) \leqslant c([1/(\beta\varepsilon)] + 1)\sqrt{\varepsilon}/\eta.$

Since $\beta = \eta/(2c_4\varepsilon)$, $\mathrm{mes}(V''(\varepsilon, \eta/2))$ is less than $c\sqrt{\varepsilon}$, which concludes the proof of the second assertion of Theorem 1. ∎

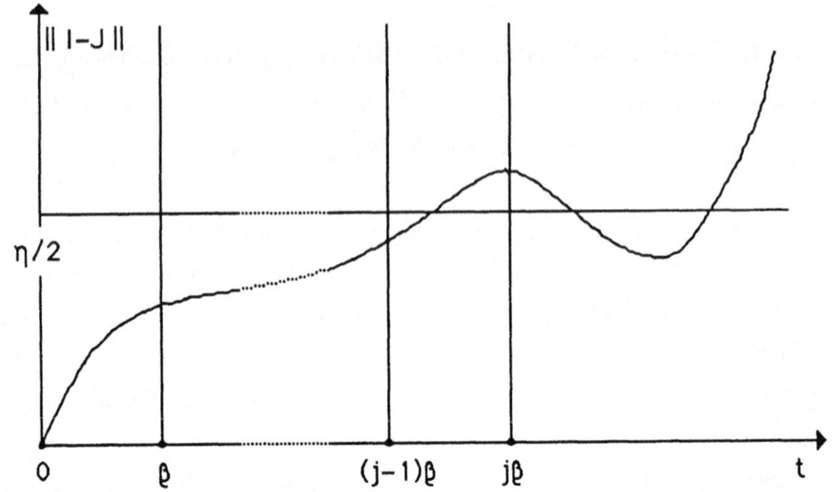

Figure 6.2: For initial conditions in the set $V'(\varepsilon, \eta/2, (j-1)\beta)$ the separation between the exact trajectory $I(t)$ and the averaged trajectory $J(t)$ remains small until time $(j-1)\beta$ but exceeds $\eta/2$ at some time between $(j-1)\beta$ and $j\beta$.

6.3 Optimality of the results of Theorem 1

Proof of Part c of Theorem 1:

The example chosen is very simple. Consider the equation:

(1) $\quad d^2\varphi/d\tau^2 = 1/2 + \sin\varphi$

equivalent to the system:

(2) $\quad dI/dt = \varepsilon(\sin\varphi + 1/2), \quad d\varphi/dt = I \quad (\tau = \sqrt{\varepsilon}\, t)$,

which we we previously encountered in Chapter 4 (pendulum with a constant driving force) and which satisfies Arnold's nondegeneracy condition since $\omega_2(I) = 1$, by adding the fictitious equation $d\varphi'/dt = 1$ and since $\omega_1/\omega_2 = I$.

The domain K is defined by $I \in [-1, A]$, where $A \geq 3/2$, and we choose $\eta < 1$. The set of initial conditions $(K - \eta) \times T^2$ then corresponds to $I \in [-1+\eta, A-\eta-1/2]$, $(\varphi, \varphi') \in T^2$. For convenience, we take $\varphi \in [-\pi/6, 11\pi/6]$.

The phase portrait of the forced pendulum (1), already presented in Chapter 4, possesses two fixed points, one elliptic ($\varphi = 7\pi/6 \mod 2\pi$), and one hyperbolic ($\varphi = 11\pi/6 \mod 2\pi$).

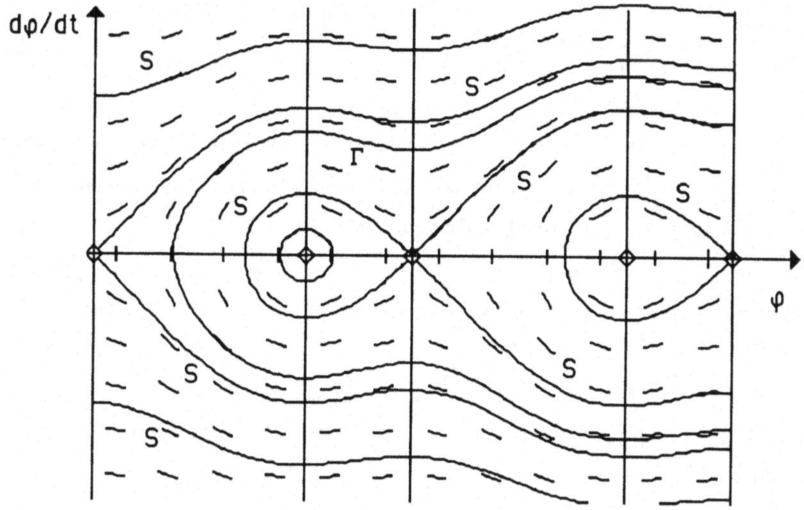

Figure 6.3: Phase portrait of the forced pendulum displaying the vectorfield together with the separatrix S and a typical trajectory Γ. In this figure φ ranges from $-\pi/6$ to $23\pi/6$.

On the closed trajectories contained inside the homoclinic loop described by the equation:

(3) $$(d\varphi/d\tau)^2 - \varphi + 2\cos\varphi = \sqrt{3} - 11\pi/6,$$

$|d\varphi/d\tau|$ remains less than $2(\sqrt{3} - \pi/3)$ for all time. Since $I = d\varphi/dt = \sqrt{\varepsilon}\, d\varphi/d\tau$, I remains less than $2\sqrt{\varepsilon}(\sqrt{3} - \pi/3)$ for trajectories with initial conditions such that $(\varphi_0, I_0/\sqrt{\varepsilon})$ is inside the homoclinic loop. We note that such initial conditions belong to $(K - \eta) \times T^2$ for ε small enough.

On the other hand, the averaged system is $dJ/dt = \varepsilon/2$, and for the initial conditions just described, $J(0)$ belongs to $[-2\sqrt{\varepsilon}(\sqrt{3} - \pi/3), 2\sqrt{\varepsilon}(\sqrt{3} - \pi/3)]$. Consequently:

(4) $$J(1/\varepsilon) - I(1/\varepsilon) > 1/2 - 4\sqrt{\varepsilon}(\sqrt{3} - \pi/3) > \eta/2,$$

provided that ε is small enough. These initial conditions thus belong to $V''(\varepsilon, \eta/2)$.

The measure of the set of initial conditions corresponding to such trajectories is equal to $2\pi\sqrt{\varepsilon}$ times the area in phase space enclosed by the homoclinic loop (φ' of course plays no role). This measure is thus of order $\sqrt{\varepsilon}$, which demonstrates the optimality of the second assertion of Theorem 1.

Using the same example, we next show the optimality of the first assertion of the theorem.

For this purpose it will suffice to find a subset C of $V'(\varepsilon, \eta/2)$ such that:

(5) $\int_C E(I_0, \varphi_0, \varepsilon)\, dI_0 d\varphi_0 d\varphi'_0 > c\sqrt{\varepsilon}.$

We examine the open separatrix branch of the hyperbolic fixed point. On the interval $[-\pi/6, 11\pi/6]$, it appears as a succession of curves superimposed on one another. The first curve, described by the equation:

(6) $I^2 + 2\varepsilon \cos\varphi - \varepsilon\varphi = a_1\varepsilon, \quad a_1 = \sqrt{3} + \pi/6,$

connects the hyperbolic point to the point $(\varphi = 11\pi/6, I = (2\pi\varepsilon)^{1/2})$. The nth curve, with equation:

(7) $I^2 + 2\varepsilon \cos\varphi - \varepsilon\varphi = a_n\varepsilon, \quad a_n = a_1 + 2n\pi,$

connects the point $(\varphi = -\pi/6, I_{n-1} = (2[n-1]\pi\varepsilon)^{1/2})$ to the point $(\varphi = 11\pi/6, I_n = (2n\pi\varepsilon)^{1/2})$.

The qth zone will be the domain lying between the $(q-1)^{th}$ and q^{th} curves (see Figure 6.4). We note that the qth zone is contained in the strip:

$$I_{q-1} \leqslant I \leqslant [I_q^2 + \varepsilon(2\sqrt{3}\pi)]^{1/2}.$$

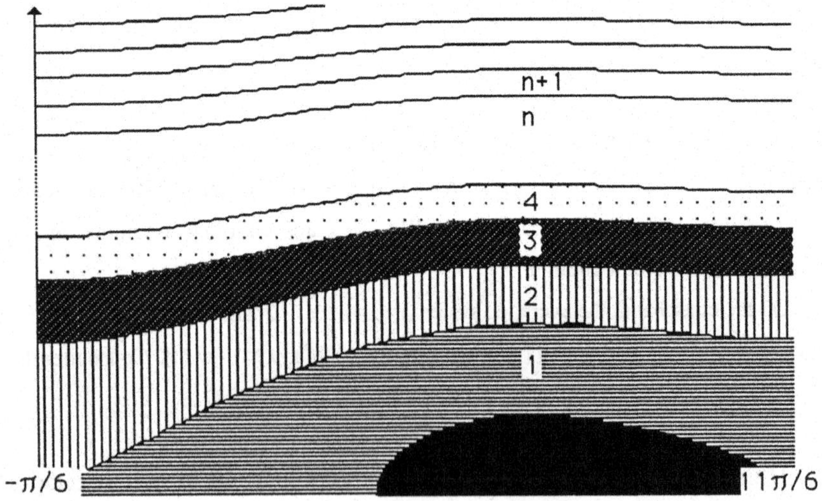

Figure 6.4: The separatrix defines disjoint zones in the domain which lies outside the homoclinic loop (black area). The nth curve separates the nth zone from the $(n+1)^{th}$ one.

The time T_q required to completely cross the qth zone lies between $2\pi/[I_q^2 + \varepsilon(2\sqrt{3}-\pi)]^{1/2}$ and $2\pi/I_{q-1}$.

Let θ be a positive number small enough to allow the exact system to be conjugated to its linear part in the portion $V(\theta)$ of the first zone defined by $|\varphi - 11\pi/6| < \theta \mod 2\pi$, and let $W(\theta)$ be the complement of $V(\theta)$ in the first zone (see Figure 6.5).

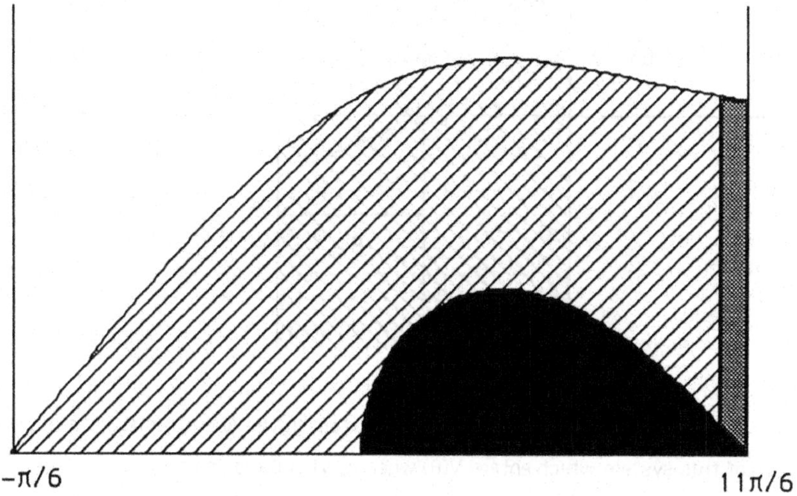

Figure 6.5: In the first zone outside the homoclinic loop (black area), two subsets must be distinguished: $V(\theta)$ (gray area) and its complement $W(\theta)$ (hatched area).

For C we choose the intersection of $W(\theta)$ with the trajectories passing through the points ($\varphi = 11\pi/6 - \theta$, $\alpha\sqrt{\varepsilon} \leq I \leq \beta\sqrt{\varepsilon}$) of the first zone. We will show that for small enough θ and ε, and for α and β correctly chosen, (5) is satisfied, then we will show that C is contained in $V'(\varepsilon, \eta/2)$.

All trajectories with initial conditions in C leave this domain in a time $T(I_0, \varphi_0)$ less than $c/\sqrt{\varepsilon}$. Since the separation at time t between exact and averaged trajectories is given by $I(t) - J(t) = \varepsilon\int_0^t \sin\varphi(u)\, du$, it is less than $c\sqrt{\varepsilon}$ for all times prior to T.

We now consider the passage of exact trajectories through the domain of linearization $V(\theta)$, where the linearized system is expressed by:

(8) $\quad dI/dt = \varepsilon\sqrt{3}/2\, \psi, \quad d\psi/dt = I \quad (\psi = \varphi - 11\pi/6).$

The hyperbolic fixed point $(0, 0)$ has eigenvalues $\lambda\sqrt{\varepsilon}$ and $-\lambda\sqrt{\varepsilon}$ with $\lambda^2 = \sqrt{3}/2$.

A trajectory which leaves the domain C enters $V(\theta)$ at the point $(\psi = -\theta, I = x)$, where $\alpha\sqrt{\varepsilon} \leqslant x \leqslant \beta\sqrt{\varepsilon}$ (see Figure 6.6).

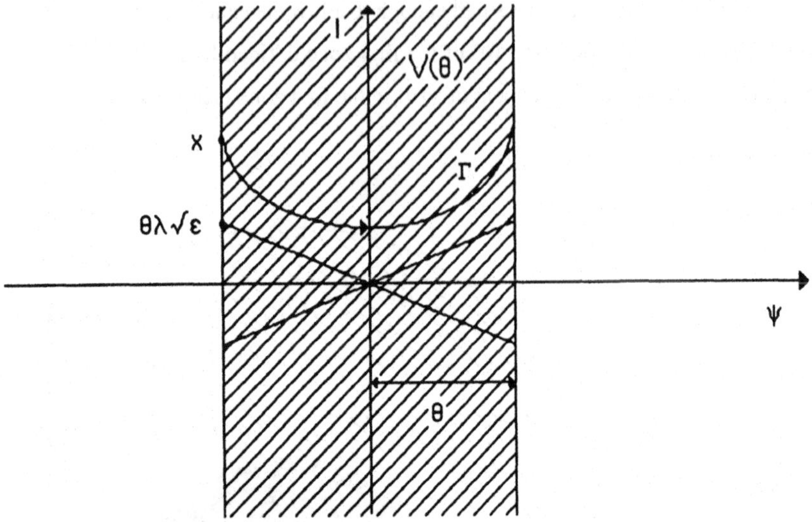

Figure 6.6: Phase portrait of the linearized system in the domain $V(\theta)$. Γ is a typical solution of this system which enters $V(\theta)$ when leaving the domain C.

The solution of the linearized system with this initial condition is simply $\psi(t) = x \sinh(\lambda\sqrt{\varepsilon}t)/(\sqrt{\varepsilon}\lambda) - \theta \cosh(\lambda\sqrt{\varepsilon}t)$. We will assume that $\alpha > \lambda\theta$ so that this trajectory is situated above the stable manifold of the linearized system. This linearized trajectory leaves $V(\theta)$ in time $T(x)$ given by the equation $\psi(T(x)) = \theta$ which, for small $\beta - \lambda\theta$, is approximated by $|\log[(x/2\theta\lambda\sqrt{\varepsilon})-1/2]|/(\lambda\sqrt{\varepsilon}) =_{def} \tau(x)$. The exact (nonlinearized) trajectory thus takes a time between $\tau(x)/2$ and $2\tau(x)$ to cross $V(\theta)$. We choose $\alpha \geqslant \theta\lambda[1 + 2\, e^{-\lambda/(4\sqrt{\varepsilon})}]$ so that this crossing time is always less than $1/(2\varepsilon)$. Since $\sin\varphi$ is close to $-1/2$ on $V(\theta)$ and since the crossing time $T(x)$ is greater than $\tau(x)/2$, the additional separation between corresponding exact and averaged trajectories during a complete passage through $V(\theta)$ is bounded below by $\varepsilon\tau(x)/4$.

We see that for every initial condition in C situated on a trajectory passing through the point $(\varphi = 11\pi/6 - \theta, I = x\sqrt{\varepsilon})$, the separation at the instant when the trajectory leaves $V(\theta)$ is bounded below by $\varepsilon\tau(x)/4 - c\sqrt{\varepsilon}$ and a fortiori by $\varepsilon\tau(x)/8$ for β chosen correctly. Since this exit takes place prior to $t = 1/\varepsilon$, for sufficiently small ε we deduce the lower bound:

(9) $$E(I_0, \varphi_0, \varepsilon) \geqslant \varepsilon\tau(x)/8 \geqslant \varepsilon|\log[(x/2\theta\lambda\sqrt{\varepsilon})-1/2]|/(8\lambda\sqrt{\varepsilon}),$$

which leads to (5) on integrating over C.

It remains to prove that C is contained in $V'(\varepsilon, \eta/2)$, which requires a bound on the separation for initial conditions in C and times less than $1/\varepsilon$. If, at such a time, the exact trajectory has not left $V(\theta)$, the separation is bounded by $c\sqrt{\varepsilon} + 1/4 - \sqrt{3}\theta/2 < 1/4$ for small enough θ, since the time to cross $V(\theta)$ is less than $1/2\varepsilon$.

If the trajectory is located in the second zone, the separation is bounded by $1/4 + (2\pi\varepsilon)^{1/2}$, since the time spent in the second zone is less than $(2\pi/\varepsilon)^{1/2}$.

Now suppose that the trajectory is located in the $(p+1)^{st}$ zone. The separation arising from complete passsage through the nth zone $(2 < n \leqslant p)$ is equal to:

$$\varepsilon\int_{-\pi/6}^{11\pi/6} \sin\varphi \; I(n, \varphi)^{-1}\, d\varphi$$

where $I(n, \varphi)$ refers to the nth zone. On integrating by parts this becomes:

$$-\varepsilon^2\int_{-\pi/6}^{11\pi/6} \cos\varphi\,(\sin\varphi + 1/2).\, I(n, \varphi)^{-3}\, d\varphi -$$
$$-\varepsilon\sqrt{3}/2\,[I(n, 11\pi/6)^{-1} - I(n, -\pi/6)^{-1}].$$

Since $I(n, \varphi)$ is greater than I_{n-1} on the nth zone, the total separation due to crossing zones 3 through p is bounded in absolute value by:

$$\varepsilon\sqrt{3}/2\,[1/I_2 - 1/I_{p+1}] + 3\pi\varepsilon^2 \sum_{n=2}^{p-1} 1/I_n^3.$$

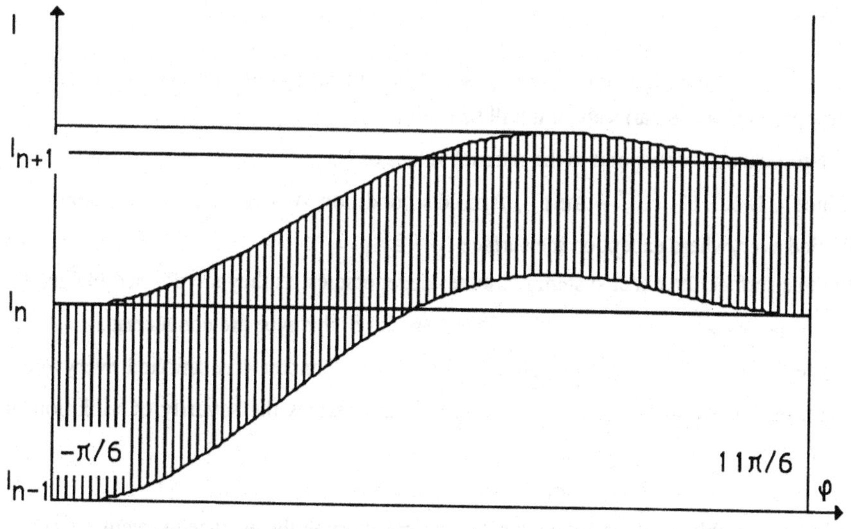

Figure 6.7: Location of the nth zone in phase space.

Since $I_n = (2\pi n\varepsilon)^{1/2}$, this expression is equal to $c\varepsilon^{1/2}$, so the total separation at the instant under consideration is less than $1/4 + (2\pi\varepsilon)^{1/2} + c\varepsilon^{1/2} + \varepsilon T_{p+1}$, where T_{p+1} is the time required to completely cross the $(p+1)^{st}$ zone; T_{p+1} is less than:

$$2\pi/I_p = (2\pi/p\varepsilon)^{1/2}.$$

We have thus established that for all initial conditions in C:

$$E(I_0, \varphi_0, \varepsilon) \leqslant 1/4 + c\sqrt{\varepsilon}.$$

By choosing $\eta > 1/2$ and ε small enough, C is contained in D. This demonstrates the optimality of the estimates in Theorem 1. ▪

6.4 Optimality of the results of Theorem 2

Proof of Part b of Theorem 2:

We refer here to the system:

$$(1) \qquad dI_1/dt = \varepsilon[I_2^r + \sin\varphi_1], \quad dI_2/dt = \varepsilon, \quad dI_3/dt = \varepsilon \cos\varphi_1$$
$$d\varphi_1/dt = I_1, \quad d\varphi_2/dt = 1,$$

where r is an arbitrary positive integer. The associated averaged system is:

$$(2) \qquad dJ_1/dt = \varepsilon I_2^r, \quad dJ_2/dt = \varepsilon, \quad dJ_3/dt = 0.$$

This example, first considered by Neistadt, is naturally suited to our purposes here. A few preliminary remarks about it will be useful:

- The resonances correspond to the rational values of I_1.
- Since $f(I).\partial\omega/\partial I = I_2^r + \sin\varphi_1$, Arnold's condition A of Chapter 3 (Section 5) is satisfied everywhere except on the surface $I_2^r + \sin\varphi_1 = 0$.
- On this surface the exact system possesses degeneracies, as they are defined in Chapter 4, at the points where $\cos\varphi_1$ vanishes. Degeneracies of this type are nongeneric.
- Since $f_0(I).\partial\omega/\partial I = I_2^r$, Neistadt's condition N of Chapter 4 is satisfied everywhere except on the surface $I_2 = 0$. The integer r characterizes the severity of violation of condition N.

The variable J_3 does not evolve along a trajectory of the averaged system. Since:

$$(3) \qquad I_3(t) - J_3(t) = \varepsilon \int_0^t \cos\varphi_1(s)\, ds,$$

if the exact trajectory spends a long time in the regions where $\cos\varphi_1 = \pm 1$, there will be a wide separation between it and the corresponding averaged trajectory. The variable I_3 plays no role in the dynamics and is introduced to characterize the separation between the trajectories, whereas the phase variable φ_2 allows us to write down Arnold's nondegeneracy condition. We may thus restrict our attention to the dynamics in the space (φ_1, I_1, I_2), given by the following equations (where the variable φ_1 is now denoted φ):

$$(4) \qquad dI_1/dt = \varepsilon[I_2{}^r + \sin\varphi], \quad dI_2/dt = \varepsilon, \quad d\varphi_1/dt = I_1,$$

or equivalently:

$$(5) \qquad d^2\varphi/d\tau^2 = [I_2(0) + \sqrt{\varepsilon}\tau]^r + \sin\varphi, \quad I_2(\tau) = I_2(0) + \sqrt{\varepsilon}\tau, \quad \tau = \sqrt{\varepsilon}t.$$

The separation between exact and averaged trajectories will be large for trajectories spending a long time near the point $(0, 0, 0)$. It is in fact possible to prove that for $2\sqrt{\varepsilon} < \rho < 1$, the measure of the set of initial conditions such that the separation is greater than ρ is at least $c(r)[\sqrt{\varepsilon} + (\sqrt{\varepsilon}/\rho)^{1+1/r}]$, where $c(r)$ is a constant depending on r. This result demonstrates the optimality of the exponents in Theorem 2 among the class of estimates in powers of ε and ρ. In fact, for arbitrarily large r, the exponent $1 + 1/r$ is arbitrarily close to 1 and for any $\alpha > 1/2$, $\beta < 1$ and $c > 0$, the measure is strictly greater than $c\varepsilon^\alpha/\rho^\beta$, for r sufficiently large.

We will not show this lower bound on the measure, which is a laborious undertaking, but we give instead the following heuristic outline: First of all, to obtain this lower bound, it will not suffice to study a single passage of a trajectory in the neighborhood of $(0, 0, 0)$; it is necessary instead to examine *successive* passages of a trajectory near this point. The desired result originates from the fact that a sufficient number of trajectories pass near $(0, 0, 0)$ sufficiently often, due to the violation of condition N at $I_2 = 0$.

Consider first a trajectory with initial condition on the surface $I_2 = 0$, which is therefore a solution of equation (5) with $I_2(0) = 0$ in the $(\varphi, d\varphi/d\tau)$ plane; in other words, a solution of the equation of a nonlinear pendulum subjected to slowly increasing force, starting at 0.

For ε small, and for an initial condition lying outside the separatrix of the unperturbed pendulum, near the hyperbolic fixed point $(\varphi=0, d\varphi/d\tau=0)$ of this system, the perturbed trajectory will slowly separate from the separatrix, *the slowness of this separation being greater with increasing* r. On the scaled time interval $[0, 1/\sqrt{\varepsilon}]$, this trajectory will pass near the hyperbolic point many times. The separation between the exact and the averaged trajectories will be close to εT, where T represents the total (unscaled) time of passage near the hyperbolic point. This may easily be seen by checking the contributions to the integral in (3) near the hyperbolic point.

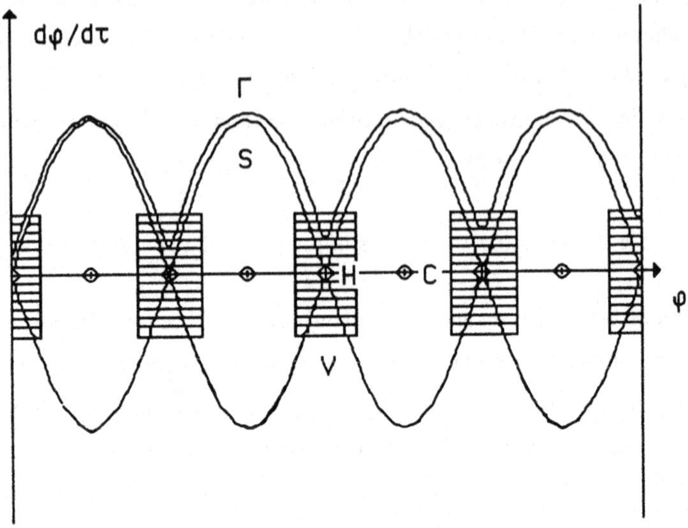

Figure 6.8: The exact trajectory Γ remains close to the separatrix S of the unperturbed pendulum for a long time. Therefore Γ visits the neighborhood V of the hyperbolic fixed point H = (0, 0) many times. C is the center fixed point (π, 0) located inside in the homoclinic loop.

The situation is hardly changed if the initial condition $I_2(0)$ is small rather than zero, as the trajectory is then the solution to equation (5), with the unperturbed pendulum equation:

$$(6) \qquad d^2\varphi/d\tau^2 = [I_2(0)]^r + \sin\varphi.$$

Since $I_2(0)$ is small, the open branch of the separatrix separates very slowly from the closed loop, more slowly with increasing r.

As we already noted in Chapter 4, condition N is not a local condition connected with the behavior of the pendulum near the hyperbolic point, but is rather a global condition connected with the asymptotic behavior of the open branch of the separatrix. For $I_2(0) = 0$, condition N is violated and the separatrix does not possess an open branch; for $I_2(0) \neq 0$, condition N is satisfied and the separatrix possesses an open branch which behaves asymptotically like a parabolic branch $[d\varphi/d\tau]^2 \sim 2[I_2(0)]^r \varphi$.

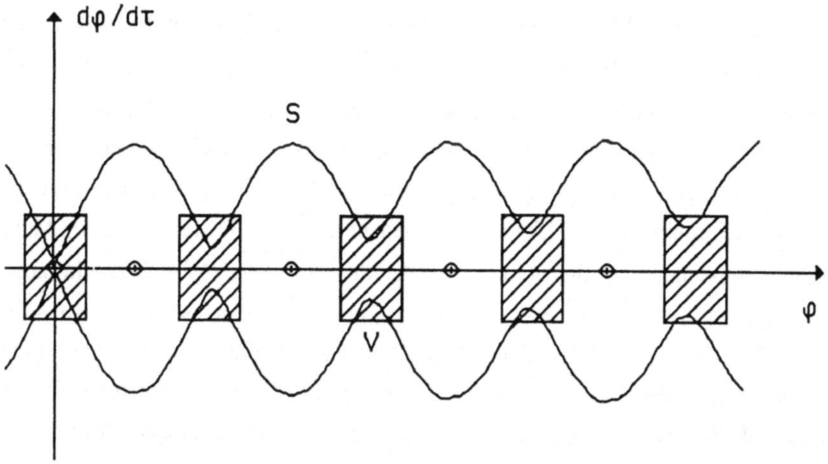

Figure 6.9: The open branches of the separatrix S visit the neighborhood V of the hyperbolic fixed point many times.

For $I_2(0)$ small, the open branch of the separatrix passes near the hyperbolic fixed point many times before definitively separating from it (see Figure 6.9). For ε small, the same is true of all perturbed trajectories with initial conditions outside the separatrix loop and near the hyperbolic fixed point. As before, the separation will be on the order of εT, where T is the total time of passage in the neighborhood of the hyperbolic point.

We thus see that to estimate the measure in $(I_1, I_2, \varphi_1, \varphi_2)$ space of initial conditions for ρ-deviant trajectories, it suffices to determine as a function of their initial conditions the total time spent by trajectories near the hyperbolic fixed point of a pendulum subjected to a force depending *linearly* on time. We note incidentally that since the system can be linearized in a small neighborhood of the hyperbolic point, it is easy to estimate the duration of one passage through this neighborhood. However, one must then precisely determine the coordinates of entry into this neighborhood on the nth passage in terms of the exit coordinates on its $(n-1)^{st}$ passage. This requires calculations too laborious to be presented here; rather than detailing the proof, we refer the interested reader to Neistadt's thesis [Nei1].

We close these two chapters about general (non-Hamiltonian) multiphase averaging for systems in the standard form with a brief but important discussion of the nondegeneracy condition which, outside the regularity hypotheses, is the main restriction to the applicability of Neistadt's results. Using Arnold's condition, the nondegeneracy of frequencies implies that the number m of slow variables is greater than the number n of phases minus one: $m \geqslant n - 1$. Kolmogorov's condition is still more restrictive, implying that $m \geqslant n$. This nondegeneracy requirement essentially provides a straightforward estimate of the volume of the resonant zones on the base space. Such a condition breaks for generic system with $m \geqslant n - 1$ on a submanifold of dimension $n - 2$. However the less stringent requirement of "finite order degeneracy" can be used ([Bac]). More precisely it is essentially sufficient to demand that the vectors $\partial\omega^j/\partial I^j$ (j is a multi-index whose length $|j|$ is less than some integer k) span the tangent space $\mathbb{R}^n$ to the frequency space. One then proves a result parallel to Theorems 1 and 2 of the present chapter, using more sophisticated lemmas only to estimate the volume of the resonant zones: In the estimates of the mean deviation (see assertion a of Theorem 1) and the measure of excluded initial conditions (see assertion a of Theorem 2) the factor $\varepsilon^{1/k+1}$ just replaces $\varepsilon^{1/2}$.

The very opposite case when no such non-degeneracy condition can ever be satisfied is of course that of strictly isochronous unperturbed systems ($\omega(I) = \omega$ a constant vector). It is simpler, as was already noticed in Sections 3.1 and 3.2. The reader may find it interesting to go back to these sections, and the experience of Chapters 5 and 6 should make it clear how the suggestions that were given there can be implemented (it should become even clearer after an acquaintance with the material of Chapter 7). As we already pointed out, this case of the perturbation of an assembly of noninteracting harmonic oscillators (a physical way of describing the unperturbed system $d\varphi/dt = \omega$, $\omega \in \mathbb{R}^n$) is quite important in applications.

CHAPTER 7: HAMILTONIAN SYSTEMS

7.1 General introduction

The present chapter is devoted to the study of perturbations of integrable Hamiltonian systems, a question which Poincaré himself termed "the fundamental problem of dynamics". Throughout this chapter, we make frequent references to the appendices, particularly Appendices 7 and 8 which introduce some essential tools in the modern theory of perturbations of Hamiltonian systems. For the fundamentals of Hamiltonian theory, we refer the reader to, e.g. [Ab], [Ar7], [Gala1], [Go] or [La2], and for an account of the more modern developments to, e.g. [Mo3-5].

We start with an n degree-of- freedom Hamiltonian H:

(1) $\quad H(I, \varphi) = h(I) + \varepsilon f(I, \varphi); \quad I \in \mathbb{R}^n, \ \varphi \in T^n.$

T^n designates the n torus; all functions are periodic with respect to φ_i $(i = 1, \ldots, n)$ with period 2π. As the unperturbed Hamiltonian h is assumed integrable, we use the action-angle variables (I, φ). Their construction is performed in Appendix 6 for one degree of freedom and the reader may consult [Ar7] or Chapter 10 for the general case; we need nothing more here than their existence, which justifies writing the perturbed Hamiltonian H as in (1). The equations of the unperturbed motion read:

(2) $\quad dI/dt = 0; \quad d\varphi/dt = \partial h/\partial I =_{def} \omega(I),$

where $\omega(I) = \nabla h$ is the vector of frequencies of h. This motion takes place on invariant tori (with equations $I = I_0$) and is characterized by the n frequencies $\omega_1(I_0), \ldots, \omega_n(I_0)$. The equations of the perturbed motion, defined by the Hamiltonian (1), may be written:

(3) $\quad dI/dt = -\varepsilon \partial f/\partial \varphi; \quad d\varphi/dt = \omega(I) + \varepsilon \partial f/\partial I.$

The Hamiltonian H is of course a constant of the perturbed motion and, as proved by Poincaré (see [Ben1]), if h and f are analytic and h satisfies a quite general condition, H is "almost surely" (i.e., for almost every choice of f and ε), the *sole* smooth constant of motion to be defined on the whole of phase space.

It is of course out of the question to survey, even quickly, the results which have been established for such systems: The restricted three body problem alone, for example, where one of the masses is assumed to be much smaller than the other two, has given rise to an enormous literature. Instead, we shall focus our attention on two complementary and relatively recent theorems which provide a rather satisfactory description of the motion and answer some of the needs of physicists: The Kolmogorov-Arnold-Moser theorem ([Ko], [Ar2], [Mo1-2]) and Nekhoroshev's theorem ([Nek1-3]).

We shall not give a complete proof of the former but merely state the theorem and recall the image of dynamics it provides. The latter, though much more recent, is less revolutionary. It completes, in a way, the classical theory of Hamiltonian perturbations as Poincaré had imagined it. As it is still far less widespread in the physics literature than the KAM theorem, we shall furnish a complete proof of this theorem in a somewhat particular case (see below) which is essentially taken from [Ben5]. Further details may be found in the original papers [Nek2-3] as well as in [Ben2].

Nekhoroshev's theorem may be considered in some sense as *the averaging theorem for Hamiltonian systems*. Indeed, if we average over angles in the first of equations (3), we obtain the trivial averaged system $dI/dt = 0$. The averaging problem therefore reduces to the study of the variations of the action variables for the system defined by (1), which is precisely the object of Nekhoroshev's theorem. In Section 7.3 we compare this result with the far simpler averaging theorem which was proved in Chapter 6 in a more general setting and which may of course be applied to the special case of system (3).

7.2 The KAM theorem

The KAM theorem originated in a stroke of genius by Kolmogorov, who, in the midst of the very intricate situation brought to light by Poincaré, Birkhoff and others (cross-sections in the vicinity of an elliptic periodic trajectory, homoclinic curves, etc.) realized that invariant tori, defined in $\mathbb{R}^n \times T^n$ by $I' = \mathrm{cst}$, where the variables (I', φ') are close to (I, φ), might be preserved in the perturbed motion. He also introduced the technique which lies at the heart of the proof and enables one to locate the invariant tori and prove their existence: the superconvergent iteration of Newton's method (see Appendix 8 for more detail) which had already been used by Siegel in his work on the conjugacy of local isomorphisms of $\mathbb{C}^n$. Contrary to the opinion which is sometimes expressed, one may therefore consider that, in his note of 1954 [Ko], Kolmogorov had really stated and essentially proved the theorem. An elementary and detailed proof following Kolmogorov's original idea can be found in [Ben6].

Arnold's paper [Ar2] provided a complete proof and introduced the notion of "ultraviolet cutoff" which allows one to prove the existence of invariant tori in a *global* way; one determines the set of preserved tori in the vicinity of a correctly chosen invariant torus $I = I_0$ of the unperturbed system instead of proving, as Kolmogorov did, the preservation of *one* given torus. This also permits the calculation of the relative measure of the set of destroyed tori (see below).

Whereas Arnold and Kolmogorov had worked with real analytic Hamiltonians, Moser uses a smoothing technique, devised by Nash (to study the isometric embedding of Riemannian manifolds), to handle the case where f and h are only differentiable ([Mo1]). Actually Moser worked on a model theorem, which deals with the existence of invariant curves for area-preserving diffeomorphisms of the plane with an elliptic fixed point located at the origin, as exemplified by the Poincaré section on the energy surface of a two degree-of-freedom Hamiltonian in the vicinity of an elliptic periodic trajectory. Note that it is still customary to first consider this theorem when one tries to improve KAM techniques and results (see [He]).

Using Moser's method and its improvements by H. Rüssman and other mathematicians, it can be shown that the theorem remains valid for systems with a finite order of differentiability. For instance the invariant curve theorem is valid and yields strong results (conjugacy to rotations on the invariant curves, positive measure of the set of invariant curves) for C^4 systems. It is clearly impossible to examine these subtle optimality questions in any further detail here and we refer to [He] for the best results available to date. Fortunately these do not affect the global picture of Hamiltonian dynamics which emerges from the KAM theorem and which we intend to describe more precisely.

We shall assume that the function h is real analytic, which implies that it may be extended to the usual complex domain $K_\rho = \{I \in \mathbb{C}^n, \mathrm{Re}(I) \in K, \| \mathrm{Im}\, I \| < \rho\}$, for ρ positive and K a ball in $\mathbb{R}^n$ (the theorem is local in I). Moreover one must assume that $h(I)$ satisfies a condition ensuring the nondegeneracy of frequencies, i.e., strict nonlinearity or strict nonisochronicity of the unperturbed motion; two different conditions are used, Kolmogorov's condition ($\det \partial^2 h/\partial I^2 \neq 0$), and Arnold's condition which demands that the determinant of the second derivative (the Hessian) of the map $(\lambda, I) \rightarrow \lambda h(I)$, from $\mathbb{R} \times \mathbb{R}^n$ to $\mathbb{R}$, be non vanishing for $I \in K$ and $\lambda = 1$ (cf. Appendix 3). After restricting K if necessary, the first condition allows the use of $\omega = \omega(I)$ instead of I as a (local) parametrization in action space; the second condition allows a similar operation on the energy surface: If, after a possible relabelling, $\omega_n(I) \neq 0$, the ratios $\omega_k(I)/\omega_n(I)$ $(1 \leq k \leq n-1)$ supply a local parametrization of the energy surface $h = \mathrm{cst}$ or also $H = \mathrm{cst}$, for ε small enough. Finally one must define the Diophantine frequencies which correspond to preserved tori. Details may be found in Appendix 4. Following the notation of this appendix, we denote by $\Omega(\tau, \gamma)$ $(\tau > n-1, \gamma > 0)$ the set of Diophantine frequencies:

$$\Omega(\tau, \gamma) = \{\omega \in \mathbb{R}^n, \forall\, k \in \mathbb{Z}^n - \{0\}, |\,\omega.k\,| \geq \gamma |\,k\,|^{-\tau}\}.$$

We may now state a first version of the KAM theorem ([Pö1]); we shall use Kolmogorov's condition which leads to a slightly simpler statement.

Theorem (K.A.M.):

Let h be a real analytic Hamiltonian defined on the ball K of $\mathbb{R}^n$ centered at I_0 and such that the map $I \to \omega(I)$ ($\omega = \nabla h$) is invertible near $I = I_0$, i.e.,

$$\det \partial^2 h/\partial I^2 \,(I = I_0) \neq 0.$$

Let εf be a C^r Hamiltonian perturbation, $r > 2\tau + 2 > 2n$. Then, for $|\,\varepsilon\,| < \varepsilon_0$ and for every $\omega \in \Omega(\tau, \gamma)$ near $\omega(I_0)$, the perturbed system defined by the Hamiltonian $H = h + \varepsilon f$ has an invariant torus on which the motion is conjugate to the linear flow (2) with frequency ω.

$\Omega(\tau, \gamma)$ is a Cantor set whose measure increases with increasing γ (see Appendix 4); $\varepsilon_0 = \varepsilon_0(\gamma, \tau, n)$ decreases with γ. For given ε, the tori in phase space corresponding to $\Omega(\gamma, \tau)$ (τ may be taken equal to n) form a set $I_\gamma \times T^n$, which of course depends on the perturbation f and is the product of a torus and a Cantor set. The arrangement of these tori has been recently studied ([Pö1-2], [Chie1-2]): The authors show that there exists an integrable Hamiltonian $H'(I')$ together with a canonical transformation: $T\colon (I', \varphi') \to (I, \varphi)$,both of them *globally* defined, such that $H'(I')$ coincides with the Hamiltonian $H(I, \varphi)$ transformed to the variables (I', φ') on the set of invariant tori of H. The precise result makes use of Whitney's theory of differentiability on closed sets. Here we shall simply say that $TH = H \circ T$ and H' coincide on the set of invariant tori of H together with all their derivatives in the sense of Whitney and give the explicit statement of Pöschel's "global" version of the KAM theorem ([Pö1-2]), which describes the foliation by the tori, in the case of an analytic perturbation f:

Theorem:

Assume that f (and therefore H) is real analytic so that it may be extended to a complex domain $K_{\rho,\sigma}$. Assumes also that h satisfies Kolmogorov's condition on this domain. Then there exists a C^∞ integrable Hamiltonian H' on $K - \eta$ ($\eta > 0$), and a near-identity canonical transformation $T\colon (I', \varphi') \to (I, \varphi)$, which is C^∞ with respect to I and analytic with respect to φ, is defined on $(K - \eta) \times T^n$, and is such that $H \circ T(I', \varphi')$

coincides with $H'(I')$ on the set:

$$\{ I' \text{ such that } T(I', \varphi') = (I, \varphi),\ \omega(I) \in \Omega(\gamma, \tau) \}$$

together with all its derivatives in the sense of Whitney.

For versions in the nonanalytic case and a short self contained exposition of Whitney's theory we recommend [Pö1-2]. We restrict ourselves here to some remarks:

1) From a technical viewpoint, this book contains all the elements required for proving the KAM theorem, at least in Arnold's version which is close to the first statement we gave above, except that it applies to the analytic case. In fact this theorem is somewhat simpler to prove than Nekhoroshev's, the only difference being in the iteration process which makes use of Newton's method (see Appendix 8 in fine for an outline of the procedure).

2) Let us recall that the tori, of dimension n, are of course located on the energy surface, of dimension $2n - 1$, which leads to two sharply different situations according to whether $n = 2$ or $n > 2$ (for $n = 1$, all systems are integrable):
a) if $n = 2$, the tori *partition* the energy surface; if an initial condition is located between two invariant tori of the perturbed Hamiltonian, the corresponding trajectory is *trapped* entirely between these tori. Therefore for *every* trajectory and for any time interval, the variation of the action variables is smaller than the maximum distance between two tori. We shall see in the following chapter (Section 8.6) that this is the key point in the proof of perpetual adiabatic invariance for one degree-of-freedom nonlinear systems.
b) When $n > 2$, the complement of the tori is a connected set and the motion may a priori be ergodic on this set with the existence of a dense trajectory. This phenomenon is called "Arnold diffusion" after the example which was given in [Ar4] ; though this point is not yet completely established, this seems to be a generic phenomenon with respect to the perturbation. Nekhoroshev's theorem aims precisely at obtaining an upper bound of the speed of Arnold diffusion.

In this sense the two theorems are complementary; the KAM theorem proves that most tori of the unperturbed motion are only slightly deformed and Nekhoroshev's theorem estimates the velocity of the action drift for trajectories located between these preserved tori. We may add that in its present form the latter theorem is not optimal; in particular, it makes no use of the existence of invariant tori and yields a uniform bound for all trajectories, including those (the most numerous) which are located on the invariant tori and undergo on any time interval only small variations of the action variables, as proved by the upper bound on the deformation of the preserved tori (see next remark). Therefore it would be of major interest to devise, if indeed it is feasible, a *common* proof of both theorems.

3) Let $\omega(I) = \omega^* \in \Omega(\gamma, \tau)$ be the frequency vector associated with a given torus (defined by I) of the unperturbed system. An invariant torus with respect to the perturbed flow then exists in the vicinity such that the perturbed motion on this torus is conjugate to a quasi-periodic motion with frequency vector ω^* (i.e., it is quasi-periodic in the new canonical variables (I', φ')). By evaluating the size of the canonical transformation $(I, \varphi) \to (I', \varphi')$ one proves ([Pöl-2], [Nei7]) that the distance between a perturbed torus and the corresponding unperturbed torus is less than $c\sqrt{\varepsilon}$. In other words, the *deformation* of the tori is at most on the order of $\sqrt{\varepsilon}$.

4) One also proves ([Pöl-2], [Nei7]) that the relative measure of the complement of the tori, i.e., the set in which Arnold diffusion takes place, is at most on the order of $\sqrt{\varepsilon}$. Thus, when ε tends to 0 the relative measure of the tori, which is greater than $1 - c\sqrt{\varepsilon}$, tends to one. Here is the origin of this result: To prove the theorem one reduces the Hamiltonian to an integrable form via an iterative process (see below Nekhoroshev's theorem), i.e., one eliminates the dependence on the angle variables in a neighborhood of the torus characterized by the frequency vector ω^*. This iterative process converges when ω^* is Diophantine, e.g. when:

$$(1) \qquad |\omega^*.k| \geqslant \gamma |k|^{-n}; \quad k \in \mathbb{Z}^n - \{0\}.$$

that is, $\omega^* \in \Omega(n, \gamma)$ (see above and Appendix 4). To estimate the measure of the set of preserved tori, one then notices that the iterative process still converges if γ depends on ε, more precisely if $\gamma = \gamma_0 \sqrt{\varepsilon}$. One then readily estimates the measure of the set of vectors ω^* which satisfy (1) with this γ, as is done in Appendix 4. The measure in action space is of the same order because if, for instance, the Hamiltonian h satisfies Kolmogorov's condition, the map $I \to \omega(I)$ is a local diffeomorphism.

5) Since the measure of the set of preserved tori is not zero and indeed has almost full measure, and since the flow leaves this set invariant, the perturbed Hamiltonian $H = h + \varepsilon f$ is not ergodic, a fact that is expressed by the following rather vague statement, the meaning of which can however be made quite precise : "The generic Hamiltonian is neither integrable nor ergodic" (see [Mar]). Soon after the KAM theorem

was first proved, some people interpreted this statement as contradicting Boltzmann's assumption of "molecular chaos". The situation is of course not that simple; in addition to the fact that the Hamiltonians considered in thermodynamics are not generic (or, still worse, regularity hypotheses are not satisfied in collision problems), the number of degrees of freedom n plays a crucial part. Only recently has the problem of the dependence on n of the various constants been tackled (starting with ε_0 which quantifies the validity threshold of the theorem), and the discrepancies between the theoretical values of constants and those numerically obtained (in the KAM as well as in Nekhoroshev's theorem) are generally of several orders of magnitude. Moreover, the questions of the dependence on n of the constants and of the existence of a thermodynamic limit $n \to \infty$ are of course meaningless unless n may be varied without modification of the Hamiltonian's specific form. To this end one works with the simplest systems of statistical mechanics, essentially chains or lattices of rotators or harmonic oscillators to which are added a weak and localized nonlinear coupling; in the case of one dimensional chains and nearest neighbor interactions the corresponding Hamiltonians may be written:

$$(2) \qquad \begin{aligned} H_{rot}(I, \varphi) &= \sum_{1 \leq k \leq n} I_k^2/2m + \varepsilon V(\varphi_{k+1} - \varphi_k), \\ H_{osc}(I, \varphi) &= \sum_{1 \leq k \leq n} \omega_k I_k + \varepsilon V(\varphi_{k+1} - \varphi_k). \end{aligned}$$

Note that the latter does not satisfy Kolmogorov's condition; however, a preliminary transformation brings it back to the standard situation. For such lattice systems and certain specific forms of perturbations, versions of the KAM theorem have now been proved for $n = \infty$ ([Vi1-2], [Way]). These generalizations and applications of KAM and Nekhoroshev's theorems in statistical physics presently constitute an active field of research (see for example [Ben3], [Galg] and references therein).

6) Attempts to clarify the situation from a theoretical viewpoint and to obtain more realistic values of the analytically predicted thresholds compared to the numerically obtained values have recently led to the publication of new proofs of the KAM theorem (or rather of the invariant curves theorem), which are based on renormalization type techniques. We shall not discuss this point further but only refer the reader to [Khan].

7) As a final but important remark, we shall now try to bring to light what extra information is provided by the KAM theorem regarding the convergence of the classical series of perturbation theory ([Lin], [Poi], [Ze]). These "Linstedt-Poincaré" series are generically asymptotic, as noticed by Poincaré, who in fact invented the very notion of asymptotic series on that occasion (see e.g. [Whitt]). For more details about these

classical tools of perturbation theory the reader may consult Appendices 5, 7 and 8 (whose notation we adopt) and for an example of their use we refer to the proof of Nekhoroshev's theorem given below. We add that we purposely do not give a very precise definition of Linstedt-Poincaré series; instead this term refers to all the classical series, some instances of which may be found below and in Chapter 8.

Let us now consider the perturbed Hamiltonian (1), where the perturbation depends analytically on ε: $f = f(I, \varphi, \varepsilon)$ may then be developed as a power series in ε. Whereas the theorems are generally proved via a recursive method (see Appendix 8 and below), in practical computations one generally builds series term by term, using for this purpose e.g. the Lie method and the formulas of Appendix 7. In fact one eliminates the angle variables order by order, via an nth order canonical transformation $T^{(n)}$ which defines the change of variables $(I_n, \varphi_n) \rightarrow (I, \varphi)$; it is then easy to obtain the inverse transformation. The canonical transformation is generated by the function $\chi^{(n)} = \sum_{k=1}^{n} \varepsilon^{k-1} \chi_k$ and $T^{(n)}$ appears as a series whose terms may be computed as functions of the coefficients χ_k, as indicated in Appendix 7. The series converges provided the χ_k's are small enough (cf. [Gio]). The determination of conditions for the convergence of these series and the study of the sequences $\chi^{(n)}$ and $T^{(n)}$ as n increases constitute a large part of classical perturbation theory as surveyed in Appendix 8. Many facts, such as the Poincaré-Fermi theorem or the existence of homoclinic manifolds show that, except in very peculiar situations, these sequences do *not* converge on an open set of action space, however small ε may be.

Now the KAM theorem proves the following fact:

Let I^* be such that $\omega(I^*) = \omega^* \in \mathbb{R}^n$ is Diophantine; it is then possible to build up a map $(\varphi, \varepsilon) \rightarrow C_\varepsilon(\varphi) = (I', \varphi')$ defined by the series:

$$(3) \qquad I'(\varphi, \varepsilon) = I^* + \varepsilon I^{(1)} + \dots ; \qquad \varphi'(\varphi, \varepsilon) = \varphi + \varepsilon \varphi^{(1)} + \dots$$

which is analytic with respect to (φ, ε) for ε small enough and which defines a parametrization of the perturbed torus of frequency ω^*, located in the vicinity of the unperturbed torus $I = I^*$.

The crucial point consists in the fact that, once the *existence* of a torus has been proved via the quadratic convergence method, this torus may be located via classical methods (i.e., non-superconvergent ones). In other words, the above series are precisely those that Poincaré considered and their convergence may be proved when ω^* is Diophantine. This stems from the fact that the canonical transformation is *uniquely* determined by the elimination of angle variables and the superconvergent method does nothing but provide a rearrangement of the series displaying its convergence.

To make this point more precise, let us recall that at the nth step, $T^{(n)}$ reduces the Hamiltonian to the form (see Appendix 8):

$$(4) \qquad H^{(n)}(I_n, \varphi_n, \varepsilon) = h^{(n)}(I_n, \varepsilon) + \varepsilon^n f^{(n)}(I_n, \varphi_n, \varepsilon)$$

where the average of $f^{(n)}$ with respect to φ_n is zero and the integrable part $h^{(n)}$ is thus uniquely defined ($h^{(n)}(I_n, 0) = h(I_n)$). Setting $\omega^{(n)} = \nabla h^{(n)}$, the equation $\omega^{(n)}(I_n, \varepsilon) = \omega^*$ may then be solved, and when $n \to \infty$, $\omega^{(\infty)} = \nabla h^{(\infty)}$ is obtained as a formal series ($\omega^{(\infty)} = \omega + O(\varepsilon)$). Now the KAM theorem ensures that the equation $\omega^{(\infty)}(I_\infty, \varepsilon) = \omega^*$ may be solved, first to obtain I_∞ ($I_\infty = I^* + O(\varepsilon)$), and in a second step to determine the torus via the equation $(I', \varphi') = T^{(\infty)}(I_\infty, \varphi_\infty)$, where $T^{(\infty)}$ is the formal limit of the sequence $T^{(n)}$ and I_∞ is kept constant. This equation is identical to (3), provided φ is changed to φ_∞, this change of notation simply corresponding to different viewpoints of transformations, as coordinate changes or as maps between two spaces.

The point to remember is the following: The formal transformation $T^{(\infty)}$, whose arguments are $(J, \psi, \varepsilon) = (I_\infty, \varphi_\infty, \varepsilon)$, converges *if* $|\varepsilon| < \varepsilon_0$ (ε_0 depends in an essential way on the Diophantine properties of ω^*) *and if* J satisfies the equation $\omega^{(\infty)}(J, \varepsilon) = \omega^*$, where $\omega^{(\infty)}$ is obtained term by term as a series.

7.3 Nekhoroshev's theorem; introduction and statement of the theorem

As mentioned above, Nekhoroshev's theorem deals with the variations of action variables during the perturbed motion. Neither the statement of the theorem nor its proof refer to the existence of invariant tori.

Once more we consider a real analytic Hamiltonian $H = h + \varepsilon f$ which may be extended to the complex domain $K_{\rho,\sigma}$, where K is a convex compact set of $\mathbb{R}^n$. The convexity of K is not essential but it does simplify some points in the geometric part of the proof and, in any case, the theorem is essentially local in the action variables. Here and below, we shall write $K_{\rho,\sigma} = D(K, \rho, \sigma)$ as in [Ben5] to account for the successive variations of K, ρ and σ; we write simply D when no confusion is possible.

<u>Theorem</u>:

Assume that h is a *convex* function, that is the quadratic form $A =_{def} \partial^2 h/\partial I^2 = \partial\omega/\partial I$ is *definite* (say positive) at every point in K. We denote by m (resp. M) its lowest (resp. highest) eigenvalue on the set K: $\forall\, I \in K$, $\forall\, v \in \mathbb{R}^n$,

$M\| v \|^2 \geq (\partial\omega/\partial I\, v, v) \geq m\| v \|^2$. We also set $E = \mathrm{Sup}(\| h \|, \| f \|)$.

Then, for ε small enough ($|\varepsilon| \leq \varepsilon_0$) and for any initial condition $I_0 = I(0)$ belonging to $K - 2\Delta$, the variation of the action does not exceed $\Delta = \delta\varepsilon^{1/c}$ for times less than T, $T = \tau\varepsilon^{-3/4}\, \exp(1/\varepsilon)^{1/4c}$, where $c = 4(n^2 + 2n + 2)$:

$$\mathrm{Sup}_{\,t\in[0,\,T]}\ \| I(t) - I(0) \| \leq \Delta.$$

δ and τ are dimensional constants (action and time respectively) and one may choose $\delta = (ME)^{1/2}/m$, which has the dimension of an action and $\tau = (EM)^{-1/2}$.

Remarks:

1) This statement of the theorem is taken from [Ben5] as well as the proof given below ([Ben2] may also be useful to the reader). It is simpler than Nekhoroshev's original proof ([Nek2-3]) but its validity is restricted to less general systems (see next remark). The proof of Nekhoroshev's theorem splits into an analytic part and a geometric part which form the subjects of the next two sections. The key idea of the analytic part consists of constructing a nth order perturbation scheme where $n = n(\varepsilon)$ is not fixed but depends on ε and tends to infinity as ε goes to zero. This can and has been exploited in other settings, where no geometry is involved. We have already alluded to such an application at the very end of Section 3.2: One can devise a perturbation scheme for general one frequency standard systems (including of course time periodic systems)) which leads to exponentially small remainders (cf. [Nei7]). Other instances will appear in the next chapter.

2) We shall restrict ourselves to convex functions h, which obviously satisfy Kolmogorov's nondegeneracy condition; only slight modifications would be required to extend the proof to *quasi-convex* functions (the function h is said to be quasi-convex when its energy surfaces are convex; in this case h, considered as a function of any variable transverse to these surfaces, is not necessarily convex; see Appendix 9).

The theorem actually applies, but with different constants, provided that h satisfies Arnold's or Kolmogorov's condition and is a *steep* function on K; steepness is precisely defined in Appendix 9 which also discusses the genericity of this property in the class of real analytic functions. All the above conditions (convexity, quasi-convexity, steepness) deal with *geometric* properties of resonances and their use is restricted to one of the lemmas in the geometric part of the proof (see Section 8.5) so that in fact not much work is required to extend the proof given below to the generic case of steep unperturbed Hamiltonians. The reader may consult [Nek1] to see the corresponding geometric lemmas.

3) A detailed reading of the proof shows that the theorem actually applies to Hamiltonians of the form $H(I, \varphi, p, q) = h(I) + \varepsilon f(I, \varphi, p, q)$, where h satisfies the above conditions and the Hamiltonian perturbation f depends on the extra canonical variables $(p, q) \in \mathbb{R}^{2m}$. Of course, one then estimates the variation of I and not that of the complete set of action variables (I, p). In contrast with this situation, we remind the reader that the KAM theorem is *not* valid for such Hamiltonians, i.e., when the perturbation brings into play extra degrees of freedom. In Section 8.6 however, we shall present a useful version of the KAM theorem in a related "degenerate" case.

In addition, Nekhoroshev's theorem may also apply to various systems that can be reduced to the above, such as Hamiltonians of the form $H(I, \varphi, t) = h(I) + \varepsilon f(I, \varphi, t)$, periodic with respect to t, which amounts to dealing with the autonomous $n + 1$ degree-of-freedom Hamiltonian $H'(I, E, \varphi, t) = E + h(I) + \varepsilon f(I, \varphi, \tau)$, with a steepness condition on $h'(I) = h(I) + E$.

4) The essential point is of course that action variables remain *confined* during an exponentially long time, on the order of $\exp(1/\varepsilon)^{1/4c}$, where c increases with n. The optimality of such estimates is briefly discussed in [Ben2]. Studies still in progress tackle this question and consider in particular the optimality with respect to n, in view of applications to statistical mechanics (see also Section 8.5 for similar problems).

5) The reader should be wary of jumping to the erroneous conclusion that I varies slowly. In fact the action variables generally vary with a speed *of the order of* ε (the variation could not be faster) but motion remains confined to a small subset of phase space. In this way, Nekhoroshev's theorem completes, in the Hamiltonian case, the much simpler and more general multiphase averaging theorem which was proved in Chapter 6, but does *not* literally improve it. Indeed, while the variation of the action variables remains small for an exponentially long time, the maximal deviation on this time interval is generally approached in a much shorter time, as pointed out by the following example.

We consider the two degree of freedom Hamiltonian:

(1) $$H(I, \varphi) = h(I_1, I_2) + \varepsilon f(\varphi_1, \varphi_2) = (I_1^2 - I_2^2) + \varepsilon \sin(\varphi_1 - \varphi_2).$$

h satisfies Kolmogorov's condition everywhere since $\det(\partial^2 h/\partial I^2) = -1$. It is everywhere *steep* and even quasi-convex except on the two diagonal lines $I_1 = \pm I_2$ (cf. Appendix 9). Nekhoroshev's theorem applies to any compact subset of the (I_1, I_2) plane outside these diagonal lines since, on such a domain, h is *uniformly* steep, which is the geometric validity condition for the result. The exact solutions are easily computed. $\lambda = I_1 + I_2$ is a first integral, and in particular for any initial condition such that $\varphi_1(0) = \varphi_2(0)$, one has:

$$(2) \qquad I_1(t) = I_1(0) - (\varepsilon/\lambda)\sin(\lambda t); \quad I_2(t) = I_2(0) + (\varepsilon/\lambda)\sin(\lambda t) \quad (\lambda \neq 0).$$

The system can of course actually be solved for any initial condition. When λ tends to 0, i.e., when the point (I_1, I_2) approaches the second diagonal line, the deviation of the action variables increases, and when λ is precisely 0, the trajectory $(I_1(t) = I_1(0) - \varepsilon t$, $I_2(t) = -I_1(t))$ is no longer confined. The first diagonal line would play a similar part for other types of perturbations (e.g. $f = \sin(\varphi_1+\varphi_2)$). $| dI/dt |$ is always on the order of ε, or at least it reaches such a value in a time interval of the order of $1/\lambda$, and the deviation of the action variables on such an interval is on the order of ε/λ. The multiphase averaging theorem, which makes no reference to geometric properties such as steepness, also applies to this example. It asserts that the variation of action variables does not exceed $\rho(\varepsilon)$ ($c\sqrt{\varepsilon} \leqslant \rho \leqslant c'$) on the timescale $1/\varepsilon$, provided the initial condition does not belong to a certain set V'' with relative measure on the order of $\sqrt{\varepsilon}/\rho$. Here we should take as V'' a strip of width $\lambda \sim \sqrt{\varepsilon}$ around the second diagonal line (intersected with, say, a large square). Actually, for $\lambda \sim \sqrt{\varepsilon}$ a time on the order of $1/\sqrt{\varepsilon}$ suffices to attain the maximal deviation $c\sqrt{\varepsilon}$.

Thus Nekhoroshev's theorem does *not* ensure that the variation of the action variables on a short interval of time is negligible; it rather bounds from above the *amplitude* of Arnold's diffusion, and not the drift velocity of the action variables.

The following two sections are devoted to the proof of Nekhoroshev's theorem. The analytic part discribes the motion near an arbitrarily given point I_0, in terms of the arithmetic properties of the frequency vector $\omega(I_0)$ (motion inside a resonant zone). The geometric part provides a global picture of the motion in action space (passage from one resonant zone to another) which allows the proof to be concluded.

7.4 Analytic part of the proof

The Hamiltonian we consider is analytic on the complex domain $K_{\rho,\sigma} = D(K, \rho, \sigma)$ with K a compact, convex set of $\mathbb{R}^n$. The local frequency $\omega(I_0)$ of the unperturbed motion (actually a vector with n components) is defined at every point I_0 in K. In the neighborhood of any such I_0, the perturbed motion essentially depends, as mentioned above, on the arithmetic properties of $\omega(I_0)$. The analytic part of the proof aims at analyzing this motion in detail. To this end the perturbed Hamiltonian H is recast into a resonant normal form, which takes into account the properties of $\omega(I_0)$, is obtained via a recursive method and describes the motion near I_0 on a certain interval of time.

The exponentially long validity time in Nekhoroshev's theorem derives from the following circumstance: Whereas classical perturbation theories limit themselves to a finite order, here the number of successive canonical transformations, though finite, depends on ε. In this way, one can make the remainder of the resonant normal form much smaller, indeed exponentially small. By contrast, in the KAM theorem, one performs an infinite number of canonical transformations, but the process converges only on a complicated Cantor set. This is summarized in the following table.

	number of steps	size of the remainder	validity time
Classical methods	finite independent of ε	ε^r	ε^{-r}
Nekhoroshev's th.	$(1/\varepsilon)^a$, $0<a<1$	$\exp(-[1/\varepsilon]^a)$	$\exp(1/\varepsilon)^a$
KAM theorem	infinite	0	infinite

All the needed results concerning resonant normal forms are embodied in the analytic lemma below which makes use of an ultraviolet cutoff $N(\varepsilon)$: High order Fourier coefficients ($|k| > N$) are never taken into account. The precise value of this cut-off is left undefined until the end of the proof when we set $N(\varepsilon) = (1/\varepsilon)^{1/c}$. This choice is easy to understand: As the perturbation f is analytic on $D(K, \rho, \sigma)$ its Fourier coefficients decrease exponentially fast (as $\| f \| e^{-\sigma|k|}$, see Appendix 1) and the remainder of its Fourier series $f^{>}_N = \sum_{|k|\geqslant N} f_k e^{ik.\varphi}$ satisfies the estimate:

$$\| f^{>}_N \|_{D(\sigma-\beta)} \leqslant c(n)\beta^{-n} \| f \|_{D(\sigma)} \, e^{-\beta N/2}$$

for $0 < \beta < \sigma < 1$ (see Proposition 3 of Appendix 1). For the choice of $N(\varepsilon)$ indicated above the ultraviolet part $f^{>}_N$ of the perturbation will be of order $\exp[-\beta/(2\varepsilon^{1/c})]$ and its influence on the motion will be negligible on intervals of time much smaller than the inverse of this quantity. This ultraviolet cutoff thus provides a Hamiltonian with a finite ε-dependent number of harmonics.

We shall now state and prove the iterative lemma which allows for one step in the process of reduction to normal form and which will be used repeatedly to prove the analytic lemma.

<u>Iterative lemma</u>:

Let M be a modulus of $\mathbb{Z}^n$. We rewrite the Hamiltonian $H(I, \varphi) = h + \varepsilon f(I, \varphi)$, which is analytic on the complex domain $D = D(K, \rho, \sigma)$ (we suppose that $\sigma \leqslant 1$ which is no restriction), in the form:

(1) $$H(I, \varphi) = h(I) + Z(I, \varphi) + R(I, \varphi)$$

where $Z(I, \varphi) = \varepsilon \sum_{k \in M} f_k(I)e^{ik.\varphi}$ is the resonant part of the perturbation εf and $R = \varepsilon f - Z$ is its nonresonant part. We assume that:

$$\| h \| \leqslant E; \quad \| f \| = \| Z + R \| \leqslant E; \quad \| R \| \leqslant \eta E \quad (\eta \leqslant 2\varepsilon).$$

We also suppose that on the compact set K:

(2) $$\forall\, k \notin M, \; | k | \leqslant N, \; | \omega(I).k | \geqslant \nu > 0.$$

Let $\alpha > 0$ and $\beta > 0$ satisfy the inequalities:

(3) $$\alpha \leqslant \rho/4, \quad \beta \leqslant \sigma/4$$

(4) $$2^{4n+4}\eta L \leqslant 1, \quad \text{where} \quad L = E/(2\nu\alpha\beta^{n+1}).$$

Then one may define on $D' = D(K-\alpha, \rho-\alpha, \sigma-\beta)$ a canonical transform $T: (J, \psi) \to (I, \varphi)$ such that:

- T is close to the identity:

(5) $$\| I - J \|_{D'} \leqslant \alpha/2, \quad \| \varphi - \psi \|_{D'} \leqslant \beta/2\,.$$

- In the new variables (J, ψ) the Hamiltonian reads:

$$H'(J, \psi) = h(J) + Z'(J, \psi) + R'(J, \psi)$$

and the resonant (Z') and nonresonant parts (R') satisfy:

(6) $$\| R' \|_{D'} \leqslant (\eta_1^{(1)} + \eta_1^{(2)})E, \quad \eta_1^{(1)} \leqslant c_1(n)E\eta e^{-\beta N/2}/\beta^n,$$
$$\eta_1^{(2)} \leqslant \varepsilon\eta L(c_2(n) + c_3(n)L),$$

with $c_1(n) = 2^{n+4}n^{n-1}$, $c_2(n) = 2^{3n+4}n$, $c_3(n) = 2^{8n+13}n^2$

(7) $$\| Z' + R' \|_{D'} \leqslant \varepsilon_1, \quad \varepsilon_1 = \varepsilon + \eta_1/2.$$

Remark:

The assumption $\eta \leqslant 2\varepsilon$ is not restrictive. Indeed the resonant part Z is obtained by projecting the perturbation $Z + R$ onto the set of resonant Fourier components: $Z = P_M(Z + R)$. Thus $\| Z \| \leqslant \| Z + R \| \leqslant \varepsilon E$, because P_M is a projection operator (of norm 1) and $\| R \| \leqslant \| Z + R \| + \| Z \| \leqslant 2\varepsilon E$. For further details we refer the reader to the last paragraph of Appendix 3.

Proof:

Following the Lie series method (see Appendix 7), we introduce an auxiliary Hamiltonian χ on $D(K, \rho, \sigma)$ to generate the canonical transformation T. The transformed Hamiltonian reads:

(8) $$H' = \exp[L_\chi] H = h + Z + R + \{\chi, h\} + \{\chi, Z+R\} + [H'-H-\{\chi, H\}].$$

and one wishes to eliminate the low-order nonresonant part. To this end we write R as the sum of an ultraviolet part $R^{>} = \sum_{|k|>N} r_k e^{ik.\varphi}$ and a low-order part $R^{\leqslant} = R - R^{>}$ and we choose for χ the zero-average solution of:

(9) $$\{ \chi, h \} + R^{\leqslant} = 0 \quad \text{or} \quad \omega.\partial\chi/\partial\varphi + R^{\leqslant} = 0.$$

The Fourier coefficients of χ are nonzero only for $k \notin M$, $| k | \leqslant N$ and may be written:

(10) $$\chi_k = i r_k /(\omega.k).$$

Therefore:

(11) $$| \chi_k | \leqslant | r_k |/\nu \leqslant \| R \|_D \nu^{-1} e^{-\sigma|k|} \leqslant \eta\nu^{-1} E e^{-\sigma|k|}$$

and Proposition 3 of Appendix 1 then allows us to estimate for any γ, $0 < \gamma < \sigma$ the norm of χ on $D(K, \rho, \sigma-\gamma)$:

(12) $$\| \chi \|_{D(K,\rho,\sigma-\gamma)} \leqslant \eta\nu^{-1} E(4/\gamma)^n.$$

Formula (12) together with Cauchy's inequality (Appendix 1, §1) will now enable us to estimate the size of the canonical transformation T, thus proving (5). We assume that (I, φ) and (J, ψ) respectively belong to $D(K-\alpha/2, \rho-\alpha/2, \sigma-\beta/2)$ and $D(K-\alpha, \rho-\alpha, \sigma-\beta)$. We have from Cauchy's inequality:

(13) $$\| \partial\chi/\partial\varphi \|_{D(K-\alpha/2,\rho-\alpha/2,\sigma-\beta/2)} \leqslant (4/\beta) \| \chi \|_{D(K,\rho,\sigma-\beta/4)}$$

(12) together with condition (4) of the theorem then yields:

(14) $$\| \partial\chi/\partial\varphi \|_{D(K-\alpha/2,\rho-\alpha/2,\sigma-\beta/2)} \leqslant \alpha/2.$$

In a similar way one obtains:

(15) $$\| \partial\chi/\partial I \|_{D(K-\alpha/2,\rho-\alpha/2,\sigma-\beta/2)} \leqslant \beta/2.$$

Lemma 1 of Appendix 7 then ensures that:

- The transformation T is defined on the whole of $D' = D(K-\alpha, \rho-\alpha, \sigma-\beta)$.

- $\| I - J \| \leqslant \alpha/2$, $\| \varphi - \psi \| \leqslant \beta/2$.

thus proving (5).

If χ is chosen as indicated above, the transformed Hamiltonian H' takes the following form on D':

$$(16) \qquad H' = h + Z + S, \quad S = R^{>} + \{\chi, Z+R\} + [H'-H-\{\chi, H\}].$$

S contains both a resonant part which we shall denote by $P_M S$, and a nonresonant part $S - P_M S$. Therefore H' may be cast into the same form as H: $H' = h + Z' + R'$, provided we define the resonant part Z' and the nonresonant part R' of the perturbation as:

$$(17) \qquad Z' = Z + P_M S; \quad R' = S - P_M S.$$

We shall now derive the estimates (6) and (7). We first estimate the norms of the three terms defining S. The first estimate is readily given by Proposition 3 of Appendix 1 (with $C = \eta E$) which assumes $\beta \leqslant 1$:

$$(18) \qquad \| R^{>} \|_{D'} \leqslant \eta_1^{(1)} E/2, \quad \eta_1^{(1)} = \eta 2^{n+4} n^{n-1} e^{-\beta N/2} / \beta^n$$

The second estimate is easily derived by using the Cauchy inequalities together with (12):

$$(19) \qquad \| \{\chi, Z+R\} \|_{D'} \leqslant \sum_{i,j} (\| \partial\chi/\partial I_i \|_{D'} \| \partial(Z+R)/\partial\varphi_j \|_{D'} + $$
$$+ \| \partial(Z+R)/\partial I_i \|_{D'} \| \partial\chi/\partial\varphi_j \|_{D'})$$
$$\leqslant (4n/\alpha\beta) \| \chi \|_{D(K,\varrho,\sigma-\beta/2)} \| Z+R \|_D$$
$$\leqslant (4n/\alpha\beta)(\eta\nu^{-1} E(8/\beta)^n)\varepsilon E \leqslant 2^{3n} n \eta \varepsilon L.$$

To evaluate the last term we notice that:

$$(20) \qquad \| H' - H - \{\chi, H\} \|_{D'} \leqslant 1/2 \| \{\chi, \{\chi, H\} \|_{D(K-\alpha/2,\varrho-\alpha/2,\sigma-\beta/2)}$$

which is readily obtained through the use of Taylor's formula at second order (Appendix 7, Lemma 1). Cauchy's inequality together with (12) then yields (as $\eta \leqslant 2\varepsilon$):

$$(21) \qquad \| H' - H - \{\chi, H\} \|_{D'} \leqslant 2^{8n+11} n^2 \eta^2 E L^2 \leqslant 2^{8n+12} n \varepsilon \eta^2 E L^2.$$

From (19) and (20) we deduce:

$$(22) \qquad \| H' - H - \{\chi, H\} + \{\chi, Z+R\} \|_{D'} \leqslant \eta_1^{(2)} E/2$$

with $\eta_1^{(2)}$ as given in the statement of the lemma. (19) and (22) in turn yield:

$$(23) \qquad \| S \|_{D'} \leqslant \eta_1 E/2$$

where $\eta_1 = \eta_1^{(1)} + \eta_1^{(2)}$. This bound will enable us to estimate $\| R' \|_{D'}$ and $\| Z'+R' \|_{D'}$.

(6) is readily derived from the definition of $R' = S - P_M S$ (see (17)) and the bound $\| P_M S \|_{D'} \leqslant \| S \|_{D'} \leqslant \eta_1 E/2$. Setting $\varepsilon_1 = \varepsilon + \eta_1/2$, we also get, by (17) :

$$(24) \qquad \| Z'+R' \|_{D'} = \| Z+S \|_{D'} \leqslant \varepsilon E + \eta_1 E/2 \leqslant \varepsilon_1 E,$$

which proves (7) and completes the proof of the lemma. ▪

It is important to realize that the size of the remainder R' is bounded by the sum of two terms (see (6)) such that:

- $\eta_1^{(1)}$ is of exponential type ($\sim \eta e^{-\beta N/2}$) and comes from the tail of the Fourier series of an analytic function with the cutoff frequency N.
- $\eta_1^{(2)}$ is of quadratic type ($\sim \varepsilon\eta$) and represents the discrepancy between the transformed Hamiltonian and the normal form, once the transformation has been chosen so as to effect the conjugacy at the linear level (i.e., by solving (9)).

We shall now state and prove the analytic lemma. As already mentioned, this lemma summarizes the local reduction to a resonant normal form and is proved via the repeated use of the above lemma.

<u>Analytic lemma</u>:

Assume that all the hypotheses of the iterative lemma are satisfied and that:

- $\alpha \leqslant \rho/8,\ \beta \leqslant \sigma/8$
- (25) $L \geqslant 1$
- (26) $c(n)\varepsilon L^2 N^{4\xi(n+2)} \leqslant 1$ where $c(n) = (2n+4)^{2n+4} 2^{4(2n+4)}$ and $1/4 \leqslant \xi < 1/2$
- (27) $\beta\sqrt{N} \geqslant 2n \log N$ (a technical condition, N must be large enough).

Then one may define on $D' = D(K-2\alpha,\ \rho-2\alpha,\ \sigma-2\beta)$ a near identity canonical transformation $T\colon (J, \psi) \rightarrow (I, \varphi)$ satisfying:

(28) $\quad \| I - J \| \leqslant \alpha, \quad \| \varphi - \psi \| \leqslant \beta,$

such that the the nonresonant part R' of the transformed Hamiltonian $H' = h + Z' + R'$ (Z' and R' are defined as before) satisfies the estimate:

(29) $\quad \| R' \|_{D'} \leqslant 8\varepsilon E e^{-N^{\xi}}.$

<u>Proof</u>:

To prove this lemma, we apply the iterative lemma $[N^{\xi}]$ times. The jth transformation T_j is defined on the domain $D_j = D(K-\sum_{k\leqslant j} \alpha_k,\ \rho-\sum_{k\leqslant j} \alpha_k,\ \sigma-\sum_{k\leqslant j} \beta_k)$ where $\alpha_k = \alpha/k^2$, $\beta_k = \beta/k^2$, and it brings the Hamiltonian to the form $H_j = h + Z_j + R_j$, with $\| Z_j + R_j \| \leqslant \varepsilon_j E$ and $\| R_j \| \leqslant \eta_j E$. As will be apparent, the factor k^{-2} is largely arbitrary and could be replaced by another sequence such that the corresponding series converges.

The proof is of course by induction; One shows recursively that:

- $[N^\xi]$ successive transformations may indeed be performed.
- The nonresonant part R_j satisfies the estimate: $\| R_j \| \leq \eta_j E$ on D_j, where $\eta_j \leq 2\varepsilon e^{-j}$ for $j \leq [N^\xi]$.

When $j = 0$, this assertion is obviously true, as $\eta \leq 2\varepsilon$; we now assume it has been proved for $j \leq p$.

The pth transformation is defined on $D_p = D(K-\rho_p, \rho-\rho_p, \sigma-\sigma_p)$ where $\rho_p = \sum_{k\leq p} \alpha_k$ is smaller than $\rho/4$ and $\sigma_p = \sum_{k\leq p} \beta_k$ is less than $\sigma/4$. As $\alpha_{p+1} \leq \alpha \leq \rho/8 \leq (\rho - \rho_p)/4$ and $\beta_{p+1} \leq \beta \leq \sigma/8 \leq (\sigma - \sigma_p)/8$, condition (3) of the iterative lemma is fulfilled. Besides, as $\eta_j \leq 2\varepsilon e^{-j}$ and $\varepsilon_j = \varepsilon_{j-1} + \eta_j/2$ $(j \leq p)$, ε_j is greater than ε and η_j is smaller than $2\varepsilon_j$. Condition (4) of the iterative lemma is still to be checked. We set $L_p = E/(2\nu\alpha_p\beta_p^{\,n+1})$. Condition (4) then reads:

$$2^{4n+4}\eta_p L_p \leq 1 \qquad (30)$$

or:

$$\eta_p L 2^{4n+4}(p+1)^{2n+4} \leq 1. \qquad (31)$$

As $L \geq 1$ (see (25)) and $\eta_p \leq 2\varepsilon e^{-p}$, this inequality is a direct consequence of (26).

We may therefore use the iterative lemma once more. The new canonical transformation T_{p+1}, which is defined on the domain D_{p+1}, brings the Hamiltonian to the form H_{p+1}, with $\| R_{p+1} \| \leq \eta_{p+1}E$ and $\| Z_{p+1} + R_{p+1} \| \leq \varepsilon_{p+1}E$ $(\varepsilon_{p+1} = \varepsilon_p + \eta_{p+1}/2)$. It only remains to show that $\eta_{p+1} \leq 2\varepsilon e^{-(p+1)}$. We write η_{p+1} as $\eta_{p+1} = \eta_{p+1}^{(1)} + \eta_{p+1}^{(2)}$ and show that each of these term is smaller than $\varepsilon e^{-(p+1)}$.

The first term, which appears in the estimate of the ultraviolet part of R_{p+1} reads:

$$\eta_{p+1}^{(1)} = \eta_p 2^{n+3} n^{n-1}/\beta_{p+1}^{\,n} \exp[-\beta_{p+1} N/2]. \qquad (32)$$

Since $\eta_p \leq 2\varepsilon e^{-p}$, and in view of the choice of the sequence (β_j), we shall require that:

$$2^{n+3} n^{n-1}(p+1)^{2n}/\beta^n \exp[-\beta N/(2(p+1)^2] \leq 1. \qquad (33)$$

For a given N this condition is satisfied for p small enough, and for the sake of simplicity, we choose a power law for the maximal possible p, requiring $p \leq [N^\xi]$, $0 < \xi \leq 1$. As the left-hand side of (33) is an increasing function of p, this condition may be replaced by:

$$2^{n+3} n^{n-1}[N^\xi]^{2n}/\beta^n \exp[-\beta N/(2[N^\xi]^2)] \leq 1. \qquad (34)$$

It is apparent from this that one must have $\xi < 1/2$. This restriction being granted, (34) is a direct consequence of (27), provided that $\xi \geqslant 1/4$; this last condition thus appears for computational reasons as does (27).

The second term $\eta_{p+1}^{(2)}$ appears when one bounds the low-order part of R_{p+1}:

(35) $\qquad \| R_{p+1} - R^{>}_{p+1} \| \leqslant \eta_{p+1}^{(2)}E.$

From the bound (see the statement of the iterative lemma):

(36) $\qquad \eta_{p+1}^{(2)} \leqslant \varepsilon_p \eta_p L_p (2^{3n+4}n + 2^{8n+13}n^2 L_p)$

one easily deduces, as $L \geqslant 1$ and $\varepsilon_p \leqslant 2\varepsilon$, since $\varepsilon_p = \varepsilon + \sum_{j<p} \eta_j/2$:

(37) $\qquad \eta_{p+1}^{(2)} \leqslant 2^{8n+14}n^2 \varepsilon L^2 \eta_p (p+1)^{4(n+2)}.$

The sought-after bound $\eta_{p+1}^{(2)} \leqslant \varepsilon e^{-(p+1)}$ then results from (26).

We have thus shown that $[N^\xi]$ canonical transformations may be successively performed and that $\eta_p \leqslant 2\varepsilon e^{-p}$ for $p \leqslant [N^\xi]$. Therefore:

(38) $\qquad \| R' \| \leqslant 2\varepsilon E e^{-[N^\xi]} \leqslant 8\varepsilon E e^{-N^\xi}$

which proves (29).

Denoting by (J, ψ) the variables defined by the sequence of canonical transformations, we also have:

(39) $\qquad \| I - J \| \leqslant \sum_{j \leqslant [N^\xi]} \alpha_j/2 \leqslant \alpha, \quad \| \varphi - \psi \| \leqslant \sum_{j \leqslant [N^\xi]} \beta_j/2 \leqslant \beta,$

which proves (28) and finishes the proof of the analytic lemma. ∎

Remarks:

1- At the end of the proof we shall fix the dependence of the ultraviolet cutoff on ε. Condition (26) precludes values of N which are too large. On the other hand, the bound (29) immediately determines the time interval on which the theorem applies, as a function of $N(\varepsilon)$. Roughly speaking, $T \sim \| R' \|^{-1}$.

2- The analytic lemma consists in the construction of a resonant normal form near a given point I of action space, and the only information we shall extract from this normal form is the fact that in the new variables the evolution of the action variables is governed by the system:

$$dJ/dt = -\, \partial Z'/\partial\psi - \partial R'/\partial\psi.$$

Since R is small and analytic, the second term of the right-hand side is also small. On the other hand, the first term is a vector $\sum_{k \in M} ikz_k(J)e^{ik.\psi}$ of the resonant module M. Therefore the evolution of action variables takes place essentially along M. Hamilton's equations thus provide a link between *action* space (I), *frequency* space (ω) and *Fourier* space (k) and this is in some sense the only place where the Hamiltonian nature of the system plays a part.

3- As already seen in the preceding chapters, one may, instead of assuming that a nondegeneracy condition is satisfied, consider the opposite case of constant unperturbed frequencies (see [Ben2], [Gala2] and Appendix 8). One then deals with perturbations of an assembly of uncoupled harmonic oscillators and the unperturbed Hamiltonian may be written $h(I) = \omega.I$ ($\omega \in \mathbb{R}^n$). Assuming that ω is Diophantine ($\omega \in \Omega(\gamma, n)$):

(40) $\quad |\omega.k| > \gamma |k|^{-n}$ for every k in $\mathbb{Z}^n/\{0\}$

the proof of Nekhoroshev's theorem reduces to its analytic part. In the statement of the analytic lemma, one sets $M = \{0\}$, $Z = 0$, $N = (1/\varepsilon)^{1/c}$ and works on the whole of action space, with (40) replacing condition (2). The analytic lemma then actually proves the theorem itself. For more details we refer the reader to the aforementioned paper [Ben2] and to the end of the second section of Appendix 8. We may add that, as was already the case for the non-Hamiltonian averaging theorems, no general theory is available in the intermediate situation when ω is neither constant nor a local diffeomorphism (see however the final remark of Chapter 6).

The analytic lemma will enable us to study the motion in a given resonant zone of action space. This point is embodied in the following lemma, called the "main lemma"by Benettin et al. in [Ben5], which is now easy to prove and concludes the analytic part. The plane which passes through I and is generated by the vectors of the resonant module M will be denoted by $\Pi_M(I)$ and will be called the plane of fast drift, for obvious reasons in view of remark 2 above.

Main Lemma:

Assume that all the assumptions of the analytic lemma are satisfied. Let $d \geq 4\alpha$ and $T > 0$ satisfy:

(41) $\quad (16\varepsilon T/\beta)e^{-N^{\xi}} \leq d.$

Then, for every $I(0)$ in $K - 3\alpha$, the deviation of the trajectory $I(t)$ from the fast-drift plane $\Pi_M(I(0))$ does not exceed d, as long as $t \leq T$ and $I(t)$ remains in $K - 3\alpha$.

Proof:

The analytic lemma permits us to reduce the Hamiltonian to a normal form $H' = h + Z' + R'$ via a canonical transformation $T: (J, \psi) \rightarrow (I, \varphi)$. T is defined on

$D' = D(K-2\alpha, \rho-2\alpha, \sigma-2\beta)$ and is close to the identity (see (28)). By Lemma 1 in Appendix 7, its image contains the domain $D'' = D(K-3\alpha, \rho-3\alpha, \sigma-3\beta)$ and the point $J(0)$, whose image is $I(0)$, belongs to $K - 2\alpha$.

Let $J'(t)$ denote the projection of $J(t)$ onto the fast-drift plane $\Pi_M(J(0))$. In view of remark 2 above, we have:

$$(42) \qquad J(t) = J'(t) + \int_0^t \partial R'/\partial\psi \, du.$$

Therefore, as long as $I(t)$ remains in $K - 3\alpha$ and $t \leqslant T$:

$$(43) \qquad \| J(t) - J'(t) \| \leqslant t \, \| \partial R'/\partial\psi \|_{D'} \leqslant T/\beta \, \| R' \|_{D''} \leqslant 8\varepsilon TE/\beta \, e^{-N^{\xi}} \leqslant d/2$$

where we have used Cauchy's inequality in the second inequality, the estimate (29) in the third, and assumption (41) in the last one.

The estimate of the distance of $I(t)$ to the plane of fast drift $\Pi_M(I(0))$ is now readily obtained:

$$(44) \qquad \text{dist}\,(I(t), \Pi_M(I(0))) \leqslant \| I(t) - J(t) \| + \text{dist}\,(J(t), \Pi_M(J(0))) + \text{dist}\,(\Pi_M(I(0)), \Pi_M(I(0))).$$

The first and third terms are smaller than α, the size of the transformation T when it acts on action variables. In view of (43) the second term is smaller than $d/2$. Since we assume $d \geqslant \alpha/4$, $\text{dist}(I(t), \Pi_M(I(0)))$ is less than d, which proves the lemma. ∎

If one defines, as will be done in the next section, a cylinder of fast drift with extension d around the fast drift plane, the lemma ensures that the trajectory either remains in that cylinder until time T or exits before that time *across the base* of the cylinder.

Now that we have gained a fairly precise understanding of what occurs locally in any zone of action space, the geometric part of the proof will provide us with a *global* picture of the motion, which will serve to finish the proof of the theorem.

7.5 Geometric part and end of the proof

Following Nekhoroshev's original construction, we must first define several simple geometrical objects on action space. In fact we shall use the terminology and notation introduced in [Ben5] which differ only slightly from those of [Nek2]. Let M be a module of dimension r $(0 \leqslant r \leqslant n)$ in $\mathbb{Z}^n$, generated by the vectors $k^{(1)}, \dots, k^{(r)}$, the norms of which are all smaller than N; we may restrict ourselves to the consideration of such modules. One then defines:

1) The *resonant manifold* $R_M = \{ I \in K, \omega(I).k = 0, \forall k \in M \} = \{ I \in K, \omega(I).k^{(j)} = 0, j = 1,\ldots, r \}$.

2) The *resonant zone* $Z_M = \{ I \in K, | \omega(I).k^{(j)} | < \nu_r, j = 1,\ldots, r \}$, the half-width of which is ν_r, which surrounds the resonant manifold. $M = \{0\}$ is the only module with dimension 0 and in this case we set $R_M = Z_M = K$. Note that, for $r = n$ ($M = \mathbb{Z}^n$), a resonant zone may exist even if ω does not vanish (nearly critical points of the unperturbed system).

3) We set $Z_r^* = \cup_{M;\dim M=r} Z_M$, with the convention that $Z_0^* = K$ and $Z_{n+1}^* = \varnothing$.

4) Let M be of dimension r. To isolate the part of the corresponding resonance with dimension precisely equal to r, one introduces the *resonant block* $B_M = Z_M \backslash Z_{r+1}^*$.

5) $\Pi_M(I)$, the *plane of fast drift* , which has been defined in the previous section, is the plane of dimension r which passes through I and is generated by M.

6) $\Lambda_M(I) = \Pi_M(I) \cap Z_M$.

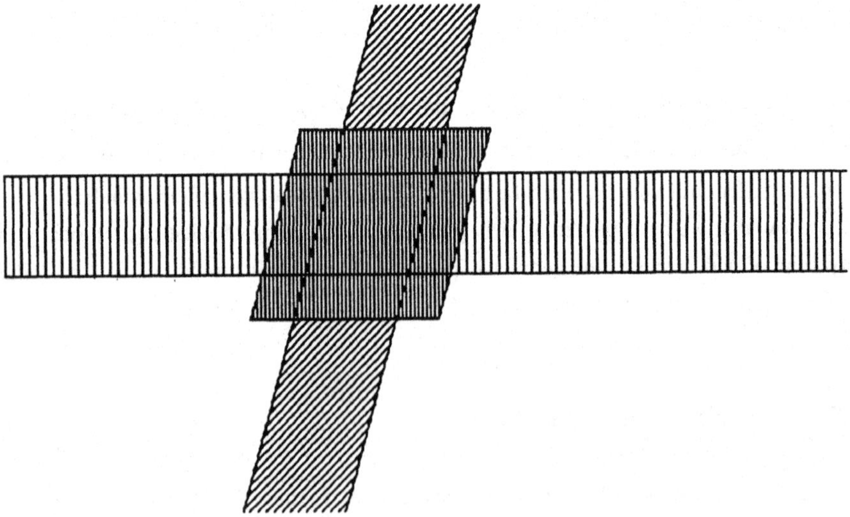

Figure 7.1: The intersection of two dimension one resonances forms a dimension two resonance. The associated resonant zone is wider than the simple resonant zones ($\nu_2 > \nu_1$; see below). A resonant block is the set of points in a given resonant zone which do not belong to any higher dimension resonant zone.

7) $C_{M,d}(I) = (\Pi_M(I)+d) \cap Z_M$ is a *cylinder* with radius d which passes through I and is included in the zone Z_M. It contains the points of Z_M which are a distance less than d from the fast-drift plane $\Pi_M(I)$. According to the "main lemma" of the previous section, motion near a given point I of action space takes place inside the cylinder with *axis* $\Lambda_M(I)$. The *bases* $C_{M,d}(I) \cap \partial Z_M$ of the cylinder, that is, the intersections with the boundaries of the resonant zone do not belong to the (open) resonant zone.

8) Finally one defines *extended blocks* $B^{ext}_{M,d} = \bigcup_{I \in B_M} C_{M,d}(I)$. $B^{ext}_{M,d}$ is the union of all the cylinders of radius d that are associated to the points of the resonant block B_M. During the passage through a resonant zone, motion remains limited to the corresponding extended block.

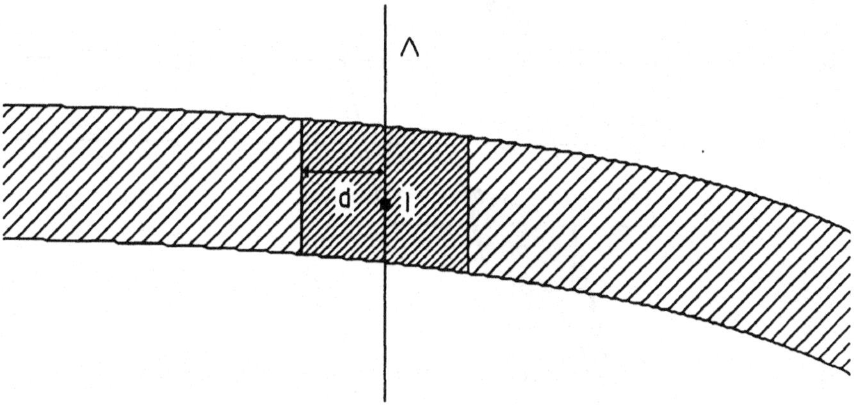

Figure 7.2: Cylinder of fast drift $C_{M,d\,(I)}$ with axis Λ. I belongs to a resonant zone Z_M.

The construction is illustrated in Figures 7.1 through 7.3. It satisfies the following assertions, (in fact it was introduced for that purpose):

a) $\bigcup_M B_M = K$.

b) If I belongs to the resonant block B_M with M of dimension r and k is any vector outside the module M with norm less than N, then $|\omega(I).k| \geqslant \nu_{r+1}$. In what follows the widths of the resonances will form an increasing sequence $\nu_0 \leqslant \nu_1 \leqslant \ldots \leqslant \nu_n$. It may thus be the case that $|\omega(I).k| \leqslant \nu_{r+p}$, $(p \geqslant 2)$, i.e., I may belong to $B_M \cap Z_{r+p}{}^*$. If M and M' are two different modules with respective dimensions r and $r+1$, then the corresponding blocks

B_M and $B_{M'}$ do not intersect; on the contrary, if the dimension of M' strictly exceeds $r+1$ then B_M and $B_{M'}$ may have a nonempty intersection.

c) Finally, and this is crucial, if a point I does not belong to any r dimensional resonant zone ($I \notin Z_r^*$), then it lies in a resonant block B_M, the dimension of M being *strictly smaller than* r. Indeed: $\cup_{\dim M=j} B_M = Z_j^*\backslash Z_{j+1}^*$ which implies that:

$$\cup_{\dim M<r} B_M = \cup_{0\leq j<r} Z_j^*\backslash Z_{j+1}^* = K\backslash \cap_{j=1,\dots,r} Z_j^*,$$

so that if $I \notin Z_r^*$, it does belong to $\cup_{\dim M<r} B_M$.

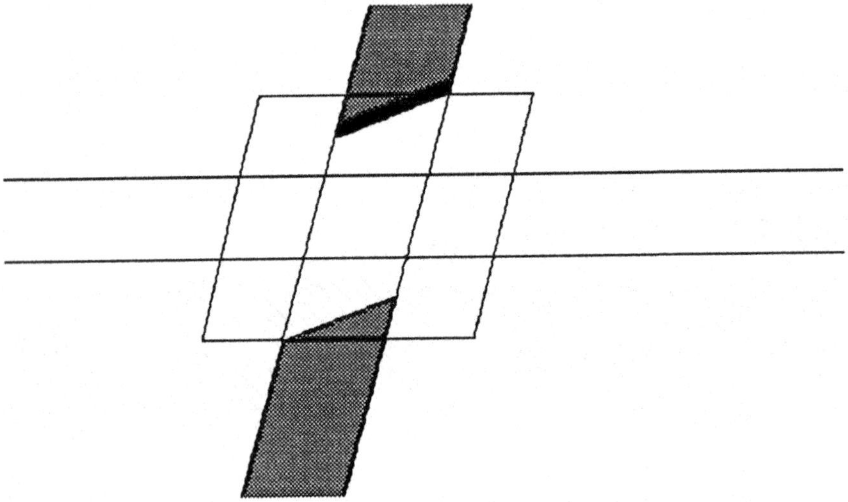

Figure 7.3: The shaded area represents an extended block $B^{ext}_{M,d}$ of dimension one which is the union of the cylinders of fast drift $C_{M,d(I)}$ for all points I belonging to a resonant block B_M. One such cylinder appears in black on the figure.

We shall now bound from above the diameter of the fast drift cylinders $C_{M,d}$ (I) from above. Because of the main lemma this will provide a bound for the variation of the action variables that may occur during the crossing of a resonant zone.

Here is the *only* place in the proof where the geometric conditions play a part. The convexity requirement ensures the *transversality* of the resonant manifold R_M and the plane of fast drift Π_M (I) (I belongs to R_M). Indeed, the plane which is *orthogonal* to R_M is spanned by the vectors $v^{(j)} = \partial\omega/\partial I.k^{(j)}$ $(j=1,\dots, r)$, where $k^{(j)}$ is a basis of M. On the other hand, Π_M itself is spanned by the vectors $k^{(j)}$. If $v = \sum_i \lambda_i k^{(i)}$ belongs to Π_M and is tangent to R_M, it is orthogonal to the vectors $v^{(j)}$, and in particular $<v, \partial\omega/\partial I.v> = 0$, in contradiction with the convexity assumption.The estimation of the diameter of the cylinders given below will rely on this transversality property.

The steepness condition is far less restrictive. It only ensures that the contact between R_M and Π_M has at most finite order and allows for an estimate of the diameter of $\Lambda_M(I) \cap Z_M$ in terms of the steepness coefficients. The corresponding lemma may be found in [Nek2]. We repeat once more that this is the only place where steepness plays a part and that the rest of the proof is the same for convex or steep Hamiltonians. However, to conclude the proof, one must then check conditions which involve the $3(n - 1)$ steepness coefficients instead of the sole coefficient m. The bound on the diameter of cylinders is given by the following lemma.

Geometric Lemma:

Let M be a resonant module of dimension r and let I belong to the resonant zone Z_M. The diameter of the fast drift cylinder $C_{M,d}(I)$ is less than:

$$d_r = 2rN^{r-1}n\nu_r/m + 2Md/m.$$

Proof:

One must first establish the following proposition which is stated in [Nek2] and proved in [Ben2] and [Ben5].

Proposition:

Let $k^{(1)},\dots, k^{(r)}$ be r independent vectors of $\mathbb{Z}^n$ and assume that $|k^{(i)}| \leqslant N$, $1 \leqslant i \leqslant r$. Let ν be a nonnegative real number. Then, if $v \in \mathbb{Z}^n$ satisfies the set of inequalities $|v.k^{(i)}| \leqslant \nu$, $1 \leqslant i \leqslant r$, one has $\| v \| \leqslant \nu r N^{r-1}$ (Euclidean norm).

We leave the easy proof to the reader (induction on r).

Consider now any two points I' and I'' in $C_{M,d}(I)$. Then for every i in $\{1,\dots, r\}$:

(1) $\qquad |(\omega(I'') - \omega(I')).k^{(i)}| \leqslant 2\nu_r,$

or:

(2) $\qquad |A^*(I'' - I').k^{(i)}| \leqslant 2\nu_r,$

where $A^* = A(I^*)$ ($= \partial\omega/\partial I(I^*)$) and the point I^* is located between I' and I''. We may write $A^*(I'' - I')$ as a sum $v + v^\perp$, where v belongs to M and $v^\perp$ to its orthogonal complement, and we likewise set $(I'' - I') = w + w^{\perp}$. Since both I' and I'' belong to $C_{M,d}(I)$, $\| w^{\perp} \| \leqslant 2d$. Hence:

$$(3) \qquad m\| I'' - I' \|^2 \leqslant | A^*(I'' - I').(I'' - I') | = | v.w | + | v^{\perp}.w^{\perp} |$$
$$\leqslant \| v \| \| I'' - I' \| + 2dM \| I'' - I' \|.$$

This entails that $\| I'' - I' \| \leqslant \| v \|/m + 2dM/m$ and the lemma then follows from the inequality $\| v \| \leqslant 2r\nu_r N^{r-1}$ that results from the above proposition. ∎

We are now in position to conclude the proof of Nekhoroshev's theorem. We must assign values to the remaining free parameters which appear in the statements of the lemmas and of the theorem: N, ξ, α, β, d, T, Δ and the sequence $\{\nu_r , 0 \leqslant r \leqslant n\}$. The use of the main lemma will require that resonances do not overlap. More precisely we demand that the extended block $B^{ext}{}_{M,d}$, which is associated to the r dimensional module M, intersect no r dimensional resonant zone (except of course Z_M): $B^{ext}{}_{M,d} \cap Z_{M'} = \varnothing$ if $M \neq M'$ and $\dim M = \dim M'$. This may be rewritten as:

$$(4) \qquad \forall I \in B^{ext}{}_{M,d}, \forall M' \neq M, \text{ such that } \dim M' = \dim M = r, \forall k' \in M' \backslash M,$$
$$| k' | \leqslant N \Rightarrow | \omega(I).k | \geqslant \nu_r .$$

The above non-overlapping condition entails the following assertion: If I belongs to the resonant block B_M and I' to the base of the cylinder $C_{M,d}(I)$, i.e., $I' \in C_{M,d}(I) \cap \partial Z_M$, then I' belongs to a resonant block $B_{M'}$, the dimension of M' being *strictly less* than the dimension of M.

Thus the "trapping mechanism", which limits Arnold's diffusion, consists in the fact that, to the level of approximation of the theorem, a trajectory always drifts from a resonant zone to a *less* resonant one (the dimension r strictly decreases) before possibly ending in the nonresonant domain ($r = 0$). Therefore, the maximal drift of the action variables cannot exceed $d_0 + ... + d_n \leqslant (n + 1)\, d_n$ where d_r is given by the geometric lemma and the d_r's form an increasing sequence, as do the ν_r ; the estimate Δ which appears in the statement of the theorem will be chosen as an upper bound of this quantity.

We must therefore fix the remaining free parameters so that:

a) The non-overlapping of resonances condition is satisfied (geometric requirement).

b) The assumptions of the main lemma are fulfilled (analytic requirement).

The drift $\| I(t) - I(0) \|$ will then be less than Δ for any initial condition in $I(0) \in K - 3\alpha$ (see the main lemma) as long as $I(t)$ remains in $K - 3\alpha$ and $t \leqslant T$. As $\Delta \geqslant 2(n + 1)d$ (because $d_r \geqslant 2d$ for every r) and $d \geqslant 4\alpha$ (see the main lemma) the estimate will actually hold at least until time T, for every initial condition in $K - 2\Delta$, thus proving the theorem.

First of all, we shall consider the non-overlapping condition. If I belongs to $B^{ext}_{M,d}$, there exists, by the very definition of extended blocks, some point $I_0 \in B_M$ such that the cylinder $C_{M,d}(I_0)$ contains I. The distance from I to I_0 will then be smaller than d_r since this quantity is an upper bound of the diameter of $C_{M,d}(I_0)$. Actually this distance is smaller than $d_r - Md/m$, as I_0 is located on the axis.

For every vector k not belonging to M and with norm less than N one has:

(5) $\quad |\omega(I).k| \geqslant ||\omega(I').k| - |(\omega(I)-\omega(I')).k|| \geqslant |\nu_{r+1} - M(d_r - Md/m)N|$.

The overlapping will thus be precluded if:

(6) $\quad \nu_{r+1} \geqslant \nu_r + MN(d_r - Md/m)$,

or, explicitly:

(7) $\quad \nu_{r+1} \geqslant [1 + 2rMN^r/m]\nu_r + M^2Nd/m$.

The easiest and almost optimal way to guarantee the validity of these inequalities is to choose the half-widths ν_r as:

(8) $\quad \nu_r = (4M/m)^r r! N^{r(r-1)/2}\nu_0$, assuming that $\nu_0 \geqslant NM^2d/m$.

We shall now fix the remaining parameters: N, ξ, α, β, d, T, Δ and ν_0. Let us recall the inequalities that must be satisfied together with the lemma in the statement of which they appear:

(9.1) $\quad \alpha \leqslant \rho/8$ (analytic lemma).

(9.2) $\quad \beta \leqslant \sigma/8$ (analytic lemma).

(9.3) $\quad L_r \geqslant 1,\ 0 \leqslant r \leqslant n$ with $L_r = E/(2\nu_r\alpha\beta^{n+1})$ (analytic lemma).

(9.4) $\quad c(n)\varepsilon L_r^{\,2} N^{4\xi(n+2)} \leqslant 1,\ 1/4 \leqslant \xi < 1/2$ (analytic lemma).

(9.5) $\quad \beta\sqrt{N} \geqslant 2n\log N$ (analytic lemma).

(9.6) $\quad d \geqslant 4\alpha$ (main lemma).

(9.7) $\quad d \geqslant 16\varepsilon T/\beta\ e^{-N^{\xi}}$ (main lemma).

(9.8) $\quad \Delta \geqslant (n+1)d_n$ with $d_n = 2nN^{n-1}n\nu_n/m + 2Md/m$ (theorem).

(9.9) $\quad \nu_0 \geqslant NM^2d/m$ (non-overlapping of resonances).

We choose $\beta = \sigma/8$, according to (9.2), and set $\alpha = d/4$, so that both (9.1) and (9.6) are satisfied for ε small enough, since d will be small with ε. As N will tend to infinity with $1/\varepsilon$, (9.5) too will be satisfied for ε small enough, and we set $\xi = 1/4$ in

(9.4). (9.7) and (9.8) are definitions of Δ and T rather than constraints, and we shall set $T = (d\sigma/2^9\varepsilon)\exp(N^{1/4})$. Taking into account the fact that ν_r is an increasing sequence, we are left with the inequalities:

(10.1) $L_0 \geqslant 1$

(10.2) $c(n)\varepsilon L_n{}^2 N^{n+2} \leqslant 1$

(10.3) $\nu_0 \geqslant NM^2 d/m$

where $L_n = L_0\nu_0/\nu_n = L_0(m/4M)^n N^{-n(n-1)/2}/n!$ and $c(n) = (2n+4)^{2n+4}2^{4(2n+4)}$ is taken from the analytic lemma.

These equations involve only the three variables d, N and ν_0, that is, the maximal transverse excursion of the action variables, the cut-off frequency and the typical resonance width. For the sake of simplicity one may choose power laws for these parameters: $d \sim \varepsilon^a$, $\nu_0 \sim \varepsilon^b$, $N \sim (1/\varepsilon)^{1/c}$. One then easily checks that all the above equations are satisfied for ε small enough and the following choice of the exponents:

$$a = 1/4, \quad b = n(n+1)/2c, \quad c = 4(n^2 + 2n + 2).$$

Δ may then be chosen as indicated in the statement of the theorem. ∎

It is obvious that in the course of the proof many more or less arbitrary choices have been made, making a discussion of the optimality of the various estimates quite difficult. As we already mentioned, the optimality problem, possibly based on a different proof, is a matter of current research.

We make one last but important remark. If the unperturbed Hamiltonian is convex, or even only quasi-convex, the geometric part of the proof can in fact be simplified: The trajectory does not visit several resonant zones but, if the sizes have been carefully chosen, it remains confined in one and the same (more or less resonant) zone for an exponentially long time. This simpler trapping mechanism is suggested in [Nek2] and implemented in [Ben2]. We shall presently say more about it but we chose to present the more complex mechanism above because it displays the "generic" behavior, which works in the case of general steep functions.

So let B_M be a resonant block, the dimension of M being r; to study the dynamics inside this block, one uses the resonant normal form of the Hamiltonian constructed in the analytic lemma. As explained in Appendix 3, one can assume that M is generated by the vectors $k^{(i)} = (0...0, 1, 0..., 0)$ with a 1 in the ith entry, $i = 1,..., r$. Indeed, one only needs to perform an appropriate canonical transformation $(I, \varphi) \to (I', \varphi') = (PI, {}^tP^{-1}\varphi)$, defined by the invertible matrix P with integer entries. $\partial\omega/\partial I$ is then changed to $P^{-1}(\partial\omega/\partial I)P$ and retains its convexity property. The normal form may be written:

$$H(I, \varphi) = h(I) + Z(I, \varphi_R) + R(I, \varphi)$$

where φ_R is an abbreviated notation for the resonant angle variables $(\varphi_1, ..., \varphi_r)$; the set of the $(n - r)$ other angle variables will be denoted by φ_{NR}, and we write similarly $I = (I_R, I_{NR})$. Z is of order ε and R is exponentially small ($\| R \| \sim \exp(-1/\varepsilon^{1/c})$). The nonresonant action variables I_{NR} evolve with a typical velocity on the order of $\| R \|/\sigma$ since $dI_{NR}/dt = -\partial R/\partial\varphi_{NR}$. On a timescale much smaller than $T \sim \exp(1/\varepsilon^{1/c})$, they may be considered as "frozen" and the evolution of the resonant action variables I_R is then governed by the r degree-of-freedom Hamiltonian:

$$H'(I_R, \varphi_R) = h(I_R, I_{NR}(0)) + Z(I_R, I_{NR}(0), \varphi_R).$$

In this approximation, the motion takes place inside the plane $\Pi_M(I(0))$, with equation $I_{NR} = I_{NR}(0)$. If the unperturbed Hamiltonian h is quasi-convex, R_M and Π_M intersect tranversally and, from the definition of Π_M, if $I(0)$ belongs to R_M and k is a vector of Π_M, $\partial h/\partial I(I(0)).k = 0$. Because of the convexity of h on the energy surface, this critical point can only be a minimum of the restriction of h to the plane $\Pi_M(I(0))$, at the point $I(0) \in R_M$. As energy is conserved, the trajectory remains trapped, to the level of approximation we consider, in a neighborhood of $I(0)$ and cannot visit any other resonant zone.

In order to show this in a quantitative way, one only needs to expand the unperturbed Hamiltonian h around a point $I_0 = I(0)$ belonging to the resonant block B_M (not necessarily to the resonant manifold R_M):

$$(11) \qquad h(I(t)) = h(I_0) + \omega(I_0).(I(t)-I_0) + 1/2\,(\partial\omega/\partial I)(I^*)(I(t)-I_0).\times \times (I(t)-I_0)$$

where $I^* \in (I_0, I(t))$ because K is convex.

We denote by a the action drift $\| I(t) - I_0 \|$. From (11) we get:

$$(12) \qquad ma^2/2 \leqslant | h(I(t))-h(I_0) | + | \omega(I_0).(I(t)-I_0) |,$$

where m is the lowest eigenvalue of the quadratic form $\partial\omega/\partial I$ on the energy surface; note that since the perturbed and unperturbed energy surfaces are ε-close, a small change in the constants will accomodate the difference, for ε small enough.

Conservation of energy $H(I(t)) = H(I(0))$ now provides an upper bound for the first term on the right-hand side of (12):

$$(13) \qquad | h(I(t)) - h(I_0) | \leqslant | Z(I(t), \varphi(t))-Z(I_0, \varphi(0)) | + 2 \| R \| \leqslant c\varepsilon.$$

The second term may be estimated as follows. Let P_M be the orthogonal projection onto M

and $P_\perp$ be the orthogonal projection onto the orthogonal complement of M. We have:

$$(14) \qquad |\omega(I_0).(I(t) - I_0)| \leqslant \| P_M[\omega(I_0)] \|.\| I(t) - I_0 \| + \| P_\perp[\omega(I_0)] \|.\| P_\perp[I(t) - I_0] \|.$$

As $|\omega(I_0).k| \leqslant \nu_r$ for every basis vector of M (since I_0 belongs to B_M), the first term on the r.h.s. is smaller than $arN^{r-1}\nu_r$ according to the proposition we used in the proof of the geometric lemma. The evolution equations of the nonresonant variables tell us that the second term is smaller than $c|t|.\| \partial R/\partial\varphi \|$. Therefore:

$$(15) \qquad |\omega(I_0).(I(t) - I_0)| \leqslant arN^{r-1}\nu_r + c\sigma^{-1}|t|.\| R \|.$$

Formulae (12),(13) and (15) yield:

$$(16) \qquad ma^2/2 \leqslant c\varepsilon + arN^{r-1}\nu_r + c\sigma^{-1}|t|.\| R \|.$$

and this last inequality allows us to bound $a = \| I(t) - I(0) \|$ from above and to complete the proof of the theorem in the case of quasi-convex Hamiltonians.

CHAPTER 8: ADIABATIC THEOREMS IN ONE DIMENSION

8.1 Adiabatic invariance; definition and examples

We first recall the definition of an adiabatic invariant at lowest order, starting from a system $dx/dt = f(x, \varepsilon t)$ with $x \in \mathbb{R}^n$ and ε a (small) parameter (f defines a slowly varying vector field). As a matter of notation, in this chapter we will write either $f(x, \varepsilon t)$, in which case $\tau = \varepsilon t$ will denote the rescaled "slow" time, or when necessary, $f(x, \lambda(\varepsilon t))$, which stresses the time dependence through an explicit real valued parameter λ. In Chapter 10, we will study the case where $\lambda \in \Lambda \subset \mathbb{R}^d$ denotes a set of d parameters and the parameter space whill then play a nontrivial role.

A function $A(x, \tau)$ is said to be an adiabatic invariant associated to the given system if its variation on the time interval $t \in [0, 1/\varepsilon]$ ($\tau \in [0, 1]$) is small together with ε, except perhaps for a set of initial conditions whose measure goes to zero with ε (this will happen in the next chapter); that is, for "most" initial conditions:

$$\lim_{\varepsilon \to 0} \operatorname{Sup}_{t \in [0, 1/\varepsilon]} \| A(x(t), \varepsilon t) - A(x(0), 0) \| = 0.$$

We shall be concerned here almost exclusively with the case of n degree-of-freedom Hamiltonian systems with evolution governed by a slowly varying Hamiltonian $H(p, q, \tau)$, and the present chapter is devoted to the case $n = 1$. The reader will find a general - but not always very effective - theory of adiabatic invariants for noncanonical systems in [Kru], which is summarized in [Lic] (see also [Lén]). Let us also mention that [Go] contains elementary but useful information on adiabatic invariants in the case to be discussed below. Much information about adiabatic invariants (and related topics) can also be found in [Bak].

In the introduction we briefly previewed the classes of systems that will be studied in this chapter and the next, but before we present some of the most important examples of adiabatic invariants, it may be useful to outline the content of this chapter, as it is an attempt to collect together the different kinds of results which have been obtained for one dimensional Hamiltonian systems. Apart from the general results, which are described in Section 2, they can be rather strictly classified as a two entry array, whith two types on each side, which gives rise to four types of results and proof techniques.

As a first criterion, the system can be linear or nonlinear; this is admittedly a poor classification, since only the harmonic oscillator (with slowly varying coefficients) is a representative of the first class, but as is well known, it has been forced upon us by the development of physics and mathematics.

The second criterion pertains to "boundary conditions", very much as is the case for PDE results: roughly speaking the Hamiltonian $H(p, q, \lambda)$ can be either periodic as a function of the parameter λ, or asymptotically autonomous, which means that the limits $\lambda_\pm = \lim_{\tau\to\pm\infty} \lambda(\tau)$ exist. Of course this distinction will enter only when we are interested in time intervals much longer than $1/\varepsilon$.

The rest of this first section will thus be devoted to the exposition of the three most classical examples of adiabatic invariants, which have largely motivated the progress of the theory. We will also demonstrate the elementary and fundamental result which says that the action is an adiabatic invariant to lowest order in (one dimensional) Hamiltonian systems, and we will point out how this is a straightforward consequence of the simplest averaging theorem (as proved in Section 3.1).

The second section introduces the higher order invariants with the aid of adiabatic Lie series, which largely clarify both the theory and the practical computations; we will give a simple example of the calculation of the second order invariant for the harmonic oscillator. This section depends heavily on the results of Appendix 7. We also show how the same results can be obtained using a noncanonical method, namely the averaging method for standard systems with one phase, as described at the end of Section 3.2.

We then come back in Section 3 to the harmonic oscillator to introduce, in the case of a periodic dependence on the parameter, the phenomenon of parametric resonance and show that it allows to study the optimality of the results obtained in Section 2. We will also summarize a technique developed by T.J.Shep ([Sh]) who uses the so-called Liouville transformation to compute the higher order invariants more quickly in the linear case.

In the next section (Section 4), we examine the "linear asymptotic case", when λ goes to limits $\lambda_\pm$ as $t \to \pm\infty$. One can then estimate the total drift in action $\Delta I = I(+\infty) - I(-\infty)$ and show how the main order term can sometimes be computed in the adiabatic limit $\varepsilon \to 0$ using the WKB method.

Section 5 is devoted to the "asymptotically autonomous nonlinear case" in which one can prove a strong result by a Nekhoroshev type technique, much as in Section 2.

Finally, in Section 6 it is shown how a version of the KAM theorem enables one to prove a result of *perpetual* adiabatic invariance in the case of a periodic dependence on the parameter for nonlinear systems satisfying a condition of frequency nondegeneracy, which serves to inhibit the mechanism of parametric resonance.

In particular, the above should make it clear how the four most important cases of adiabatic invariance (linear or nonlinear systems, periodic or asymptotically vanishing dependence on the parameter) are treated using, to a certain extent, four different techniques (Hill's theory, WKB method, Nekhoroshev perturbation technique, KAM theory) which stresses the difference in the various phenomena involved.

Let us now start with the classical examples of adiabatic invariance.

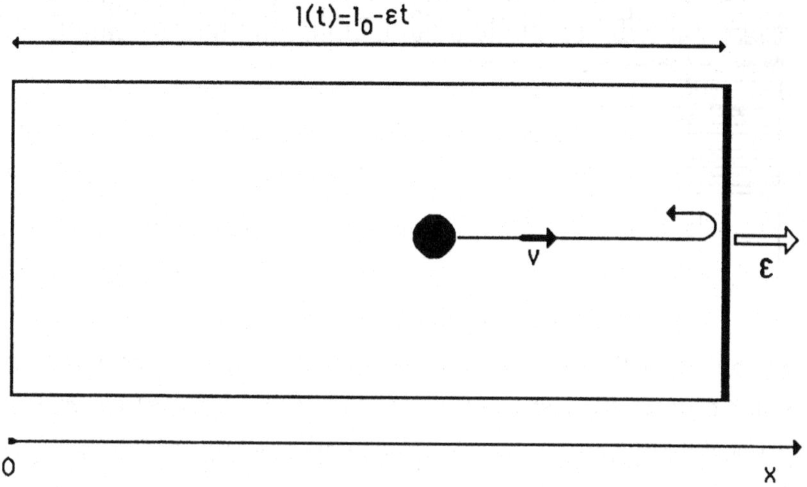

Figure 8.1: A ball bouncing in a box enclosed by a slowly moving wall (speed ε). Collisions are supposed to be perfectly elastic.

The first and perhaps simplest example is the following: Consider a box closed by a slowly moving wall (with constant velocity ε) and a ball of unit mass undergoing perfectly elastic reflections (see Figure 8.1); at time $t = 0$, it is at $x = 0$, with a positively oriented velocity v. Let t_n denote the time of the nth bounce against the wall and l_n and v_n the length of the box and the velocity of the ball at that time. It is easy to see that the following recursion formulas hold:

$$t_n = t_{n-1} + 2(l_0+\varepsilon t_{n-1})/(v_{n-1}-2\varepsilon)$$

(1) $$l_n = l_{n-1} + 2\varepsilon(l_0+\varepsilon t_{n-1})/(v_{n-1}-2\varepsilon)$$

$$v_n = v_{n-1} - 2\varepsilon$$

which prove the invariance of $A = l_n v_n$; within a constant factor this is simply the action of the system, or the area enclosed by the trajectory in phase space (see Figure 8.2), which in the case of constant velocity turns out to be an exact invariant (not only adiabatic).

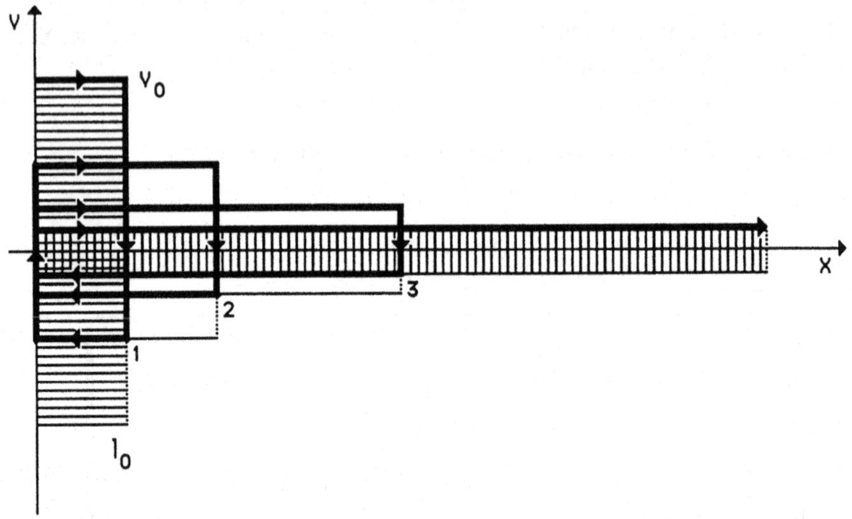

Figure 8.2: The ball's velocity v and the length l of the box change with time (the system's trajectory appears as a bold curve) while the action (hatched areas), is an invariant of the motion. Three successive collisions are shown.

However simple it may be, this example calls for a few remarks:

- This is a nonsmooth - nondifferentiable - system which therefore does not exactly meet with the requirements of this chapter; it is possible however, as is nearly always the case, to construct a theory dealing with symplectic maps (or area preserving, in dimension one) which parallels the one for smooth Hamiltonian flows.
- In the case where the velocity is not a constant, the invariance of action is no longer exact but is instead adiabatic (holds for $\varepsilon \to 0$) and, what is more important, things are much more complicated for time intervals much longer than $1/\varepsilon$. If in particular the motion of the wall is periodic with respect to $\tau = \varepsilon t$ (say $= \sin\tau$) one is led to the Fermi model, which has been studied extensively (see [Gu] and [Lic] along with their bibliographies) in relation with the so-called standard - or Chirikov - mapping (cf. [Chir]). The KAM theorem then permits one to show (more directly than in Section 6 below) that for any ε, not necessarily small, the ball cannot be accelerated without bound (starting from a slow velocity), a result which is not immediately intuitive; in fact, this is

in strong contrast to the corresponding original non deterministic model which, using a random phase approximation, predicts an unbounded acceleration. It is useful to notice that here the natural parameter is the ratio of the flight time ($\sim 1/v$) to the period of the motion of the wall ($\sim 1/\varepsilon$), that is, it goes like $\varepsilon l/v$ and gets small when v increases, whatever ε, so that, as is immediately clear from the physics, the random phase approximation ceases to be valid.

- If instead of a single ball, one considers a huge number of them (typically 10^{23}), the system will form a "hard sphere gas" and according to results due to Y. Sinaï (published only in the case of 2 spheres) it is ergodic on the energy surface. If one then admits that the results of Chapter 9 may be extended to this non smooth system, they show the adiabatic invariance of the volume in phase space, which is related to physical entropy via the Boltzman formula as will be discussed in more detail at the end of Chapter 9.
- The above discussion hinges upon the thermodynamical meaning of "adiabaticity", that is, "absence of heat transfer". As F. Fer notes in [Fer], it would in fact be more accurate to simply speak of "reversibility" as far as one stays in the framework of analytical mechanics (as we do here) where the question of heat is not even raised, and where from a thermodynamical point of view all phenomena are (implicitly) supposed to take place adiabatically. For these subtle questions half way between mechanics and thermodynamics we direct the reader to more specialized books, e.g. to Chapter 8 of the famous book by L. Brillouin ([Bri]), to [Bak] and to the lucid exposition of F. Fer.

We now come to the most famous and seminal example of an adiabatic invariant, namely the action of a harmonic oscillator with slowly varying frequency, which in that case is the ratio E/ω of the energy to the pulsation. As mentioned in the introduction, this was first suggested by Einstein [Ei1] in the days of the "old" quantum mechanics, in an attempt to understand the appearance of Planck's constant (one should think of the Bohr-Sommerfeld formula $E = (n + 1/2)\hbar\omega$ resulting in the quantization of the oscillator). A detailed account of these discussions, including quotations from the Solvay congress proceedings can be found in [Bak].

The proof of the adiabatic invariance is very simple and is valid for all one degree-of-freedom Hamiltonian systems with a Hamiltonian $H(p, q, \tau)$ ($\tau = \varepsilon t$). The system is integrable for fixed τ so with the help of a generating function $S(q, \varphi, \tau)$ one can then construct, as is done in Appendix 6, the action-angle variables (I, φ) which depend on τ. In these variables the system is governed by the Hamiltonian:

$$(2) \qquad K(I, \varphi, \tau) = H(I, \tau) + \varepsilon \partial S/\partial \tau$$

where H is independent of φ. The equations of motion read:

$$dI/dt = -\varepsilon\partial^2 S/\partial\tau\partial\varphi$$

$$(3) \qquad d\tau/dt = \varepsilon$$

$$d\varphi/dt = \omega(I,\tau) + \varepsilon\partial^2 S/\partial\tau\partial I, \quad \omega = \partial H/\partial I$$

which is in the standard form used in Chapter 3 (with I and τ as the slow variables). One can then apply Theorem 1 of that chapter, noticing that $\partial^2 S/\partial\tau\partial\varphi$ has zero mean value (with respect to φ) and find the invariance of I as:

$$(4) \qquad \text{Sup}_{t\in[0,\,1/\varepsilon]} \mid I(t) - I(0) \mid = O(\varepsilon).$$

The intermediate variable, ε-close to I and evolving with $O(\varepsilon^2)$ velocity, will be $P = I + \varepsilon\omega^{-1}\partial S/\partial\tau$. The theorem however requires that ω be bounded from below $\omega(I, \tau) \geqslant \omega_0 > 0$ for $\tau \in [0, 1]$, which here is the same as requiring that the trajectory keep away from the separatrices, that is, the curves on which the period of the motion increases to infinity. We thus obtain the following

Elementary adiabatic theorem:

Let $H(p, q, \tau)$ be a one dimensional C^2 Hamiltonian slowly depending on time. We suppose that for a certain region of the phase space and for any τ ($0 \leqslant \tau \leqslant 1$) there exists an action-angle transformation $(p, q) \rightarrow (I, \varphi)$ (not necessarily one to one), which is the same as supposing that the trajectories of the "frozen" (fixed τ) systems stay away from the separatrices. This also implies that $\mid \omega(I, \tau) \mid \geqslant \omega_0 > 0$ in the given region.

Then the action is an adiabatic invariant in the sense that it satisfies (4).

Let us mention the formulas for the harmonic oscillator:

$$H = 1/2\,(p^2 + \omega(\tau)^2 q^2)$$

$$(5) \qquad I = H/\omega; \quad \varphi = \text{Arctg}(\omega(\tau)p/q)$$

$$K(I, \varphi, \tau) = \omega I + \varepsilon\omega'/(2\omega)\, I \sin(2\varphi) \quad (\omega' = d\omega/d\tau).$$

The third example of adiabatic invariants we shall present, albeit in an incomplete way, is much more complex. Because of its physical and practical importance, it has motivated much of the work devoted to the elaboration of a noncanonical formalism in the theory of adiabatic invariants (see especially [Kru]; more recent references are given below). It deals with the motion of a charged particle in an electromagnetic field that is slowly varying - in space and time - with respect to the Larmor radius and Larmor frequency (see below). The theory of adiabatic invariance in this case was largely motivated by the development of magnetic confinement machines in controlled fusion experiments (magnetic mirrors, Tokamak type toroidal machines). A similar situation is encountered in the motion of electrons in the Van Allen belts of the terrestrial magnetosphere (see e.g. [Bak]).

The most striking feature of this example lies in the appearance of a *hierarchy* of adiabatic invariants. To understand this phenomenon, we need a more general definition of adiabatic invariance that allows for slow *local* variations. One starts with a Hamiltonian function $H(I, \varphi, \varepsilon p, \varepsilon q, \varepsilon t)$ which includes r *fast* degrees of freedom (I, φ) and $s = n - r$ *slow* degrees of freedom (p, q), along with a possible slow time dependence. Notice that there is in fact no difference between time and the other slow degrees of freedom, as can be shown by considering t and H as conjugate variables and writing down the equivalent *autonomous* Hamiltonian in one more degree of freedom. If the motion is integrable when ε, p, q and $\tau = \varepsilon t$ are fixed, a canonical transformation will bring the Hamiltonian into the form:

$$H = H_0(I, \varepsilon p, \varepsilon q, \varepsilon t) + \varepsilon H_1(I, \varphi, \varepsilon p, \varepsilon q, \varepsilon t).$$

If $r = 1$, the action I will be adiabatically invariant (on the time scale $1/\varepsilon$). The invariance of I allows the "elimination", so to speak, of one of the degrees of freedom, as it is essentially frozen on the timescale $1/\varepsilon$. If the dynamics of the remaining variables (p, q) occur on various timescales and if one degree of freedom is faster than the others, it will be possible to iterate the same operation and reduce the number of degrees of freedom again by one. The best one can hope for in the end is thus to find parameters ε_1, ε_2,... each of which describes the motion on a different time scale, with all but one degree of freedom frozen. The many degree of freedom motion is then described as a "cascade" of (integrable) one degree of freedom systems. On a given timescale, all degrees of freedom but one are "frozen": The faster ones are adiabatically conserved, while the slower ones do not have time to evolve. Let us stress that this description, rarely implemented except on model problems, has nothing to do with the adiabatic invariance of action in multidimensional systems to be discussed in the next chapter.

The application of the above program to the motion of a charged particle with mass m and charge e in a slowly varying electromagnetic field is far from obvious. When the magnetic field B is uniform and constant and no electric field E is present, a particle with velocity v moves with *parallel* velocity $v.b$ (b is the unit vector B/B) and *angular frequency* $\omega = qB/m$ (called the Larmor frequency) on a helix with axis B and radius

$\rho = mv/eB$ (called the Larmor radius). This fast motion is usually called *Larmor gyration.* We are thus already faced with two timescales corresponding to the fast Larmor gyration and the slower parallel motion. If an electric field is present and both fields varies slowly in space (on a characteristic length $L >> \rho$) and time (on a characteristic time $T >> 1/\omega$), the particle also *drifts* across field lines. For instance if an electric field is superimposed on an uniform and static magnetic field, a transversal drift occurs at the speed

$$u = cE \times B/B^2.$$

Two new timescales then obviously appear: The timescale of the drift motion and the characteristic time T. A small parameter clearly plays a part in this problem:

$$\varepsilon = \rho/L = v/\omega L = mv/qBL.$$

Unfortunately, it does not show up in the Hamiltonian:

$$(6) \qquad H(p, r) = 1/2m \, \| p - e\, A(r, t)/c \|^2 + e\varphi(r, t)$$

where r denotes the position of the particle, the velocity v of which is given by $mv = p - e/c\, A$, and (A, φ) are the potentials of the electromagnetic field. This explains why the problem is not easily treated in the Hamiltonian framework, in other words by performing canonical transformations.

A specially adapted non-Hamiltonian method, referred to as the guiding center theory, has been devised. It essentially consists of successive averagings of the Lorentz equation of motion:

$$(7) \qquad m\,dv/dt = eE + v/c \times B.$$

We shall give a brief account of it below in the nonrelativistic case, following the clear exposition of T.G. Northrop in his book [No1]. The same author - in collaboration with J.A. Rome - has more recently given (in [No2]) a lucid account of the higher order results which can be found in the literature and usually involve intricate vector algebra. This algebraic complexity explains why results appear under various guises (see [No2] for a detailed discussion).

The guiding center theory starts with an explicit separation of the Larmor gyration and the slower motion. To this end one introduces an average position R of the particle (the actual position of which is r), known as the *guiding center* and defined as follows:

$$(8) \qquad \rho = (mc/eB^2)B \times (v-u), \quad R = r - \rho$$

(throughout this section capital letters will be used for quantities related to the guiding center).

As the electric field is not necessarily weak, the drift speed is not always small. That is why we introduce the drift velocity u in this definition: The transverse motion of the particle is approximately circular only in the frame of reference moving at speed u. We assume that the fields in (8) (recall that $u = cE \times B/B^2$) are evaluated at the actual position r of the particle. This choice is arbitrary and we could for instance evaluate the fields at the guiding center position R, the difference being of order ε^2 in the equation for ρ. Part of the trouble one encounters when dealing with guiding center theory stems from this arbitrariness. It explains why the equations obtained after averaging are usually non-Hamiltonian and accounts for some of the discrepancies in the results derived by various authors at the next order of the theory.

We shall now perform the first averaging process (averaging on the Larmor gyration). The equation of motion (7) may be cast in the nondimensional form:

$$(9) \qquad \varepsilon d^2r^*/dt^2 = e/c\ dr^*/dt \times B^* + eE^*$$

where $r^* = r/L$, $t^* = v_0t/L$ (v_0 is a typical particle's velocity), $B^* = B/B_0$ (B_0 is a typical strength of the magnetic field) and $E^* = cE/v_0B_0$. Formally one deals with the same perturbation problem when one considers equation (7) and assumes m/e to be a small parameter ε. We shall stick to that latter approach in what follows. Setting $r = R + \rho$ in (7) and averaging over the gyrophase of the fast Larmor motion, one obtains after rather lengthy calculations (which may be found in [No1]) the following set of guiding center equations of motion:

$$(10.1) \qquad m/e\ dV_{||}/dt = E_{||} - M/e\ \partial B/\partial s + m/e\ u.(\partial b/\partial t + V_{||}\partial b/\partial s + u.\nabla b) + O(\varepsilon^2)$$

$$(10.2) \qquad V_\perp = b/B \times [-cE + Mc/e\ \nabla B + mc/e\ (V_{||}\partial b/\partial t + V_{||}^2\partial b/\partial s + V_{||}\ u.\nabla b + \partial u/\partial t + V_{||}\ \partial u/\partial s + u.\nabla u)\ ] + O(\varepsilon^2).$$

In (10) s denotes the distance along the field line, $E_{||} = E.b$ is the parallel electric field, $V_{||} = dR/dt.b(R)$ the parallel velocity of the guiding center, $V_\perp = dR/dt - V_{||}b$ its perpendicular velocity and $M = \rho^2\omega e/2c$ the magnetic moment. Equations (10) describe the average motion of the guiding center on a timescale which is slow with respect to the Larmor gyration: (10.1) is the equation of motion along the lines of force, whereas (10.2) gives the speed of drift across the field lines (it is obtained by inserting (8) into (7) and then crossing the resulting guiding center equation of motion with b); this drift phenomenon originates in the existence of a transverse electric field, the inhomogeneity of B and the time-dependence of both fields. They are derived under the following assumptions:

- No small parameter other than ε plays a part in this problem. Indeed one assumes that the ratio $1/\omega T$ of the characteristic times of Larmor gyration and field evolution is of order ε. Formally this amounts to setting $\partial/\partial t \sim \varepsilon$ in the equations and it greatly simplifies the ordering.

- the parallel electric $E_{\parallel}$ is at most of order ε. In the opposite case, the guiding center would move along the field line during one Larmor period, over a distance sufficient to feel the effects of the fields' variations.

We shall now prove that the magnetic moment $M = m/2B\ (v_\perp - u)^2$ is an adiabatic invariant of the guiding center motion ($v_\perp$ is the transverse velocity of the charged particle and should not be confused with $V_\perp$). We shall restrict ourselves to the time-independent case where the proof relies on energy conservation. The mean kinetic energy of the particle (averaged over the Larmor gyration) reads in the general time-dependent case:

$$(11) \qquad \langle mv^2/2\rangle = mV_{\parallel}{}^2/2 + mu^2/2 + MB + O(\varepsilon^2).$$

The sum of the first two terms is the kinetic energy associated with guiding center motion, while the third term is the kinetic energy of Larmor rotation (equivalent to the energy of a magnetic dipole M in a field B). Using (10) one may compute its time derivative:

$$(12) \qquad d/dt\ \langle mv^2/2\rangle = eV.E + M\partial B/\partial t + O(\varepsilon^2).$$

The first term on the r.h.s. is the work of the electric force and the second one is an induction effect originating in the time dependence of the magnetic field. In the static case $E = -\nabla\Phi$ and (12) becomes:

$$(13) \qquad d/dt\ \langle mv^2/2\rangle = -\,ed\Phi/dt + O(\varepsilon^2).$$

(11) then yields:

$$(14) \qquad d/dt\ (mu^2/2 + MB) = -d/dt\ (mV_{\parallel}{}^2/2 + e\Phi).$$

We may use (10.1) to estimate $d/dt\ (mV_{\parallel}{}^2)$. Then:

$$(15) \qquad d\Phi/dt = V_{\parallel}\partial\Phi/\partial s + V_\perp.\nabla\Phi + O(\varepsilon^2).$$

Or, taking (10.2) into account:

$$(16) \qquad d\Phi/dt = V_{\parallel}\partial\Phi/\partial s + u.\nabla\Phi + b/B \times (Mc/e\ \nabla B + m/e\ V_{\parallel} db/dt + $$
$$+ m/c\ du/dt).\nabla\Phi + O(\varepsilon^2).$$

As $\partial\Phi/\partial s = -E_{\parallel}$ and $u.\nabla\Phi = 0$, we have:

$$(17) \qquad d\Phi/dt = -V_{\parallel}E_{\parallel} - u.(M/e\ \nabla B + m/e\ V_{\parallel}\ db/dt + m/e\ du/dt) + O(\varepsilon^2).$$

This enables us to evaluate the r.h.s. of (14) and obtain:

$$(18) \qquad 1/e\ d(MB)/dt = M/e\ (V_{\parallel}\partial B/\partial s + u.\nabla B) + O(\varepsilon^2) = M/e\ dB/dt + O(\varepsilon^2)$$

that is:

$$(19) \qquad dM/dt = O(\varepsilon^2),$$

thus proving the adiabatic invariance of the magnetic moment M.

The most important consequence of the adiabatic invariance of M is certainly the existence of ***magnetic mirrors***. Consider the magnetic configuration with rotational symmetry of Figure 8.3.

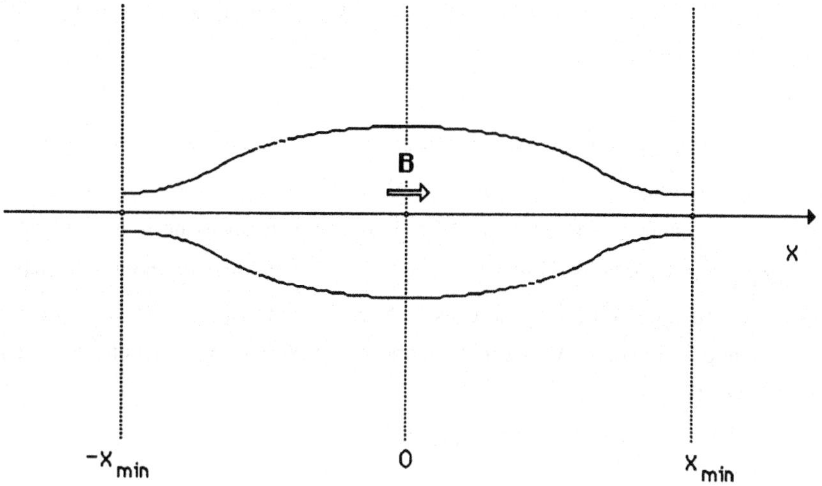

Figure 8.3: A magnetic configuration with rotational symmetry constitutes a "magnetic bottle" where particles with low parallel velocity remain trapped.

On a given line of force, the strength of the magnetic field presents a minimum B_{min} (at $x = 0$). Its maximal value B_{max} is attained at the two symmetric points $x = \pm 1$. Charged particles are trapped in this *magnetic bottle* provided the parallel velocity $V_{||}$ is not too large. This stems from the adiabatic invariance of M ($1/e\ dM/dt = O(\varepsilon^2)$) and the quasi-conservation of the average energy $W = mV_{||}^2/2 + MB + O(\varepsilon^2)$ (we assume the electric field is weak, that is, of the order of ε). Let $V_{||}^*$ denote the value of the parallel velocity of the guiding center at the time when the particle crosses the plane $x = 0$, and let $v_{\perp}^*$ be the perpendicular velocity of the particle at this same time. We have:

$$(20) \qquad mV_{||}^2/2 + MB \simeq mV_{||}^{*2}/2 + MB_{min}$$

and, provided that $mV_{||}^{*2}/2 < M(B_{max} - B_{min})$, the parallel velocity vanishes in the limit $\varepsilon \to 0$ near the points of the line of force where the field strength is:

$$(21) \qquad B^* = B_{min} + mV_{||}^{*2}/2M = B_{min}(1 + [V_{||}^*/v_{\perp}]^2).$$

Thus, provided their parallel velocity is not too large, charged particles are trapped and bounce between the mirror points with a period T_{bounce} on the order of T_{Larmor}/ε.

We shall henceforth consider only such particles and shall study the average drift motion of the guiding center across the field lines, the average being taken over the periodic bounce motion. In what follows, M will be considered as an invariant of the motion. To study the drift motion it is convenient to introduce curvilinear coordinates (α, β, s) such that $A = \alpha \nabla \beta$ (then $B = \nabla\alpha \times \nabla\beta$). Let us introduce the quantity:

$$(22) \qquad K = mV_{\parallel}^2/2 + MB + e\Phi + e\Psi$$

where $\Psi = \alpha/c \; \partial\beta/\partial t$.

In the static case, $K = W$ and is exactly conserved. In the general case of time-dependent fields, K is no longer conserved but we shall prove that the action-like quantity $J = \oint p_{\parallel}\, ds = \oint 2m[K - e(\Phi+\Psi) - MB]^{1/2}\, ds$ is an adiabatic invariant of the drift motion (the integration is performed over the period T_{bounce} of the bounce motion). J may be considered a function of K, α, β and t, as B, Φ and Ψ are functions of (α, β, s, t) and M is a constant of the motion. After simple but tedious algebraic manipulations one obtains the set of equations:

$$(23.1) \qquad < d\alpha/dt > = c/(eT_{bounce}) \; \partial J/\partial\beta$$

$$(23.2) \qquad < d\beta/dt > = - c/(eT_{bounce}) \; \partial J/\partial\alpha$$

$$(23.3) \qquad < dK/dt > = - 1/T_{bounce} \; \partial J/\partial t$$

$$(23.4) \qquad 1 = 1/T_{bounce} \; \partial J/\partial K.$$

The first two equations govern the average transverse drift motion of the guiding center (the notation $< >$ means the average over the bounce period T_{bounce} has been performed). One easily derives from (23) that, at lowest order in ε:

$$\begin{aligned}(24) \qquad dJ/dt &= \partial J/\partial\alpha \; d\alpha/dt + \partial J/\partial\beta \; d\beta/dt + \partial J/\partial K \; dK/dt + \partial J/\partial t \\ &= eT_{bounce}/c \, [<d\alpha/dt> d\beta/dt - <d\beta/dt> d\alpha/dt] + \\ &+ T_{bounce} \, [dK/dt - <dK/dt>]\end{aligned}$$

and $<dJ/dt>=0$. A more rigorous derivation, which takes into account the higher order terms, actually yields: $dJ/dt = O(\varepsilon)$, $<dJ/dt> = O(\varepsilon^2)$. This proves the adiabatic invariance of J on the timescale of the transverse drift motion (which is slow when compared to the bounce motion).

As T_{bounce} is a function of (α, β, K, t) the equations (23) are *not* Hamiltonian. However, if one considers instead K as a function of (α, β, J, t) one easily derives from

(23):

(25.1) $\langle d\alpha/dt \rangle = -c/e\ \partial K/\partial\beta$

(25.2) $\langle d\beta/dt \rangle = c/e\ \partial K/\partial\alpha$

(25.3) $1 = T_{bounce}\ \partial K/\partial J.$

α and β are conjugate canonical variables for the nonautonomous Hamiltonian K. As expected, K depends, at the order under consideration, on the action variable J, but not on its conjugate angle variable. Note also that the corresponding angular frequency is, quite naturally, $1/T_{bounce}$. From (23.3) and (23.4) one also deduces the expected relation:

(25.4) $\langle dK/dt \rangle = \partial K/\partial t.$

We shall now study the third and last invariant of the hierarchy. In some physical situations (Van Allen belts, fusion mirror devices, ...) the center guide slowly precesses on a magnetic surface during its drift motion across the field lines. The corresponding period is $T_{prec} \sim T_{bounce}/\varepsilon$. Each such magnetic surface is perfectly characterized by the three parameters M, J and K. Indeed the equation $J = J(\alpha, \beta, K, M, t)$ implicitly defines a cylinder in the space (α, β, s). We shall henceforth denote by Φ the flux across any section of the magnetic surface (no confusion should arise as the electrostatic potential will never be used in what follows). In the static case, the energy K is conserved, the magnetic surfaces do not vary in time and the flux Φ is trivially a constant of motion. If the electromagnetic field slowly evolves (with a characteristic time T much longer than the precession period: $T \sim T_{prec}/\varepsilon$), K is no longer conserved but Φ remains an adiabatic invariant of motion on this timescale. The proof of this assertion is easy. The flux across any section of the "frozen" magnetic surface (M, J, K) on which the guiding center lies at time t is:

(26) $\Phi(M, J, K, t) = \oint A.dl = \oint \alpha\nabla\beta.dl = \oint \alpha(\beta, M, J, K, t)\ d\beta.$

As M and J are adiabatic invariants, the evolution of Φ on the time-scale of the precession motion is given by:

(27) $d\Phi/dK = \partial\Phi/\partial K \langle dK/dt \rangle + \partial\Phi/\partial t$

where, as before, <> stands for the average over the bounce motion. In view of (25.2) and (26) we have:

(28) $\partial\Phi/\partial t = \oint \partial\alpha/\partial K\ d\beta = \oint d\beta/(\partial K/\partial\alpha) = c/e \oint d\beta/\langle d\beta/dt\rangle = T_{prec}.$

Likewise we deduce from (25.2) and (25.4):

(29) $\partial\Phi/\partial t = \oint \partial\alpha/\partial t\ d\beta = -\oint (\partial K/\partial t)/(\partial K/\partial\alpha)\ d\beta$

$= -\ c/e \oint \langle dK/dt\rangle/\langle d\beta/dt\rangle\ d\beta = -\ c/e \oint \langle dK/dt\rangle\ dt$

$= -\ cT_{prec}/e \langle\langle dK/dt\rangle\rangle$

where the notation <<>> denotes a double average over the bounce and precession motions. Therefore:

(30) $d\Phi/dt = cT_{prec}/e\,[< dK/dt > - << dK/dt >>]$

and we see that the average of $d\Phi/dt$ over the precession motion vanishes at the order considered. This proves the adiabatic invariance of the flux Φ on a timescale which is slow with respect to the precession motion.

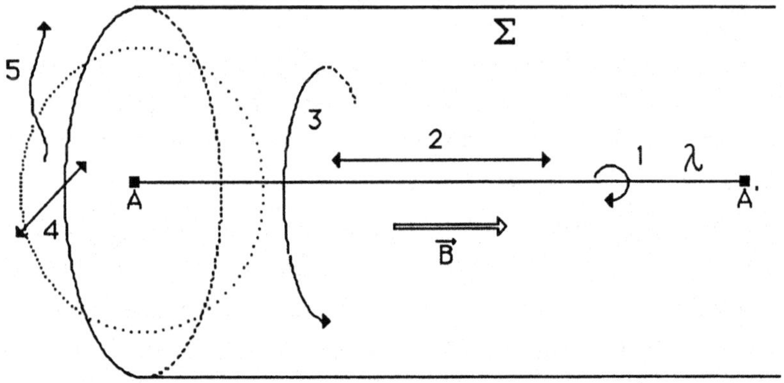

Figure 8.4: A charged particle in a time-varying magnetic field B may undergo motions on different time scales: Rapid Larmor gyration (1) around a field line λ, bounce motion (2) between the turning points A and A', slow precession (3) on the surface Σ across the field lines, slow motion (4) transverse to the magnetic surface Σ and finally a very slow drift motion (5).

To summarize, the motion of a charged particle in the situation just considered involves five different timescales (see Figure 8.4):

- On the fastest timescale, the particle rotates around the magnetic field with the frequency ω_{Larmor}.
- On a slower timescale, the guiding center bounces along a field line with the frequency $\omega_{bounce} \sim \varepsilon\omega_{Larmor}$. This motion preserves the adiabatic invariant M (the magnetic moment).
- On a still slower timescale, the guiding center drifts across the field lines and slowly precesses on a closed magnetic surface with the frequency $\omega_{prec} \sim \varepsilon\omega_{bounce}$. This

transverse motion is described by the equations (25.1) and (25.2) which are obtained by averaging the guiding center motion equations (10) over the bounce motion. The action J is an adiabatic invariant of this precession motion.

- On the next timescale, the guiding center drifts across the magnetic surfaces (because of the time-dependence of the magnetic field). The flux Φ is adiabatically invariant on this timescale.

- Finally, the adiabatic invariance breaks down at large times and the particle slowly drifts in configuration space.

To conclude this discussion, note that the hierarchy of invariants must not be confused with the adiabatic series of a *given* invariant to be discussed in the next section. As regards this last notion, only the second term of the adiabatic series of M has actually been computed (we refer the reader to [No2] for further details on this subject).

Apart from the above specialized computations pertaining to plasma physics, attempts have been made to make the treatement of this example "as canonical as possible", the most elegant one problably being Littlejohn's (see [Litt1-4]). This approach makes use of symplectic forms and is based on the classical Darboux theorem (see §43 of [Ar7]), which may be stated as follows:

Darboux's theorem:

Let ω be a closed non-degenerate two-form on phase space and z an *arbitrary* coordinate system. Then one can find a new coordinate system z' in which ω takes the standard form: $\omega = dz'_1 \wedge dz'_2 +\ldots+ dz'_{2n-1} \wedge dz'_{2n}$. Moreover one of the new coordinates may be chosen at will.

The proof of the theorem is iterative and relies on the repeated use of a lemma which ensures that an adequate change of variables $z \to z'$ brings ω into the form $\omega = dz'_1 \wedge dz'_2 + \omega'$ where $\omega' = \sum_{i\notin\{1,2\},j\notin\{1,2\}} \omega_{i,j}(z')\, dz'_i \wedge dz'_j$ (the matrix $\omega_{i,j}$ of the components of ω in the basis z' is then block-diagonal). Let us now assume that ω is the symplectic form associated with the problem of the motion of a charged particle in an electromagnetic field and the original coordinates are the usual cartesian coordinates (x, y, z) (together with the conjugate momenta). The lemma then allows us to uncouple the fast degree of freedom corresponding to Larmor gyration from the other ones. This is the heart of Littlejohn's method, which we shall now describe.

The four degree of freedom *autonomous* Hamiltonian of the problem reads:

(30) $$H(x, p, t, h) = 1/2\,[p - A(x, t)/\varepsilon]^2 + \Phi(x, t)/\varepsilon - h$$

x stands as usual for (x, y, z). We have set $m = c = 1$ for the sake of simplicity and we write as before $e = 1/\varepsilon$ (see above). Two successive changes of variables will enable us to introduce more physical coordinates:

1) $(x, p, t, h) \rightarrow (x, v, t, k_0)$

where $v = p - A(x, t)$ is the particle's velocity and k_0 its kinetic energy. In these coordinates, the Hamiltonian reads $H(x, v, t, k_0) = v^2/2 - k_0$.

2) $(x, v, t, k_0) \rightarrow (x, t, k, v_{||}, \theta, w)$

where $v_{||} = b.v$ is the parallel velocity of the particle, $w=[(v-u)^2 - u^2]^{1/2}$ the perpendicular velocity in the frame of reference which drifts at the speed $u = cE \times B/B^2$ (see above), $k = k_0 - v.u + u^2/2$ the kinetic energy in this same frame of reference and θ the gyrophase. θ may be defined as follows: $\theta = \psi + \mathrm{Arctg}[\tau_1.(v - u)/\tau_2.(v - u)]$ where $\tau_1(x, t)$ and $\tau_2(x, t)$ are two given orthonormal vectorfields perpendicular to the magnetic field and $\psi(x, t)$ is a scalar field defining the origin of angles in the plane perpendicular to $B(x, t)$. ψ is chosen so that the final expressions will be as compact as possible. In these coordinates, the Hamiltonian reads:

$$H(x, v, t, k_0) = 1/2\,(u^2+v_{||}^2) - k.$$

One may then apply the lemma alluded to above. It yields a new set of coordinates (X, t, K, U, θ, M) and the Hamiltonian takes the form:

$$H(X, t, K, U, \theta, M) = MB + U^2/2 - K + \varepsilon H_1(X, t, K, U, \theta, M).$$

At lowest order, M is the magnetic moment, whereas the variables X, K, U coïncide with x, k, $v_{||}$. The Darboux transform uncouples the fast and slow degrees of freedom. Indeed the Poisson brackets of the coordinates are then:

(32) $$\{Z, \theta\} = 0, \quad \{Z, M\} = 0, \quad \{\theta, M\} = 1/\varepsilon$$

where the notation Z designates any coordinate other than M and θ (in other words the Poisson tensor is block-diagonal).

However the variables (M, X, K, U) still contain rapidly oscillating terms (at a higher order than $(x, k, v_{||})$). X for instance is the guiding center position only at lowest order and displays order ε oscillations. One must perform a new change of variables to get rid of these oscillations. This "averaging transform", which is symplectic, yields the new variables $(X', t, K', U', \theta, M')$.

In these variables the Hamiltonian reads:

$$H'(X', t, K', U', \theta, M') = M'B(X', t) + U'^2/2 - K' + \\ + \varepsilon H'_1(X', t, K', U', \theta, M').$$

Actually one may perform a final transformation:

$(X', t, K', U', \theta, M') \to (X'', t, K'', U'', \theta, M')$ to eliminate the first order part in the Hamiltonian which then becomes:

$$H'(X'', t, K'', U'', \theta, M') = M'B + U''^2/2 - K'' + O(\varepsilon^2).$$

This yields a precise definition of the guiding center at order ε. However there is a price to pay: The symplectic two-form no longer has block-diagonal components.

To conclude this section, we note that, though to our knowledge it has not been done, Littlejohn's method could a priori be used to prove the adiabatic invariance of J and Φ.

8.2 Adiabatic series

In the last section we proved the elementary but fundamental result of the adiabatic invariance of the action for one dimensional Hamiltonian systems on the timescale $1/\varepsilon$. We will now introduce higher order invariants I_k ($I = I_0$) to improve this estimate. We will work with a Hamiltonian $H(p, q, \tau)$ which is smooth. (C^∞) in the canonical variables (p, q) and of class C^{k+1+} ($k \geqslant 0$) in the slow time variable $\tau = \varepsilon t$ ($0 \leqslant \tau \leqslant 1$). As in the elementary adiabatic theorem of last section, we consider a region D of phase space which is a union of trajectories, where action-angle variables for the frozen (τ fixed) systems exist, and which keeps away from the separatrices of these systems, so that the frequency is bounded away from zero. Then:

Theorem 1:

There exists a kth order invariant $I_k(p, q)$ such that for any trajectory with initial condition in $D - \delta$ ($\delta > 0$), one has:

(1) $\quad |I - I_k| = O(\varepsilon)$, $\sup_{t \in [0, 1/\varepsilon]} |I_k(t) - I_k(0)| \leqslant c_k \varepsilon^{k+1} \quad c_k > 0.$

The construction runs as follows; beginning with the Hamiltonian $H(p, q, \tau)$, one first transforms to action-angle variables for any fixed τ and so obtains the new

Hamiltonian:

$$(2) \qquad H(I, \varphi, \tau) = H_0(I, \tau) + \varepsilon H_1(I, \varphi, \tau)$$

which is still denoted by H for the sake of simplicity (see Appendix 8 and formula (2) in Section 1 above). One is then faced with the system of equations (3) of last section ($H_1 = \partial S/\partial\tau$), which can be considered as a standard system with *one* phase φ and *two* action (slow) variables I and τ, so that one can use the scheme devised in Section 3.2 for these systems. Although we did not state there what precise regularity assumptions are necessary to go to a given order, the mixed conditions we have adopted above (and which are rather natural in the present context) do allow for the above stated result. In fact, $H(I, \varphi, \tau)$ is of class $C^{k+}([0, 1], C^\infty(K \times T^1))$, i.e., it is C^∞ in $(I, \varphi) \in K \times T^1$ (where K is the domain corresponding to D in action space) and C^{k+} with respect to τ. The reader may check, looking through Sections 3.1 and 3.2, that this permits to derive an order $(k + 1)$ result.

Because the system is Hamiltonian, the vector field of the averaged system for the slow variables which are constructed recursively vanishes to a certain order (the equation for τ remains of course unchanged), and one can construct a transformation $(I, \varphi) \rightarrow (I_k, \varphi_k)$ such that:

$$(3) \qquad I_k = I + \varepsilon I^{(1)} + \varepsilon^2 I^{(2)} + \dots + \varepsilon^k I^{(k)},$$

$$dI_k/dt = \varepsilon^{k+1} X'_{k+1}(I_k, \tau, \varphi_k) + O(\varepsilon^{k+2})$$

where X'_{k+1} has zero average with respect to the phase. Using lemma 1' of Section 3.1 leads to the estimate (1). ■

We repeat here what we already noted in Chapter 7, namely that the sequence $\{I_k\}$ is generally asymptotic, i.e., the constants c_k in the statement of the theorem may rapidly increase with k. In particular, for a given $\rho > 0$ only finitely many invariants will be constant within ρ on the timescale $1/\varepsilon$.

We can easily extend the timescale in the following way:

Amplification:

Suppose that $H(I, \varphi, \tau) \in C^{k+1}(\mathbb{R}^+, C^\infty(K \times T^1))$, and that all its derivatives, up to order $k + 1$ are *uniformly* bounded on $K \times T^1 \times \mathbb{R}^+$; then one can choose the invariant I_k such that:

$$(4) \qquad |I - I_k| = O(\varepsilon), \quad dI_k/dt = O(\varepsilon^{k+2})$$

these estimates being *uniform* for $\tau \geqslant 0$ (or equivalently $t \geqslant 0$). In particular:

$$(5) \qquad \sup_{t\in[0,\ 1/\varepsilon^r]} | I_k(t) - I_k(0) | = O(\varepsilon^{k-r+2}), \quad 0 \leqslant r \leqslant k+1.$$

This is easily proved by inspection. Here and above, we have identified the invariants with what we called the intermediate variables (denoted as P) in Chapter 3, because anyway, I_k is only defined within $O(\varepsilon^{k+1})$, although for the first invariant, the choice of the action itself is of course most natural. Also, the regularity condition has been made slightly more stringent as, working on a longer timescale, one cannot use lemma 1' of Section 3.1 to gain one order of differentiability. We are here in the simplest case allowing for timescale extension, namely that where the averaged vector field for the slow variable (except again that for τ) vanishes; the only peculiarity of the present situation is that τ being itself one of the variables, the domains cannot be bounded as soon as soon as $t >> 1/\varepsilon$, so that one needs *uniform* estimates on the quantities involved. ■

Let us notice that the case where the Hamiltonian is periodic with respect to τ and that where it is asymptotically autonomous (independent of τ) are two important instances in which uniform estimates do hold true.

Although the above proof is simple and reduces the problem to the case of standard systems with one phase, it is not very elegant, since it almost leaves aside the fact that we are dealing with canonical equations, and uses it only to notice that the averaged system is trivial at any order. It is of course desirable to devise a purely Hamiltonian scheme, this being also useful for the effective computation of the higher order invariants, and we now proceed to decribe it.

Starting from the Hamiltonian (2); one performs as usal a canonical transformation, using e.g. Lie series, with a generating auxiliary Hamiltonian χ according to the formulae (40) of Appendix 7 (in this Appendix the meaning of the letter τ is different from what it is here). To obtain the kth invariant I_k , one can restrict oneself to the first k terms and this leads to a well-defined transformation $(I, \varphi) \to (I_k , \varphi_k)$ which brings the Hamiltonian into the form:

$$(6) \qquad H^{(k)}(I_k , \varphi_k , \tau) = H_0(I_k , \tau) + \varepsilon H_1(I_k , \tau) + \ldots + \varepsilon^k H_{k-1}(I_k , \tau) + \\ + \varepsilon^{k+1} R_k(I_k , \varphi_k , \tau, \varepsilon)$$

with a bounded remainder R_k if I_k , φ_k and τ also remain bounded. By writing Hamilton's equations and applying the first averaging theorem of Chapter 3, one immediately shows the validity of estimate (1) for I_k . The extension of the timescale, as stated in the amplification, requires one more step in the algorithm.

It remains to determine the χ_n's (see formula (40) in Appendix 7) and the expression of I_k as a function of the original variables. We shall work formally and obtain formulas which are valid at any finite order. The series which determines χ is generated as in Appendix 7, whose formula (44) holds within a slight modification; in fact, we have:

$$(7) \qquad \varepsilon\partial\chi/\partial\tau + \{\chi, H'\} = \partial H'/\partial\varepsilon - T\partial H/\partial\varepsilon$$

in which H' is the new Hamiltonian, obtained after the transformation T generated by χ. Because of the slow time variation, refering to Appendix 7, we need only replace $\partial\chi/\partial t$ by $\varepsilon\partial\chi/\partial\tau$ on the l.h.s.. To order n we obtain, instead of equation (45) in Appendix 7, the following formula:

$$(8) \qquad \{\chi_n , H_0\} = n(H'_n - H_n) - F_n - \partial\chi_{n-1}/\partial\tau, \quad \chi_0 = 0$$

with F_n defined by formula (46) in Appendix 7. Let us write down the first three formulas, using (47) in the same Appendix:

$$\{\chi_1 , H_0\} = H'_1 - H_1$$

$$(9) \qquad \{\chi_2 , H_0\} = 2(H'_2 - H_2) - L_1(H'_1 + H_1) - \partial\chi_1/\partial\tau$$

$$\{\chi_3 , H_0\} = 3(H'_3 - H_3) - L_1(H'_2 + 2H_2) - L_2(H'_1 + 1/2\, H_1) -$$

$$- 1/2\,(L_1)^2 H_1 - \partial\chi_2/\partial\tau.$$

The system (8) is solved order by order, chosing the H'_n so as to eliminate the secularities; the new action variable I' is found as $I' = T^{-1}I$ with the series for T given in formula (42) of the appendix where it is also shown how the series for T^{-1} is immediately constructed from the series for T (because the transformations are unitary).

We shall give as an example the computation of the second order invariant for a harmonic oscillator with slowly varying frequency. To this end we return to formulas (5) in Section 1, keeping in mind that the computation is expressed in the new variables (see Appendix 7); in fact as is pointed out in this appendix, the variables can be treated as dummy variables, provided the final result is correctly interpreted. We shall denote (I_2 , φ_2) by (J, ψ) for the sake of clarity, and for any function χ, we have $\{\chi, H_0\} = \omega(\tau)\partial\chi/\partial\psi$, provided that $\omega(\tau) \geqslant \omega_0 > 0$. Since $H_1 = \omega'/(2\omega)\, J \sin(2\psi)$ has zero mean value (with respect to ψ), we can set $H'_1 = 0$ in the first of equations (9), which gives:

$$(10) \qquad \chi_1(J, \psi, \tau) = \omega'/(4\omega^2)\, J \cos(2\psi)$$

$$\{\chi_1 , H_1\} = - \omega'^2/(4\omega^3)\, J.$$

In the second equation (9), $H'_1 = H_2 = 0$ and $\partial\chi_1/\partial\tau$ has zero mean value, whereas $L_1(H_1) = \{\chi_1 , H_1\}$ is constant in J; the nonsecularity condition imposes:

$$(11) \qquad H'_2 = 1/2\, \{\chi_1 , H_1\} = - \omega'^2/(8\omega^3)\, J$$

which makes it possible to write down the new Hamiltonian to second order in the new variables:

$$(12) \qquad H' = (\omega - \varepsilon^2\omega'^2/(8\omega^3))J + O(\varepsilon^3)$$

and this is independent of ψ as it should be. Notice that this is again an oscillator with slowly varying frequency which, however, has been renormalized and, due to the linearity of the equations, this is in fact true at any order. Working "to second order" means setting $\chi_n = 0$ $(n > 2)$ in equations (8), which imposes the value of H'_n. Once H'_2 has been determined, χ_2 follows from equations (9):

$$(13) \qquad \chi_2(J, \psi, \tau) = - 1/(8\omega)\, (\omega'/\omega^2)'\, J \sin(2\psi).$$

It remains to compute J as a function of (I, φ) as indicated above; to second order:

$$(14) \qquad J = T^{-1}I = I - \varepsilon\, \{\chi_1 , I\} - \varepsilon^2/2\, \{\chi_2 , I\} + \varepsilon^2/2\, \{\chi_1 , \{\chi_1 , I\}\} + O(\varepsilon^3)$$

$$= I + \varepsilon\omega'/(2\omega^2)\, I \sin(2\varphi) + \varepsilon^2/(4\omega)\, (\omega'/\omega^2)'\, I \sin(2\varphi) +$$

$$+ \varepsilon^2/8\, (\omega'/\omega^2)^2\, I + O(\varepsilon^3).$$

At this point of the book, it should come as no surprise that a result with an exponential estimate can be proved in the analytic case. We state it as

Theorem 2:

Let the Hamiltonian $H(I, \varphi, \tau)$ be analytic on the domain

$$K_{\rho,\sigma} \times \{\tau \in \mathbb{C}, |\operatorname{Im} \tau| < \zeta\}, \ \rho > 0, \ \sigma > 0, \ \zeta > 0,$$

and bounded on its closure. Then, for $I_0 = I(0) \in K$,

$$(15) \qquad \sup_{|t|<T} |I(t) - I(0)| = O(\varepsilon), \ T = O(e^{c/\varepsilon}), \ c > 0.$$

Again this may be considered as a particular case of Theorem 2 in Section 3.2., but we sketch however a purely Hamiltonian proof which consists of nothing but a much simplified version of the iteration procedure which forms the analytic part of the proof of Nekhoroshev's theorem (see Chapter 7 and [Nei6]). By now the reader should have no trouble in writing the general, non-Hamiltonian proof of Theorem 2 (Section 3.2), which as we mentioned there, can be found in [Nei7].

The idea consists, as usual, in constructing (recursively) the invariants I_k with $k = k(\varepsilon)$ tending to infinity as ε tends to zero. One performs canonical transformations which eliminate the angle variable from the Hamiltonian order by order, while gradually shrinking the analyticity domain. Because the dimension is one and the frequency is bounded away from zero, no small denominators can arise, which permits a larger number of normalizing transformations than in the general version of the theorem: More precisely this number may be taken to be of order $1/\varepsilon$ rather than $1/\varepsilon^\alpha$ $(0 < \alpha < 1)$ as in Chapter 7. As in the general version, every transformation divides the size of the nonintegrable part of the perturbation by a constant factor c_1 $(c_1 > 1)$ so that the angle is finally relegated to a remainder of order $(1/c_1)^{c_2/\varepsilon} \sim e^{-c/\varepsilon}$

The scheme thus runs as follows. Start from $H = H_0 + \varepsilon H_1$ and suppose (which is not restrictive) that $\| H_1 \| = 1$ (sup norm) on the analyticity domain defined above. After k steps one finds the Hamiltonian (6) and the following facts are checked recursively:

- $H^{(k)}$ is well-defined on the closure and bounded on the domain

$$| I_k - I_0 | \leqslant \rho/2\,(1 + c_3 2^{-k}), \quad | \mathrm{Im}\, \tau | \leqslant \zeta_k = \zeta - c_4 k\varepsilon/\omega_0 ,$$

$$| \mathrm{Im}\, \varphi_k | \leqslant \sigma/2\,(1 + c_5 2^{-k})$$

(one can work locally in the action variable) where in the determination of ζ_k we have stressed the role of the smallest intervening denominator, namely ω_0, defined as the infimum of $| \omega(I,\tau) | = | \partial H_0/\partial I |$.

- On this domain, $| R_k | \leqslant (1/c_1)^k$ $(c_1 > 1$; one may chose $c_1 = 4)$.

It is possible in this way to perform $N = [c_2/\varepsilon]$ canonical transformations, with e.g. $c_2 = \zeta\omega_0/2c_4$, so that in the end $| H_N | \leqslant (1/c_1)^{c_2/\varepsilon}$. This proves the theorem, and also that the constant c in its statement can be taken proportional to ζ, the width of the strip on which the Hamiltonian can be analytically continued in the time variable. ▪

Theorems 1 and 2 do not make any difference between linear and nonlinear systems; moreover, the extension of the timescale requires uniform estimates in time but there is no need to distinguish between the various systems which may satisfy these estimates, especially between periodic (in τ) and asymptotically autonomous systems. Thirdly, no small denominators have shown up in the above proofs and computations.

We would like to comment briefly on these three related facts.

First of all the rescaled time variable is treated rather differently in the non canonical and in the canonical schemes. In the former, it definitely appears as an *action* (slow) variable,.whereas in the latter, it is treated more as an *angle* variable, with conjugate action εH. The "small denominators" never arise in this scheme simply because in a way they are always pushed up to higher orders, as a result of the ordering that has been chosen, which can be called the "natural adiabatic ordering". In fact, in the case where the Hamiltonian is τ-periodic (with period 2π), forgetting about the ordering, one should solve equations like:

$$(16) \qquad \varepsilon \partial\chi/\partial\tau + \{\chi, H_0\} = F$$

with F a 2π periodic function in τ with zero mean value. F can be expanded as:

$$(17) \qquad F(I, \varphi, \tau) = \sum_{(l,m)\neq(0,0)} F_{l,m}(I)e^{i(l\,\varphi+m\tau)}$$

and the nonsecular solution of (16) reads:

$$(18) \qquad \chi(I, \varphi, \tau) = \sum_{(l,m)\neq(0,0)} - iF_{l,m}(I)[l\omega(I, \tau) + \varepsilon m]^{-1}e^{i(l\,\varphi+m\tau)}$$

In the above expression the intervening denominators can be small only when the ratio m/l is large, which corresponds to a *resonance* between the local frequency $\omega/2\pi$ and the high order harmonics of the slow frequency ($= \varepsilon/2\pi$). More precisely, $|l| + |m|$ is then at least on the order of $1/\varepsilon$ and, according to the regularity of F, this yields an estimate on the Fourier coefficient $F_{l,m}$. This explains why we could completely disregard the effects of this possible resonance and prove results on timescales too short for it to become important. We shall presently see that, for linear systems, it does play a crucial role, giving rise to the phenomenon called parametric resonance, which proves that the results of the present section are in fact (almost) optimal and precludes *perpetual* adiabatic invariance for linear periodic systems. For truly nonlinear sytems, on the contrary, a slightly elaborate version of the KAM theorem, to be given in Section 6, allows for a proof of adiabatic invariance for all times. These systems being nonisochronous, a change in I will produce a change in ω, which introduces a detuning and hinders the development of parametric resonance and the corresponding divergence of the series. However, as we shall see, this intuition is not so easily turned into mathematics.

8.3 The harmonic oscillator; adiabatic invariance and parametric resonance

The last section was devoted to the construction of higher order invariants, and we stressed the connection between the regularity of the system and the timescale on which these invariants are valid, regardless of the fact that the system is linear or not. Here we explore in some sense the optimality of these results, through a more detailed study of the linear case.

The equations under investigation read:

$$(1) \qquad d^2y/dt^2 + \omega^2(1+\eta f(\varepsilon t))y = 0$$

where ε and η are parameters, f is periodic with period T, of zero mean value and with unit (sup) norm. We assume that $0 \leqslant \eta < 1$, so that the term between brackets does not vanish; we are thus dealing with the equation of a modulated harmonic oscillator with average pulsation ω. To introduce parametric resonance, we set $\varepsilon = 1$ and consider the Poincaré map P, defined as:

$$(2) \qquad P(y_0, y_1) =_{def} (y(T), y'(T))$$

where $y(t)$ is the solution of (1) with initial conditions (y_0, y_1). P is a *linear* map of the phase plane to itself, which preserves area because the system is Hamiltonian, hence divergence free, and also orientation, because it is defined as the time T map of a flow. This entails that $\det P = 1$ and the spectrum of P is a function of the trace only. More precisely:

- If $|\operatorname{Tr} P| < 2$, one can set $\operatorname{Tr} P =_{def} 2\cos\theta$ and the eigenvalues are $\lambda_\pm = e^{\pm i\theta}$.
- If $|\operatorname{Tr} P| = 2$, P may be either diagonalizable, in which case $P = \pm 1$ and all solutions are periodic (with period T if $P = 1$ and $2T$ if $P = -1$) or it may be cast into a reduced Jordan form (only one of the solutions is periodic).
- If $|\operatorname{Tr} P| > 2$ (say $\operatorname{Tr} P > 0$) then $\operatorname{Tr} P =_{def} 2\operatorname{ch}\sigma$ and the eigenvalues read $\lambda_\pm = e^{\pm\sigma}$.

In the first case the origin is an elliptic fixed point; it is hyperbolic in the third case (with a reflection if $\operatorname{Tr} P < 0$); the second case determines a (linear) sheared flow if $P \neq \pm 1$. The system will be termed unstable if there exists a solution y(t) such that:

$$\sup \lim_{t\to\infty} |y(t)| = \infty.$$

Otherwise it will be called stable. It is obvious that:

- If $|\operatorname{Tr} P| < 2$ the system is stable and so is any sufficiently close (area preserving) system; it will then be said to be strongly (or structurally) stable.
- If $|\operatorname{Tr} P| = 2$ the system is stable (but not strongly stable) if and only if $P = \pm 1$.
- If $|\operatorname{Tr} P| > 2$ the system is strongly unstable.

One easily computes P and $\operatorname{Tr} P$ for the unperturbed system, that is, when $\eta = 0$;

one finds $\mathrm{Tr}\, P = 2\cos(\omega T)$, so that the system is strongly stable (in particular it will be stable for η small enough) except when $T = n\pi/\omega = n/2.T_0$ where n is an integer and $T_0 = 2\pi/\omega$ is the period of the unperturbed motion. This corresponds to the familiar fact that a swing can be set in motion only by a force whose period is an integer multiple of half the proper period.

For a given function f, one can draw the stability diagram as a function of T/T_0 and η, the boundary of the instability domains ("Arnold's tongues") being given by the equation $|\,\mathrm{Tr}\, P\,| = 2$. A very schematic representation of the result can be seen on Figure 8.5. When the parameters lie inside a hatched region, one of the eigenvalues of P has modulus larger than unity, implying the existence of solutions with arbitrarily small initial conditions (because of the linearity) which increase to infinity (along with their derivatives).

We shall need some more notation: For a fixed value value of η and a nonzero integer n, we denote by $\Omega_n = [\omega_{2n-1}, \omega_{2n}]$ (resp. $\Omega'_n = [\omega'_{2n-1}, \omega'_{2n}]$) the interval of instability which shrinks to the point $\omega = n2\pi/T$ (resp. $\omega = (n - 1/2)2\pi/T$) as η tends to zero (see Figure 8.5). Let us notice parenthetically that these intervals have received different names according to the domains of physics in which they arise, such as instability intervals, zones, bands, gaps; in solid state physics the names of Brillouin and Bloch are attached to them or closely related quantities. We shall denote by $\delta_n = \omega_{2n} - \omega_{2n-1}$ (resp. $\delta'_n = \omega'_{2n} - \omega'_{2n-1}$) the length of Ω_n (resp. Ω'_n). These are functions of η, and for special functions f, they may vanish for some n and nonzero η (the corresponding interval then collapses to a point). Finally we set $\mathrm{Tr}\, P = \Delta(\omega)$; it is also a function of the parameter η and is called the *discriminant* of equation (1). One has $\Delta(\omega) \geqslant 2$ for $\omega \in \Omega_n$ (resp. $\Delta(\omega) \leqslant -2$ for $\omega \in \Omega'_n$).

The following remarks are important and partly supported by physical intuition:

- As n increases the instability intervals get ever narrower as a function of η, near the axis $\eta = 0$. If in particular f is a trigonometric polynomial of order s, δ_n and δ'_n decrease as $\eta^{[2n/s]}$ (see [Ar9] and [Mag] Section 5.6).
- The linear model usually ceases to be physically relevant for very large solutions and the larger n, the more easily parametric resonance is swamped by the addition of small dissipative terms.
- Third, for fixed η and increasing n, δ_n and δ'_n decrease, together with the instability rate σ (defined as above by $|\,\Delta(\omega)\,| = 2\,\mathrm{ch}\sigma$ inside an instability interval).

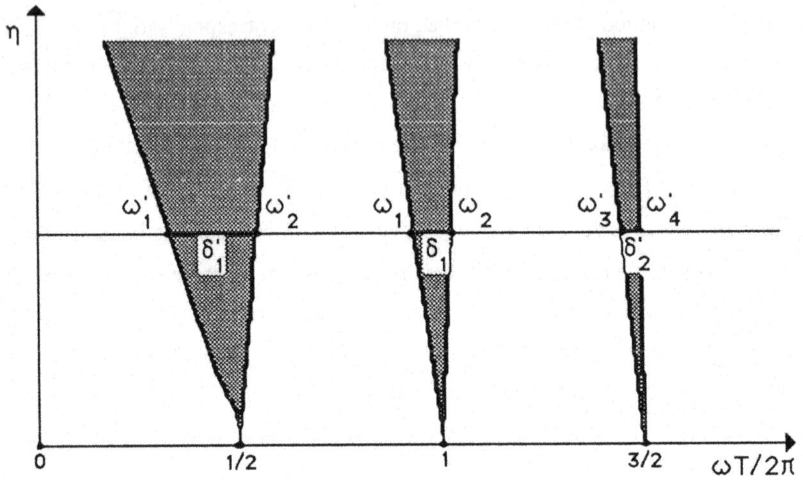

Figure 8.5: Schematic representation of instability regions in parameter space. $\omega T/2\pi$ is the ratio T/T_0 of the modulation period T to the proper period $T_0 = 2\pi/\omega$, while η denotes the modulation amplitude.

The limit of validity of adiabatic invariance for linear systems (and thus the optimality of the results obtained in the last section) is intimately connected with this last property, and we will describe it more precisely below, as it has been the object of much study. But let us first explain the connection with adiabatic invariance.

Let ε be again arbitrary in (1). Using $\tau = \varepsilon t$ as the new (rescaled) time variable, one finds, looking at equation (1), that this is equivalent to changing ω into ω/ε, considering the Poincaré map at time $\tau = T$ ($t = T/\varepsilon$). The bifurcation diagram still looks as on Figure 8.5. Keeping η fixed, let now ε tend to zero; the point representative of the system will move on a horizontal line to the right, and whatever small ε may be, it will sometimes lie in an instability interval. More precisely, this happens for $\omega T/(2\pi\varepsilon)$ close to $m/2$ ($m \in \mathbb{N}$), i.e., $\varepsilon \sim c/m$ with $c = \omega T/\pi = 2T/T_0$.

Working in the original time variable t, we denote by $\theta(\varepsilon)$ the maximum time until which adiabatic invariance holds. The unstable eigensolution $y(t)$ at time $t = \theta(\varepsilon) >> 1/\varepsilon$ has been multiplied by a factor $\pm \exp(\sigma\varepsilon\theta(\varepsilon)/T)$ (+ for m even, - for m odd) and so has been its derivative; the adiabatic invariant I, being quadratic in y and y', has been multiplied by the square of this same factor. Taking I to be of order unity, adiabatic invariance will break precisely when:

$$\exp(2\sigma\varepsilon\theta(\varepsilon)/T) - 1 = O(\varepsilon)$$

which is the same as $\theta(\varepsilon) = O(1/\sigma(\varepsilon))$.

This is the criterion we were aiming at, and it is rigorous. The meaning of $\sigma(\varepsilon)$ should be clear: One has $m \sim c/\varepsilon$, $m \in \mathbb{N}$, and we set:

$$\sigma(\varepsilon) =_{def} \sup_{\omega\in\Omega_n} \sigma(\omega), \quad \text{if } m = 2n,$$

$$\sigma(\varepsilon) =_{def} \sup_{\omega\in\Omega'_n} \sigma(\omega) \quad \text{if } m = 2n - 1.$$

It now only remains to study the maximum instability rate as a function of m, and this is connected with the asymptotic width of the instability intervals. To this end we return to equation (1) with $\varepsilon = 1$ and f at least of class C^2, and we rewrite it as:

(1') $\qquad d^2y/dt^2 + \omega^2p^2(t)y = 0, \quad p^2(t) =_{def} 1 + \eta f(t).$

It can now be transformed into Hill's equation:

(3) $\qquad d^2z/dx^2 + [\lambda\gamma^2 + q(x)]z = 0,$

via the so-called Liouville's transformation (see [Mag] Section 3.2):

(4) $\qquad x = \gamma^{-1}\int_0^t p(s)\,ds, \quad z(x) = p^{1/2}(t)y(t),$

$$q(x) = p^{-1/2}d^2/dx^2(p^{1/2}) = -\gamma^2p^{-3/2}d^2/dt^2(p^{-1/2}).$$

$q(x)$ is periodic with period π, $\lambda = \omega^2$ and γ is the unimportant constant:

$$\gamma =_{def} 1/\pi \int_0^T p(s)\,ds.$$

Hill's equation (3) (with $\gamma = 1$), which is an eigenvalue problem in standard form, has been extensively studied, starting from the work of Hill (or even before, see [Mag]). There has recently been a revival of interest in its properties, when it was recognized that isospectral deformations are connected with the integrability of the so-called Korteweg-de Vries hierarchy with periodic boundary value conditions (see [McK1-2]).

We shall only need some results from the classical theory: The instability intervals are defined as above, mutatis mutandis, and their lengths decrease with n like the Fourier coefficients of the potential $q(x)$ (this is of course no coincidence): Actually if $q \in C^k((0, \pi))$, they are of size $o(n^{-k})$ (cf. [Hoc]). Taking into account the fact that $\lambda = \omega^2$, this implies that δ_n and δ'_n are of size $o(n^{-k+1})$. However, for q to be of class C^k, one p must belong to $C^{k+2}((0, T))$, because the Liouville's transformation (4) uses up two derivatives.

We wish to estimate $\sup_{\omega\in\Omega_n} \sigma(\omega)$ (the case of Ω'_n being identical). Now

$\sigma^2(\omega) \sim \Delta(\omega) - 2$ and $\Delta(\omega) \sim 2\cos(\omega T)$, together with its derivatives, so that:

$$\Delta''(\omega) \sim -2T^2$$

when ω is close to n. This implies:

$$\sup_{\omega\in\Omega_n} \Delta(\omega) - 2 \sim -1/2\, \Delta''(\delta_n/2)^2 \sim T/4\, \delta_n{}^2.$$

Hence:

$$\sup_{\omega\in\Omega_n} \sigma(\omega) \sim \sqrt{T/2}\, \delta_n = c\delta_n\, .$$

Let us now look back at the Amplification of Theorem 1 in the last section. To ensure that $H(I, \varphi, \tau) \in C^{k+1}(\mathbb{R}^+, C^\infty(K \times T^1))$, f must belong to $C^{k+2}((0, T))$ in equation (1) (or equivalently $p \in C^{k+2}((0, T))$ in equation (1')) because one loses one derivative by passing from the coordinates (y, y') to the action-angle variables (I, φ) (the explicit formulas for the harmonic oscillator appear in (5) of Section 8.1). $q(x)$ is then in $C^k((0, \pi))$, which from the above implies that $\delta_n = o(n^{-k+1})$, $\sup_{\omega\in\Omega_n} \sigma(\omega)$ being of the same order. If we finally use the connection between $\theta(\varepsilon)$ and the instability rate σ, we find that $\theta(\varepsilon) = O(1/\varepsilon^{k+1})$ which is the prediction of the Amplification of Theorem 1 (Section 8.2). In as much as the estimates on the length of the instability intervals are known to be optimal, this shows the optimality of the estimate (5) in Section 8.2 (we have really treated only the case $r = k + 1$ but the case of a higher precision on a shorter timescale is obviously equivalent).

The analytic case is analogous: If f can be analytically continued to the strip $|\operatorname{Im} t| < \zeta$ and is bounded on its closure, δ_n and δ'_n will be of size $O(e^{-cn\zeta})$ like the Fourier coefficients of f, and the optimality of this estimate demonstrates the optimality of Theorem 2 of the last section.

To end this section, we sketch another method to construct the higher order adiabatic invariants in the linear case. This is connected with the well-known WKB method (which plays a prominent role in the next section) and is due to T.J. Shep, whose original paper ([Sh]) we follow here.

The equation of the oscillator is rewritten in the form:

$$(5) \qquad d^2y/dt^2 + \omega^2(\lambda)y = 0$$

where for purely typographical reasons we have set $\lambda = \tau = \varepsilon t$ and the C^∞ function $\omega(\lambda)$ is still bounded from above and below. Now to any strictly positive smooth function $F(\lambda, \varepsilon)$, we can associate a Liouville transformation $(Y, T, \Lambda) = \Phi(y, t, \lambda)$ through the formulas (which are a slight rewriting of formulas (4)):

(6) $$Y = yF^{1/2}(\lambda, \varepsilon); \quad T = \int_0^t F(\varepsilon s, \varepsilon)\, ds; \quad \Lambda = \int_0^\lambda F(\mu, \varepsilon)\, d\mu.$$

One then checks that in these new variables equation (5) for the oscillator becomes:

(7) $$d^2 Y/dT^2 + \Omega^2(\Lambda, \varepsilon)Y = 0$$

where $\Lambda = \varepsilon T$ and Ω is given as:

(8) $$\Omega^2 = \omega^2 F^{-2} + \varepsilon^2 F^{-3/2}(d^2 F^{-1/2}/d\lambda^2).$$

In particular, if F solves the following equation:

(9) $$\varepsilon^2(d^2 F^{-1/2}/d\lambda^2) - \omega^2 F^{-1/2} - F^{3/2} = 0$$

$\Omega = 1$ and the oscillator admits the exact invariant $I = 1/2\, [Y^2 + (dY/dT)^2]$ which in terms of the original variables reads:

(10) $$I(y, t, \varepsilon) = y^2 F + [F^{-1/2} dy/dt + \varepsilon y F^{-1} dF^{1/2}/d\lambda]^2.$$

Although equation (9) is not any simpler to solve than the original linear equation (5), it sometimes lends itself more easily to perturbation expansions and the time-dependent invariant (10) has been put to use to some extent in the practical study of equation (5) (cf. [Lew] and [Sy]). In fact the solution F of equation (9) is nothing but the phase of the exact WKB solution of (5), which, with two arbirary constants α and β reads:

(11) $$y(t) = \alpha F^{-1/2} \exp(i\int_0^t F(\varepsilon s, \varepsilon)\, ds) + \beta F^{-1/2} \exp(-i\int_0^t F(\varepsilon s, \varepsilon)\, ds)$$
$$= \alpha F^{-1/2} e^{iT} + \beta F^{-1/2} e^{-iT}.$$

Shep's idea consists in iterating the Liouville transform defined by equations (6) and (8) starting from the initial values $\omega_1 = \omega$, $F_1 = \omega$. Rewriting equation (8) slightly, one is led to the following recursive formulas:

(12) $$\omega_n^2 = \omega_{n-1}^2 F_{n-1}^{-2} + \varepsilon^2 F_{n-1}^{-3/2}(d^2 F_{n-1}^{-1/2}/d\lambda^2)$$
$$F_n^2 = \omega_{n-1}^2 + \varepsilon^2 F_{n-1}^{1/2}(d^2 F_{n-1}^{-1/2}/d\lambda^2).$$

The product of the transformations $\Phi_1, \ldots, \Phi_n$ defines the variables (y_n, t_n, λ_n) and one can show that $(\omega_n^2 - 1)$ is of order ε^{2n}. The oscillator with frequency ω_n has the action:

$$I^{(n)} = 1/(2\omega_n)\, [\omega_n y_n^2 + (dy_n/dt_n)^2]$$

as an adiabatic invariant. By reexpressing $I^{(n)}$ in terms of the original variables, one obtains an invariant of order $(2n - 1)$ (with a deviation of order ε^{2n} over a time interval $1/\varepsilon$). These invariants are of course equivalent in this (linear) case to those we found for a general Hamiltonian system in the previous section by using the scheme of Section 3.2 or the method of Lie series.

8.4 The harmonic oscillator; drift of the action

We again consider a harmonic oscillator obeying the equation:

$$d^2y/dt + \omega^2(\varepsilon t)y = 0,$$

whose frequency ω is a *real analytic* function which can be continued as a meromorphic function to the complex plane (or rather a strip around the real axis) and is bounded from below by $\omega_0 > 0$ on the real axis. We shall assume in this section that the limits $\omega_\pm = \lim_{\tau\to\pm\infty} \omega(\tau)$ exist and we shall estimate, using the WKB method, the total drift of the action $\Delta I = I(+\infty) - I(-\infty)$ along a given trajectory (but not the difference $| I(t) - I(0) |$ for any t). In the next section, we shall examine the same quantity for generic (nonlinear) systems.

As t goes to $\pm\infty$, $y(t)$ - which is real on the real axis - is equivalent to functions of the form:

$$(1) \qquad y_\pm = c_\pm e^{i\omega_\pm t} + c_\pm^* e^{-i\omega_\pm t}.$$

This holds true assuming a sufficiently fast decrease of $| \omega - \omega_\pm |$ at infinity. Precisely, we assume that $\omega(\tau)$ satisfies the following conditions:

$$(2) \qquad \int^{+\infty} | \omega(\tau) - \omega_+ | d\tau < \infty, \quad \int_{-\infty} | \omega(\tau) - \omega_- | d\tau < \infty,$$

$$\int_{-\infty}^{+\infty} (\omega'^2 + | \omega'' |)\, d\tau < \infty$$

in which case (1) holds and can be differentiated once. The associated asymptotic values of the action are $I_\pm = 1/2\ \omega_\pm | c_\pm |^2$ and we wish to evaluate:

$\Delta I = I_+ - I_- = 1/2\ (\omega_+ | c_+ |^2 - \omega_- | c_- |^2)$. Now the equation of the oscillator is identical to the stationary one-dimensional Schrödinger equation; switching to more familiar notation in this context, it reads:

$$(3) \qquad -\varepsilon^2 d^2\psi/dx^2 - \omega^2(x)\psi = 0$$

with $x = \tau = \varepsilon t$ and ψ instead of y. ψ is the zero energy solution of the equation with the negative potential $V = -\omega^2$, whereas the small parameter ε plays the role of Planck's reduced constant $\hbar$.

We are thus brought back to the quantum reflection problem above a potential barrier and the exponential smallness of the drift will be equivalent to that of the corresponding reflection coefficient in the semiclassical limit $\hbar \to 0$. Let us first make more precise the correspondence between adiabatic invariance and semiclassical scattering theory, this being perhaps the most important point of the present section (cf. for example [La1]).

One first defines the Jost solution ψ_0 by requiring the following asymptotic forms:

(4) $$\psi_0(x) \sim_{x\to-\infty} e^{i\omega_- x/\varepsilon} + Re^{-i\omega_- x/\varepsilon}$$

$$\psi_0(x) \sim_{x\to+\infty} Te^{i\omega_+ x/\varepsilon}.$$

R and T are defined to be the reflection and transmission coefficients, and the equivalences (4) can be differentiated once. The Schrödinger equation thus admits a one parameter family of *real* (for real x) solutions whose asymptotic expressions are:

(5) $$\psi(x) \sim_{x\to-\infty} \lambda(1+R^*)e^{i\omega_- x/\varepsilon} + \lambda^*(1+R)e^{-i\omega_- x/\varepsilon}$$

$$\psi(x) \sim_{x\to+\infty} \lambda Te^{i\omega_+ x/\varepsilon} + \lambda^* T^* e^{-i\omega_+ x/\varepsilon}$$

with $\lambda \in \mathbb{C}$. An identification with (1) produces $c_- = \lambda(1 + R^*)$ and $c_+ = \lambda T$. Moreover R and T satisfy the energy or probability conservation relation:

(6) $$\omega_+ |T|^2 = \omega_-(1 - |R|^2)$$

which allows the variation of action ΔI along a trajectory to be written:

(7) $$\Delta I = -2I_+|R|(\cos\alpha + |R|)/(1 - |R|^2), \quad \alpha =_{def} \mathrm{Arg}(c_+{}^2 R/T^2).$$

The reader may notice that ΔI does not depend on the origin of time, i.e., in these coordinates of the x axis. If x^* is taken as a new origin, R and T are changed into $R' = Re^{-2i\omega_- x^*/\varepsilon}$ and $T' = Te^{i(\omega_+ - \omega_-)x^*/\varepsilon}$ (only phases can be affected), whereas λ changes to $\lambda' = \lambda e^{i\omega_- x^*/\varepsilon}$ and it is easily checked that α and thus ΔI remain unchanged.

It now only remains to evaluate the amplitude of the reflection coefficient R when $\varepsilon \to 0$ and this will give:

(8) $$|\Delta I/I| \sim 2|R|\cos\alpha \qquad (\cos\alpha \neq 0)$$

$$\sim |R|^2 \qquad (\cos\alpha = 0).$$

We do not know of any interpretation of the (intrisically defined) angle α.

The determination of R is a typical problem of semiclassical scattering (with $\hbar$ as the small parameter) that we shall only sketch, refering the reader to [Frö] and especially to [Ber5] for an elementary but comprehensive review of the theory which goes far beyond the scope of this book. A more recent review can be found in [Kn] and a deep but clear mathematical exposition may be found in [Vo] which presents, besides original results, the present "state of the art". Among older papers, [Dy] and [Fed] deal with the problem in a more or less formal way; let us finally mention that the latter paper by Fedoryuk is in the context of adiabatic invariance and also treats the case of the adiabatic perturbation of an ensemble of oscillators described by the system:

(9) $$d^2y/dt^2 + A(\varepsilon t)y = 0$$

with $y \in \mathbb{R}^n$ and $A(\tau)$ a strictly positive symmetric matrix with distinct eigenvalues (for any $\tau \in [0, 1]$) and asymptotic limits at both ends. It is easily understood that this case does not differ very much from the case $n = 1$; one finds n adiabatic invariants on the time-scale $1/\varepsilon$, whose drifts are exponentially small under suitable analyticity conditions, and which correspond to the n projections onto the normalized eigenvectors of A.

Let us now come back to the determination of R, partially following [Be5] and without complete attention to rigor. The WKB phase is defined as:

$$(10) \qquad w(x) = \int_0^x \omega(x')\, dx'$$

x being possibly complex; the lower bound which determines a "phase reference level" can be chosen arbitrarily on the real axis. ψ may be written as:

$$(11) \qquad \psi = \psi_+ + \psi_- = b_+(x)\omega^{-1/2} e^{iw(x)/\varepsilon} + b_-(x)\omega^{-1/2} e^{-iw(x)/\varepsilon}$$

ψ_+ (ψ_-) represents the right (left)-travelling part of the wave and the coefficients $b_\pm$ satisfy a first order coupled system:

$$(12) \qquad b_+' = \omega'/(2\omega)\, b_- e^{-2iw/\varepsilon}, \quad b_-' = \omega'/(2\omega)\, b_+ e^{2iw/\varepsilon}$$

('= x-differentiation). The Jost solution (4) corresponds to the asymptotic boundary conditions $b_+(-\infty) = 1$, $b_-(+\infty) = 0$ and it thus satisfies the following integral version of (12):

$$(13) \qquad b_+ = 1 + \int_{-\infty}^{x} \omega'/(2\omega)\, b_- e^{-2iw/\varepsilon}\, dx$$

$$b_- = -\int_x^{+\infty} \omega'/(2\omega)\, b_+ e^{2iw/\varepsilon}\, dx.$$

Now this system can be solved recursively starting from $b_+^{(0)} = 1$, $b_-^{(0)} = 0$. To first order this produces:

$$(14) \qquad b_-^{(0)}(-\infty) = R^{(1)} = -\int_{-\infty}^{+\infty} \omega'/(2\omega)\, e^{2iw/\varepsilon}\, dx$$

leading to a classical problem of integrals with strongly oscillating phase. In this context, this was first tackled by Pokrovski et al. in [Pok1-2] and [Vo] provides a very adequate discussion of the important phenomena in the context of semiclassical theory. One can take w as a new (possibly many valued) variable and rewrite:

$$(15) \qquad R^{(1)} = -\int_{-\infty}^{+\infty} \omega'/(2\omega)\, e^{2iw/\varepsilon}\, dw$$

('= w-differentiation) which is adapted to the stationary phase or better saddle point method.

To be more specific, assume that $\omega'/(2\omega)$ is analytic in $|\operatorname{Im} w| < c_0$ and notice that, all quantities being real on the real x axis, the problem is symmetric ($w(x^*) = w^*(x)$...) with respect to reflection in the real w axis. Choosing the correct side of the real axis and shifting the contour of integration we find that under certain mild conditions, the exact statement of which we leave to the reader (asymptotic vanishing of the contributions of the vertical segments):

$$| R^{(1)} | = O(e^{-2c/\varepsilon}) \quad \text{for any c, } 0 \leqslant c < c_0 .$$

The next step consists in showing that the phase oscillating integrals that appear in the iteration of (13) are at most of the same size, leading to the same estimate for the exact reflection coefficient R. (8) then leads to the exponential smallness of the drift in action under the above mentioned conditions, that is: There exists a reflection symmetric domain in x on which $\omega(x)$ is analytic and nonzero and such that its image under the transformation $x \to w$ contains the strip $| \mathrm{Im}\ w | < c_0$ $(c_0 > 0)$.

There are cases where a more precise result is possible, namely one may find the principal part of $| R |$ as a function of ε. This is achieved by means of the so-called connection formulas which link the two asymptotic behaviors ($x \to \pm\infty$) of the solutions of the Schrödinger equation when there are "turning points", that is, places on the axis where the potential changes sign, determining the "classically forbidden" regions. We shall not enter this subject, which pertains both to elementary quantum theory (tunnel effect...) and to the study of linear analytic differential equations (Stokes phenomenon...), but only examine the simplest case, refering the interested reader again to [Be5], [Vo] and the extensive bibliographies of these papers (one may also consult [Was] on these matters).

We thus restrict ourselves to the situation where:

- ω is strictly positive on the real axis and analytic in a strip $| \mathrm{Im}\ x | < a$ $(a > 0)$.
- ω^2 has only two simple zeros at $x = x_0$ and the conjugate point in the strip.

Then the following evaluation holds:

$$(16) \qquad | R | \sim \exp(- 2/\varepsilon\, | \mathrm{Im}\ w(x_0) |).$$

For a concrete example one may come back to what is sometimes termed the Einstein pendulum, i.e., the case where $\omega^2 = g/l$ with g the acceleration of gravity and $l = l(\tau)$ the length of the pendulum; this does not in fact exactly correspond to the equation of a physical pendulum with slowly varying length $l(\tau)$ but a simple transformation brings it back to this form. To meet the above requirements, set $l(\tau) = l_0(\tau^2 + \beta)/(\tau^2 + \alpha)$ $(0 < \alpha < \beta)$ which describes a slow variation from l_0 to $l_0\beta/\alpha$ and back, with two arbitrary parameters.

We are now in the case of a problem on the line $\mathrm{Im}\ x = \mathrm{Im}\ x_0$ with only one simple turning point and one can show that this simplest caustic corresponds to a reflection coefficient equal to $-i$, that is, a phase shift of $-\pi/2$ (see e.g. [Be 5] and [Was]). In other words, there exists a solution ψ such that:

$$(17) \qquad \psi(x) \sim_{\mathrm{Re}\, x\to-\infty} \omega^{-1/2}(e^{iw/\varepsilon} - ie^{-iw/\varepsilon})$$

$$\psi(x) \sim_{\mathrm{Re}\, x\to+\infty} \omega^{-1/2}e^{iw/\varepsilon}.$$

In these formulas however, attention must be paid to the fact that w is defined according to (10) but with a phase reference level (lower bound of the integral) situated on the line $\mathrm{Im}\, x = \mathrm{Im}\, x_0$, say at the point x_0. If we now wish to compute the reflection coefficient on the real axis, the phase reference level has to be shifted to this axis, say at the origin, and that produces a phase difference $w(x_0)$. It is now easy to check, just as in the discussion below formula (7), that the reflection coefficient on the real axis is:

$$R = -\, ie^{-2i/\varepsilon\, w(x_0)}$$

and this leads to (16), perhaps changing x_0 to the conjugate point (in fact, as was already mentioned, one has to choose the "right" side of the real axis from the start).

Let us only add that the above case corresponds to the simplest caustic with the model problem of a linear potential ($\omega^2(x) = x$) which is exactly solved by the Airy function; the case of higher order turning points can be found e.g. in [Hea] using the model problems $\omega^2(x) = x^n$, $(n > 0)$. Some formulas are also available in the case of several simple turning points.

8.5 Drift of the action for general systems

We return to a general Hamiltonian system with Hamiltonian $H(p, q, \lambda(\tau))$ $(\tau = \varepsilon t)$ and do not make any assumption regarding the dependence of the frequency on the action for fixed τ; that is, the system may be linear or not. We suppose however that the two limits $\lambda_\pm = \lim_{\tau\to\pm\infty} \lambda(\tau)$ exist and we are again interested in an evaluation of the total drift $\Delta I = I(+\infty) - I(-\infty)$. We distinguish two cases according to whether $\lambda(\tau)$ is differentiable (finitely or infinitely many times) or analytic; in this way, we obtain in the analytic case, for general systems, and using the usual averaging methods of Section 2, a result close to that of the last section. The results presented in this precise form are due to A.I.Neistadt (in [Nei6]) but the C^∞ and analytic cases were essentially folk wisdom.

Let us start with the case where only finitely many derivatives exist, that is:

- Hamilton's function $H(p, q, \lambda)$ is jointly analytic with respect to all the variables.
- $\lambda(\tau) = \lambda_+$ for $\tau \geqslant 1$ and $\lambda(\tau) = \lambda_-$ for $\tau \leqslant -1$.

- λ is C^{p+2} ($p \geqslant 1$) on the open interval $(-1, 1)$.
- λ and the derivatives $\lambda^{(k)}$, $k = 1, 2, \ldots, p-1$ are continuous - and thus vanish - at $\tau = \pm 1$.
- $\lambda^{(p)}$ is discontinuous at one of the two points ± 1, or at both.

A possible $\lambda(\tau)$ is sketched in Figure 8.6 (with $p = 1$). As always (see Appendix 6 and Section 1 above) one starts by transforming the family of "frozen" (τ fixed) systems to action-angle variables, using a generating function $S(q, I, \lambda)$ leading to the new Hamiltonian (still denoted as H):

(1) $$H(I, \varphi, \lambda) = H_0(I, \lambda) + \varepsilon H_1(I, \varphi, \lambda)$$

where $H_1 = \lambda'(\tau)\partial S/\partial\lambda$ and S has been reexpressed as a function of (I, φ, λ). H_1 vanishes for $|\tau| > 1$ and possesses one less τ-derivative than λ. The drift ΔI is of course given as $\Delta I = I(1+0) - I(-1-0)$, where for any function F, $F(1+0)$ ($F(-1-0)$) denotes the right (left) limit at $+1$ (-1), so that we are in fact working on a $1/\varepsilon$ timescale.

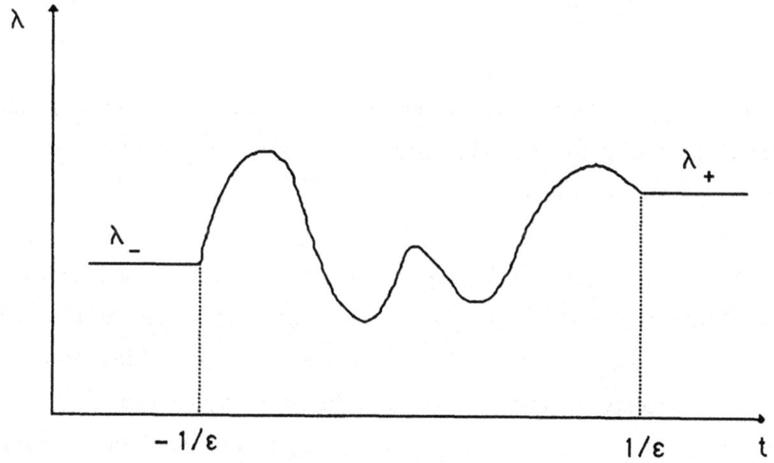

Figure 8.6: The control parameter is a function of time of finite differentiability and assumes constant values outside the t-interval $[-1/\varepsilon, 1/\varepsilon]$.

We now proceed as in Section 2, performing a canonical transformation that eliminates the angles to order p, using formulas (7) and (8) of that section. We have:

(2) $$\chi = \sum_{k=0}^{k=p-1} \varepsilon^k \chi_{k+1}(I, \varphi, \tau)$$

where $\{\chi_k, H_0\} = \omega.\partial\chi_k/\partial\varphi$ and $\omega = \omega(I, \tau) =_{def} \partial H_0/\partial I$ is assumed to be (say) positive and bounded from above and below (note that τ here does not have the same meaning as in Appendix 7). A transformation $T: (I, \varphi) \to (I_p, \varphi_p)$ is thus generated, where I_p is nothing but the pth adiabatic invariant of Section 2. To simplify notation somewhat we write $(I_p, \varphi_p) = (J, \psi)$.

We aim at showing that ΔI is of order ε^p and at computing its principal part. Now J is a constant within order ε^{p+1} on the timescale $1/\varepsilon$ and here it is in fact nearly constant on any time interval ($J = I$ when $|\tau| > 1$), whence in particular $\Delta J \sim \varepsilon^{p+1}$. Also $I = TJ$ is found according to the ε-expansion of formulas (42) in Appendix 7. It remains to show that T exists, given the assumed smoothness of λ (and thus of H_1), and to compute its principal part at $\tau = 1 + 0$ and $\tau = -1 - 0$.

For this purpose we return to formula (8) in Section 2 in which we set $H_n = 0$ $(n \geq 2)$ and $\chi_n = 0$ $(n \geq p+1)$; in particular H'_n $(n \geq p+1)$ is then completely determined. Explicitly:

$$(3) \qquad \omega\partial\chi_1/\partial\varphi = H'_1 - H_1$$

$$\omega\partial\chi_n/\partial\varphi = nH'_n - T_{n-1}H_1 - \textstyle\sum_{m=1}^{n-1} L_{n-m}H'_m - \partial\chi_{n-1}/\partial\tau \quad (n \geq 2)$$

with $L_k f = \{\chi_k, f\}$. We will use the operator Φ (see Notation, Section 1.2) which to a periodic function φ associates the integral of its fluctuating part; if $f(I, \varphi, \tau)$ is 2π-periodic in φ with Fourier coefficients f_k:

$$(4) \qquad \Phi[f] = \textstyle\sum_{k\neq 0} - ik^{-1}f_k(I, \tau)e^{ik\varphi}.$$

Notice that Φ does not alter the regularity with respect to τ and commutes with $\partial/\partial\tau$. To order 1, we thus find $\chi_1 = -1/\omega\,\Phi[H_1]$. On the other hand H_1 is of class C^{p+1} in τ on $(-1, 1)$ and $\partial^n H_1/\partial\tau^n(\tau = \pm 1) = 0$ for $n = 0, 1, \ldots, p-2$. Using formula (3) it is easy to show recursively that χ_n and H'_n have the same τ regularity as $\partial^{n-1}H_1/\partial\tau^{n-1}$ for $1 \leq n \leq p-1$ and vanish at ± 1 for $n \leq p-1$, as do the first three terms on the r.h.s. of (3), including $n = p$. This shows that the transformation T exists and satisfies:

$$(5) \qquad \chi_n(\tau = \pm 1) = -1/\omega\,\partial\Phi[\chi_{n-1}]/\partial\tau(\tau = \pm 1), \quad n = 2, \ldots, p.$$

Whence

$$\chi_n(\tau = \pm 1) = (-1/\omega)^n \partial^{n-1}\Phi^n[H_1]/\partial\tau^{n-1}(\tau = \pm 1),$$

which is useful only for $n = p$. Thus:

$$(6) \qquad I = TJ = J + \varepsilon^p/p\, L_p(J) + O(\varepsilon^{p+1}) = J + \varepsilon^p/p\, \partial\chi_p/\partial\psi + O(\varepsilon^{p+1})$$

with χ_p expressed as a function of (J, ψ), this being valid at $1 + 0$ and $-1 - 0$.

In other words all the higher order invariants I_n are equal to I for $n \leqslant p-1$ and $|\tau| \geqslant 1$. In fact the following slightly more general assertion is easily seen to hold: If the perturbation H_1 satisfies $\partial^n H_1/\partial\tau^n(\tau = \tau^*)$, $n = 0, 1, \ldots, k$ then $I_n(\tau^*) = I(\tau^*)$ for $n \leqslant k+1$. Here, given the expression of H_1 we have, again at $1+0$ and $-1-0$:

$$(7) \qquad \begin{aligned} \partial\chi_p/\partial\psi &= (-1/\omega)^p \partial^{p-1}\Phi^{p-1}[\lambda'(\tau)\partial S/\partial\lambda]/\partial\tau^{p-1} \\ &= (-1/\omega)^p \lambda^{(p)}\Phi^{p-1}[\partial S/\partial\lambda] \end{aligned}$$

since the singular (nonzero) part at ± 1 is given by the $\lambda^{(p)}$ term. To finish the evaluation of ΔI using formula (6) we need only make the arguments of the function S a little more precise in (7); according to (6) it is enough to evaluate $\partial\chi_p/\partial\psi$ to lowest order and one checks that the derivative $\partial S/\partial\lambda$ must be computed at $(I_\pm^*, \varphi_\pm^*, \lambda_\pm)$ with:

- at $-1-0$: $(I_-^*, \varphi_-^*, \lambda_-) = (I_-, \varphi_-, \lambda_-)$.
- at $1+0$: $(I_+^*, \varphi_+^*, \lambda_+) = (I_-, \varphi_*(1), \lambda_+)$ where:

$$(8) \qquad \varphi^*(\tau) = \varphi_- + 1/\varepsilon\int_{-1}^{\tau} \omega[I_- + \varepsilon\omega^{-1}(I_-, -1)H_1(I_-, \varphi_-, -1+0), \sigma]\, d\sigma$$

if $p = 1$

$$\varphi^*(\tau) = \varphi_- + 1/\varepsilon\int_{-1}^{\tau} \omega(I_-, \sigma)\, d\sigma \quad \text{if } p \geqslant 2.$$

These formulas determine the phase to lowest order and the correction introduced in the first formula vanishes as soon as $p \geqslant 2$. Hence the final result:

$$(9) \qquad \begin{aligned} \Delta I &= \varepsilon^p/p\,[R(1+0) - R(-1-0)] + O(\varepsilon^{p+1}) \\ R &= (-1/\omega)^p \lambda^{(p)}\Phi^{n-1}[\partial S/\partial\lambda] \end{aligned}$$

and R must be evaluated at $(I_\pm^*, \varphi_\pm^*, \lambda_\pm)$; (I_-, φ_-) may represent the initial conditions.

This in particular implies that if λ is C^∞ (with support in $-1 \leqslant \tau \leqslant 1$) then $\Delta I = O(\varepsilon^p)$ for any p, which is often denoted by $\Delta I = O(\varepsilon^\infty)$ and expressed by saying that ΔI is "beyond any perturbation order". Because here $J = I$ for $|\tau| > 1$, this had in fact already been proved in Section 2.

Passing to the case where λ is analytic (and so with noncompact support) one can obtain, using again the nonresonant Nekhoroshev scheme of Section 2, an exponential estimate for the drift. The precise assumptions are as follows (cf. [Nei6]):

- λ can be analytically continued to the strip $|\operatorname{Im}\tau| \leqslant \zeta$ ($\zeta > 0$) and the integrals:

$$\int_{-\infty}^{\infty} |\lambda'(s)|\, ds$$

computed on the horizontal lines $\operatorname{Im} s = \xi$, $|\xi| \leqslant \zeta$ are uniformly bounded. Also, $\lambda'(s) \to 0$ when $\operatorname{Re} s \to \pm\infty$ in that strip.

- $H(I, \varphi, \lambda)$ is analytic and bounded in:

$|I - I_0| \leqslant \rho$, $|\mathrm{Im}\,\tau| \leqslant \zeta$, $|\mathrm{Im}\,\varphi| \leqslant \sigma$, $\rho > 0, \sigma > 0, \zeta > 0$,

and $|\omega(I, \tau)| = |\partial H_0/\partial I| > \omega_0 > 0$ on this domain; this repeats the assumptions of Theorem 2 in Section 2 ($I_0 = I(0)$).

The convergence of the integral of $|\lambda'(s)|$ on the real line ensures the existence of the limits $I_\pm =_{def} \lim_{\tau\to\pm\infty} I(\tau)$ (ε small enough), since $\| H_1(\tau) \| \sim |\lambda'(\tau)|$. The quantity to be evaluated $\Delta I = |I_+ - I_-|$ therefore also exists. Under the above conditions, the following estimate holds:

(10) $\qquad \Delta I = |I_+ - I_-| = O(e^{-c/\varepsilon}) \quad (c > 0).$

The main part of the proof is identical to that of Theorem 2 in Section 2, to which we refer the reader; having constructed the coordinates $(J, \psi) =_{def} (I_N, \varphi_N)$ one may estimate ΔI. Because $\lambda'(\tau)$ - hence H_1 - goes to zero as $\tau \to \pm\infty$ on the real axis, the transformation $(I, \varphi) \to (J, \psi)$ approaches the identity and $J_\pm = \lim_{\tau\to\pm\infty} J(\tau) = I_\pm$. This implies:

(12) $\qquad \Delta I = J_+ - J_- = - \int_{-\infty}^{\infty} \partial R_N/\partial\psi(J, \psi, \tau, \varepsilon)\, d\tau.$

As R_N is defined and bounded for $|\mathrm{Im}\,\psi| \leqslant \sigma/2$, one has:

(13) $\qquad |\partial R_N/\partial\psi| \leqslant 2/\sigma \sup_{|\mathrm{Im}\,\psi| \leqslant \sigma/2} |R_N|.$

Using the analyticity with respect to τ, one conludes that the integrals:

$$\sup_{|\mathrm{Im}\,\tau|<\zeta_n} \int_{-\infty}^{\infty} \sup_{(I_n, \varphi_n)} |R_n|\, d\tau$$

(with the domain in (I_n, φ_n) defined as in Section 2), decrease like the norm of R_n on this same domain. This yields the sought after estimate for ΔI. ∎

This is the fifth time in the course of this book that we mention an exponential estimate in the simple, one phase, nonresonant situation. In Sections 3.2 (in fine), 8.2 and in the present case, which are essentially identical, we used the Nekhoroshev iteration technique to derive a result which is valid both for nonlinear *and* linear systems. Indeed since resonances are excluded a priori from the perturbation scheme, the nondegeneracy condition, which precludes linearity (isochronicity) becomes aimless (though resonances do play a role when one goes to very long time intervals, as has been already explained, and will be detailed in the next section). For linear systems on the other hand, we derived specific results, using the study of Hill's equation in the periodic case (Section 8.3) and the WKB method for asymptotically autonomous systems (Section 8.4).

The final estimate for the remainders always reads $e^{-c/\varepsilon}$ but it arises in two essentially different ways. In the Nekhoroshev scheme, it is really $(1/c_1)^N$ with $c_1 > 1$ and $N = [c_2/\varepsilon]$, whereas for linear systems it looks rather like $|e^{ia/\varepsilon}|$, where $c = \mathrm{Im}\, a$. Note that in both cases however, this can be made more precise in that c is always proportional to the width of the strip around the real axis on which the important object can be analytically continued (this quantity is denoted by σ or ζ). But in any case, the estimate in the linear cases appears as much more intrinsic; in particular it allows for a proof of optimality (within a constant).

There is at present a growing interest in these "exponentially small remainders", which show up in *analytic* systems, and whose very existence rules out the possibility of convergence of the "classical perturbation series", like those we somewhat loosely called the Linstedt-Poincaré series in the canonical case; in this respect, one should also mention the well-known exponentially small splitting of the separatrices in the neighborhood of the resonances of a near integrable Hamiltonian system, a phenomenon already known to Poincaré. The main goal of contemporary studies is to really estimate these remainders, not only to obtain upper bounds. This can be achieved only if one understands their intrinsic meaning, which in turn seems to require a "complexification" of the problem, although it is not always easy to determine what is the object that should be analytically continued to the complex values of the variable. Therefore one can venture to state that we may be witnessing the birth of truly complex methods in nonlinear pertubation theories.

As a final remark, we add that the situation is much worse for multidimensional systems. More precisely, in dimension n the remainders take the form ae^{-b/ε^c} where the dependence of $a = a(\varepsilon, n)$ on ε is algebraic; b and c also vary with n but only a and b are expected to depend of the specific problem at hand.We have listed here a, b and c in order of increasing importance (and decreasing precision). For *one dimensional* problems, c is known and is usually unity (for the "natural" choice of ε); b can often be determined in the *linear* problems (e.g. via the WKB method) and is the object of studies in progress in the *nonlinear* case, whereas a (the so-called preexponential factor) is still a matter of controversy in many *linear* problems (where it is independent of ε). By contrast, in the *multidimensional* case, as we already pointed out in Chapter 7, one has not even determined hitherto the optimal exponent $c(n)$ whose intrinsic meaning remains largely obscure.

8.6 Perpetual stability of nonlinear periodic systems

The story begins in the now usual way with a Hamiltonian $H(p, q, \tau)$ ($\tau = \varepsilon t$) which here is assumed to be 2π-periodic with respect to τ. As in the previous sections, a canonical transformation using a generating function $S(q, I, \tau)$ (see Appendix 6) transforms the family of "frozen" systems (fixed τ) to action-angle variables (I, φ) whose evolution is governed by the Hamiltonian:

(1) $$H = H_0(I, \tau) + \varepsilon H_1(I, \varphi, \tau), \quad H_1 = \partial S/\partial \tau$$

and S has been reexpressed as a function of (I, φ, τ). We still write $\omega(I, \tau) = \partial H_0/\partial I$ and:

(2) $$\Omega(I) = 1/2\pi \int_0^{2\pi} \omega(I, \tau)\, d\tau,$$

is the *average* frequency of the unperturbed system. We assume that:

- $H(I, \varphi, \tau)$ is analytic and bounded on the domain:

(3) $$D = \{ (I, \varphi, \tau), I_1 \leqslant \mathrm{Re}\, I \leqslant I_2 , |\mathrm{Im}\, I| \leqslant \rho, |\mathrm{Im}\, \varphi| \leqslant \sigma, |\mathrm{Im}\, \tau| \leqslant \sigma \}$$

where ρ and σ are two (strictly) positive constants and $\mathrm{Re}(\varphi, \tau) \in T^2$.

- $|\omega(I, \tau)| \geqslant \omega_0 > 0$ on D.

- $|d\Omega/dI| \geqslant m > 0$ on D.

This last condition ensures, as mentioned at the end of Section 2, that the system is not sensitive to parametric instability, because a drift in action induces a variation of the mean frequency which precludes a possible resonance condition needed for instability. The theorem is then as follows (cf. [Ar1], [Ar3] and [Nei5]):

Theorem:

Under the conditions described above, and given $\eta > 0$, there exist $\varepsilon_0 > 0$ and $c > 0$ such that for any $\varepsilon \leqslant \varepsilon_0$ and any initial condition $(I(0), \varphi(0))$ with $I_1 + \eta \leqslant I(0) \leqslant I_2 - \eta$:

(4) $$\sup_{t \in \mathbb{R}} |I(t) - I(0)| \leqslant c\varepsilon.$$

Under the assumption that the system is in some sense strictly nonisochronous, the action variable is thus a *perpetual* adiabatic invariant. The most useful reference on this problem may still be the paper by Arnold [Ar3] which also presents a few applications of this result (with a slightly less precise estimate). In particular, one can show the perpetual invariance of the magnetic moment associated with a charged particle in a slowly varying electromagnetic field (the first invariant in the third of the examples presented in Section 1) when it is confined inside an axisymmetric magnetic mirror device.

The proof of the theorem relies on a version of the KAM theorem adapted to a "degenerate" case, which we shall state below. Before it can be applied however, one has to perform a string of canonical transformations so as to recast the original Hamiltonian in a suitable form, following an idea of Arnold ([Ar1]).

The first transformation is as usual the passage to action-angle variables $(p, q) \to (I, \varphi)$. By a slight abuse of notation H will still denote the Hamiltonian in these variables. Next, the integral curves of Hamilton's equations are also, as is well known (see Appendix 6), the characteristic curves of the so-called Poincaré-Cartan differential form (or "relative integral invariant") and these are of course unchanged by multiplication by a constant. This motivates the interpretation of the following identity:

(5) $\qquad I d\varphi - H dt = -1/\varepsilon \, (H d\tau - \varepsilon I d\varphi).$

One can in fact perform the transformation $(I, \varphi, H, t) \to (H, \tau, L = \varepsilon I, T = \varphi)$ viewing H as a new action variable, τ as an angle variable, $T = \varphi$ as the new "time" and $L = \varepsilon I$ as the new governing Hamiltonian; this essentially trivial transformation may serve to illustrate the flexibility of the canonical formalism. L can be expressed in terms of H, τ and $\varphi = T$ by inverting the definition (1) of the Hamiltonian; this is at least locally possible because of the hypothesis on ω and we shall assume that it can be done globally (otherwise one would have to patch together a finite number of estimates). Explicitly:

(6) $\qquad \varepsilon I = L(H, \tau, T) = \varepsilon I_0(H, \tau) + \varepsilon^2 I_1(H, \tau, T, \varepsilon).$

$I_0 = H_0^{-1}$ is the inverse function of H_0 : $I_0 [H_0(I_0, \tau)] = I_0$. The "time" T appears only in the perturbed part, according to the purpose of this transformation, and it is now natural to change to the action-angle variables of the new unperturbed part $I_0(H, \tau)$, which is integrable, being one dimensional and autonomous. Letting (P, Q) be these new variables, the transformation $(H, \tau) \to (P, Q)$ leads to the Hamiltonian:

(7) $\qquad k = k(P, Q, T, \varepsilon) = \varepsilon k_0(P) + \varepsilon^2 k_1(P, Q, T, \varepsilon)$

with $k(P, Q, T) = L(H, \tau, T)$ and we shall shortly see that k_0 can be identified as the inverse of the function h_0 that is defined as the average of H_0 :

(8) $\qquad h_0(P) =_{def} 1/2\pi \int_0^{2\pi} H_0(P, \tau)\, d\tau.$

The inversion is possible because $dh_0/dP = \Omega(P)$ has constant sign since ω does not approach zero. To identify k_0 we proceed as follows; the transformation $(H, \tau) \to (P, Q)$

is effected via a generating function $F(\tau, P)$ such that:

(9) $\quad H = \partial F/\partial\tau, \quad Q = \partial F/\partial P$

and by definition of the action-angle variables:

(10) $\quad I_0(H, \tau) = I_0(\partial F/\partial\tau, \tau) = k_0(P)$

or, since $I_0 = H_0^{-1}$:

(11) $\quad \partial F/\partial\tau = H_0(k_0(P), \tau), \quad F(\tau, P) = \int^{\tau} H_0(I, \sigma)\, d\sigma$

to within an inessential constant and with $I = k_0(P)$. To determine k_0 we then use another property of the action-angle variables, namely: $Q(H, \tau+2\pi) = Q(H, \tau) + 2\pi$ or, using (9) and (11):

(12) $\quad d/dP\, [1/2\pi \int_0^{2\pi} H_0(k_0(P), \tau)\, d\tau] = 1$

that is:

(13) $\quad 1/2\pi \int_0^{2\pi} \omega(k_0(P), \tau) dk_0/dP\, d\tau = 1$

or:

(14) $\quad \Omega(k_0(P)) dk_0/dP = 1.$

This confirms that $k_0 = h_0^{-1}$, again within a constant that can be chosen to make this equality exact. We also note that $dQ/dt = dQ/dT.dT/dt = \omega/\Omega$ for the unperturbed system. There remains a fourth and last transformation to be performed that will bring the system with Hamiltonian k into the form of a two degree of freedom autonomous system. This is classical; considering k and T as conjugate variables, the Hamiltonian may be written:

(15) $$K(P, k, Q, T) = k - \varepsilon k_0(P) - \varepsilon^2 k_1(P, Q, T, \varepsilon)$$
$$= K_0(k, P, \varepsilon) - \varepsilon^2 k_1(P, Q, T, \varepsilon).$$

On the energy surface $K = 0$, this is equivalent to the nonautonomous system governed by $k(P, Q, T, \varepsilon)$. The version of the KAM theorem stated below will be applied to K.

This Hamiltonian is non generic in three ways: first and foremost, the two action variables (k, P) do not both enter at lowest order (P is missing); second, the principal part is linear in k, and finally, this same variable is absent at the next order. One can check that this makes Kolmogorov's determinant (the Hessian of K_0) vanish, whereas Arnold's determinant, which ensures the isoenergetic nondegeneracy (see Appendix 3), equals $\varepsilon k_0''(P) = -\varepsilon\Omega'/\Omega^2(I) = \varepsilon(\Omega^{-1})'(I)$ $(P = h_0(I))$. This is nonzero by the hypothesis $\Omega' \neq 0$, which is essential since we are really interested only in the energy surface $K = 0$. It is interesting to notice that the frequencies associated to K_0 are 1 and $-\varepsilon/\Omega(I)$ so that, after multiplication by $\Omega(I)$, the intervening small denominators will be of the form $| m_1\Omega(I) + m_2\varepsilon |$ ($(m_1, m_2) \in \mathbb{Z}^2$), which is the τ average of those we encountered at the end of Section 2 (see formula (18) of that section) while briefly discussing the validity of adiabatic Lie series.

The problem of perpetual adiabatic invariance of the action has finally been reduced to a particular case of a version of the KAM theorem which applies to a once degenerate two degree of freedom - but otherwise generic - Hamiltonian.

Theorem ([Nei5]):

Let $H(I, \varphi, \varepsilon)$ be a Hamiltonian (2π periodic in φ) of the form:

(16) $$H = H_0(I_0) + \varepsilon H_{10}(I) + \varepsilon H_{11}(I, \varphi) + \varepsilon^2 H_2(I, \varphi, \varepsilon)$$

where $I = (I_0, I_1)$, $\varphi = (\varphi_0, \varphi_1)$ and H_{11} has zero mean value in φ_0 (this is not restrictive). It is assumed that for small enough ε, H is analytic inside and bounded on the domain $K_{\rho,\sigma}$, where K is a compact convex region of the plane and, as usual:

(17) $$K_{\rho,\sigma} = \{ (I, \varphi) \in \mathbb{C}^2, |\mathrm{Re}\, I| \in K, |\mathrm{Im}\, I| \leqslant \rho, |\mathrm{Im}\, \varphi| \leqslant \sigma \}.$$

We also assume that there exist α and β such that on the same domain:

(18) $$|\partial H_0/\partial I_0| \geqslant \alpha > 0, |\partial H_{10}/\partial I_0| \geqslant \alpha > 0, |\partial^2 H_{10}/\partial I_0 \partial I_1| \geqslant \beta > 0.$$

We shall write $\omega_0 = \partial H_0/\partial I_0$, $\omega_1 = \partial H_{10}/\partial I_1$.

Then there exist $\eta = \eta(\varepsilon) = O(\varepsilon)$ and $\varepsilon_0 > 0$ such that for any $\varepsilon \leqslant \varepsilon_0$ and any real E, the intersection of the (real) energy surface $H(I, \varphi, \varepsilon) = E$ with $I \in K - \eta$ is divided into two sets V' and V'' such that:

- $\mathrm{mes}\, V'' = O(e^{-c/\varepsilon})$ $(c > 0)$.
- V' is the union of tori T_I which are invariant under the flow defined by H (I belongs to a Cantor set in $K \subset \mathbb{R}^2$).

Moreover:

- The torus T_{I*} is ε-close to the invariant torus $I = I^*$ of the unperturbed Hamiltonian $H_0(I_0) + \varepsilon H_{10}(I)$ (the distorsion of the tori is of order ε).
- On T_{I*} the motion is conjugate to a rectilinear flow with frequency ω^* through a diffeomorphism of the 2-torus which is ε-close to the identity and $\omega^* = (\omega'_0, \varepsilon\omega'_1)$ satisfies:

(19) $$\omega'_0 = \omega_0(I^*) + O(\varepsilon), \omega'_1 = \omega_1(I^*) + O(\varepsilon).$$

Corollary:

For any trajectory with initial conditions $(I(0), \varphi(0))$ on the energy surface $H = E$ with $I(0) \in K - \eta$, one has:

(20) $$\sup_{t \in \mathbb{R}} |I(t) - I(0)| = O(\varepsilon).$$

The corollary is a straightforward consequence of the theorem and more precisely of the evaluation of the distorsion of the tori and their partitioning of the energy surface for two degree of freedom systems. It finishes the proof of perpetual adiabatic invariance of the action, when applied to the Hamiltonian K which satisfies the hypotheses of the theorem. In that case, the action variable P is equivalent to the original I variable ($P = h_0(I)$); the exponential smallness of the volume of the destroyed tori does not however lend itself to an easy translation in terms of the original variables.

We shall briefly describe the ingredients and the scheme of the proof of the theorem. Once again, one has first to bring the Hamiltonian into a more manageable form by considering it as a perturbation of H_0 with fixed (I_1, φ_1). Then since ω_0 stays away from zero, one can apply, as was done in the last section, the Nekhoroshev scheme to push the φ_0 dependence into an exponentially small term. Indeed, this is effected by a canonical transformation $(I, \varphi) \to (J,\psi)$ $((J_1, \psi_1) = (I_1, \varphi_1))$ of size $O(\varepsilon)$ that is obtained as the product of an $O(1/\varepsilon)$ number of transformations and brings H to the form:

$$K(J, \psi) = H_0(J_0) + \varepsilon H_{10}(J) + \varepsilon^2 K_2(J, \psi_1, \varepsilon) + R(J, \psi, \varepsilon) \tag{21}$$

with $\| R(J, \psi, \varepsilon) \| = O(e^{-c/\varepsilon})$. The fact that H_{11} has zero mean value in φ_0 has here been used to push the ψ_1 dependence into the second order term.

Now if R is neglected, the first three terms on the r.h.s. in (21) define an integrable Hamiltonian (J_0 being fixed) for which one can find action-angle variables that define a transformation $(J, \psi) \to (J', \psi')$ $((J'_0, \psi'_0) = (J_0, \psi_0))$ again of order ε, leading to a new Hamiltonian:

$$\begin{aligned} H'(J', \psi') &= H_0(J'_0) + \varepsilon H'_1(J', \varepsilon) + R'(J', \psi', \varepsilon) \\ &=_{def} H'_0(J', \varepsilon) + R'(J', \psi', \varepsilon) \end{aligned} \tag{22}$$

with $\| R'(J', \psi', \varepsilon) \| = O(e^{-c/\varepsilon})$ and $H'_1(J', 0) = H_{10}(J')$.

At this point, it remains to apply the usual quadratically convergent (Newton-like) algorithm that is used in the proof of the KAM theorem (see Chapter 7 and Appendix 8 in fine) to the Hamiltonian H'. Notice that the Arnold's determinant of $H'_0(J', \varepsilon)$ is equal to:

$$- \varepsilon\omega_0{}^2 \partial^2 H_{10}/\partial I_1{}^2 + O(\varepsilon^2)$$

and so is nonzero by hypothesis. The small denominators that arise will be of the form:

$$| \omega'_0 m_0 + \varepsilon\omega'_1 m_1 |, \ (m_0, m_1) \in \mathbb{Z}^2$$

with $\omega'_0 = \omega_0 + O(\varepsilon)$, $\omega'_1 = \omega_1 + O(\varepsilon)$. The "good " frequencies are still of Diophantine type (see Appendix 4) and here satisfy the inequalities:

$$| \omega'_0 m_0 + \varepsilon\omega'_1 m_1 | \geqslant \gamma(\varepsilon)|m|^{-2}, \quad m = (m_0, m_1) \in \mathbb{Z}^2 - \{(0, 0)\}, \tag{23}$$
$$| m | = | m_0 | + | m_1 |.$$

More importantly, as we saw in Chapter 7, $\gamma(\varepsilon)$ may be chosen on the order of the square root of the size of the perturbation. Since R' is of order $e^{-c/\varepsilon}$, we may set $\gamma(\varepsilon) = \gamma_0\, e^{-c/2\varepsilon}$ ($\gamma_0 > 0$), and this provides an exponential estimate for the volume of the destroyed tori (with the substitution $c \to c/2$ in the statement of the theorem). The rest of the proof runs along the same lines as the usual proof of the KAM theorem in the analytic case (as it was first presented in [Ar2]).

CHAPTER 9: THE CLASSICAL ADIABATIC THEOREMS IN MANY DIMENSIONS

9.1 Invariance of action, invariance of volume

In the preceding chapter we gave the most important known results on adiabatic invariance for systems of one degree of freedom, in which case there is no need to distinguish between ergodic and integrable systems, and to lowest order (Section 8.1) we needed only the simple one phase averaging theorem of Section 3.1. On the other hand, the adiabatic theorems considered in this chapter have proofs very similar to those of the multiphase averaging results of Chapters 5 and 6. Despite their obvious physical significance, these theorems are rarely correctly stated, and even more rarely correctly proved. We will distinguish the only two cases for which precise theorems exist, though it would be interesting to consider cases intermediate between "integrable" and "ergodic":

i) We begin with a family $H(p, q, \lambda)$ of integrable Hamiltonians of n degrees of freedom with slowly varying parameter $\lambda = \varepsilon t$. As an almost immediate consequence of the results of Chapter 6, we have the adiabatic theorem which ensures that, for λ fixed, on the timescale $1/\varepsilon$, the variation of the actions I_i $(i = 1,..., n)$ is less than $\rho(\varepsilon)$ for all initial conditions not belonging to a certain set with measure of order $\sqrt{\varepsilon}/\rho(\varepsilon)$.

From the physical standpoint, this generalizes the original example of the pendulum discussed in the preceding chapter, and one may similarly ponder the quantization of the action variables and the "validity" of the Bohr-Sommerfeld rules. Einstein ([Ei2]) was the first to stress the intrinsic nature of these rules (their invariance under coordinate change), for which he argued using the concepts of deformation and homotopy in phase space, quite new ideas at the time.

We stress the fact that the actions are not adiabatically constant on *all* trajectories (as is sometimes claimed) but only on those trajectories lying outside a certain set of small measure. The first proof of this result may be found in an article by P.A.M. Dirac ([Di]), where it is discussed in connection with quantum mechanics. Dirac's proof is difficult to read and the result is less explicit than the one given here (which, we repeat, follows readily from Neistadt's theorem), but he clearly states that the result is not uniformly valid on all of phase space.

Finally, there have recently appeared results, due to J.H. Hannay and M. Berry (see below), which describe the evolution of angles in adiabatic transport and establish a connection with the analogous theorem in quantum mechanics, as we shall see in the next chapter.

ii) The second result on adiabatic invariance for classical multidimensional systems has its origins in statistical thermodynamics (from which the term "adiabaticity" is derived, as explained in Section 8.1) particularly in Boltzman's work. This result provides for the conservation of volume in phase space for an ergodic system undergoing a slow variation. One may think of the physical example of a perfect gas enclosed in a cylinder with a movable piston, though the theorem is not immediately applicable here because of the nondifferentiability of the hard sphere interaction potentials, and also because the proof of ergodicity for this system is not considered complete (see the discussion of the first example in section 8.1). It is also possible to connect the conservation of phase space volume to the conservation of entropy via Boltzman's formula (see below for some details).

Although this theorem has been well known in the physics community for nearly a century, no rigorous statement or proof was given, to our knowledge, until Kasuga's work in 1961 ([Kas]), which we reproduce here in a somewhat clarified version. We have already mentioned that this work contains the idea later adapted to integrable systems by Neistadt in the proof of the theorem presented in Chapter 6.

9.2 An adiabatic theorem for integrable systems

We consider a family $H(p, q, \lambda)$ of integrable Hamiltonians defined on $D \times [-1, 2]$ where D is a domain in $\mathbb{R}^{2n}$; we shall see below why λ varies in this interval. We assume that for all λ, there exist n independent first integrals $F_1 = H, F_2, \dots, F_n$ which are C^∞ and in involution ($\{F_i, F_j\} = 0$, where $\{.,\}$ is the Poisson bracket), and that every level surface $M_f =_{def} \{(p, q)$ such that $F_i(p, q) = f_i$ for $1 \leq i \leq n\}$ is compact and connected.

We first recall some well known facts pertaining to Hamiltonian integrable systems in more than one dimension (see e.g. [Ar7]), which also serves to establish notational conventions. Each level surface M_f of the Hamiltonian $H(p, q, \lambda)$ (λ fixed) is diffeomorphic to a torus of dimension n, and by choosing the closed curves $\gamma_1, \dots, \gamma_n$, which form a basis for the fundamental group of the torus, one may define n action variables $I_1, \dots, I_n$ (depending on λ) via the formula: $I_i = 1/2\pi \oint_{\gamma_i} p\, dq$ ($p\, dq = \sum_j p_j\, dq_j$).

We denote by $D(\lambda)$ the domain in phase space defined as the union of tori of the Hamiltonian $H(p, q, \lambda)$ entirely contained in D and at each point of which $\det(\partial I_i/\partial F_j)$ is nonzero (nondegenerate tori). Each torus $T^* = T^*(f_1^*, \dots, f_n^*; \lambda)$ contained in the interior of $D(\lambda)$ has an open neighborhood $O(T^*)$ in $\mathbb{R}^{2n}$, contained in $D(\lambda)$, such that:

1) $O(T^*)$ is a union of invariant tori of $H(p, q, \lambda)$.

2) $O(T^*)$ is connected.

3) On $O(T^*)$, the tori are determined by the given action variables $I_1, \dots, I_n$; in other words, the implicit function theorem allows one to label the tori locally by the values of the I_k rather than the F_k. Henceforth tori in $O(T^*)$ will be denoted by $T(I, \lambda)$.

Let $K(\lambda)$ be the domain in action space consisting of the image of $D(\lambda)$; that is, $K(\lambda)$ is the set of I such that the torus $T(I, \lambda)$ is in the domain D and is nondegenerate. Finally, we define the closed subset of $\mathbb{R}^n$ $K = \bigcap_{\lambda\in[-1, 2]} \mathrm{clos}(K(\lambda))$ as the intersection of the closures of the $K(\lambda)$ corresponding to those values of action such that, for all λ in $[-1, 2]$, the torus $T(I, \lambda)$ is nondegenerate and is contained in the domain D. Henceforth we will restrict our attention to a closed connected component of K which we still denote by K and for sufficiently small positive η, we work on $K - \eta$ which, for each fixed value of λ, defines a set of invariant tori of the Hamiltonian $H(p, q, \lambda)$ situated a finite distance from the degeneracies. We will take $V = (K-\eta) \times T^n$ as the set of initial conditions on $\mathbb{R}^{2n}$.

There are n frequencies $\omega_i(I, \lambda)$ associated with motion on the torus $T(I, \lambda)$, and we must impose a nondegeneracy condition on them. Here we assume that the map $(I, \lambda) \to \omega(I, \lambda)$ is of maximal rank on the domain considered (Kolmogorov's nondegeneracy condition). It should be noted that this is not the same as demanding that, for each fixed value of λ, the Hamiltonian $H(p, q, \lambda)$ satisfy Kolmogorov's condition of nondegeneracy of the frequencies. One could also adopt a condition similar to Arnold's by demanding that the map, which associates to the point (I, λ) the point in real projective space of dimension $n - 1$ corresponding to the half line in $\mathbb{R}^n$ with direction vector $\omega(I, \lambda)$, be of maximal rank at every point in the domain.

We may now state the adiabatic theorem in a precise way:

Theorem:

Let $\rho(\varepsilon)$ be a continuous positive function satisfying the inequality $c_1\sqrt{\varepsilon} \leqslant \rho(\varepsilon)$.

Then there exists, for sufficiently small ε, a partition $V = V' \cup V''$ such that:

a) For each initial condition in V', the variation of action along the corresponding trajectory of the time dependent Hamiltonian $H(p, q, \varepsilon t)$ is less than $\rho(\varepsilon)$ on the time interval $[0, 1/\varepsilon]$:

$$\operatorname{Sup}_{t \in [0, 1/\varepsilon]} \| I(t) - I(0) \| \leqslant \rho(\varepsilon),$$

b) the measure of V'' is of the order $\sqrt{\varepsilon}/\rho(\varepsilon)$.

To prove this theorem, it is first necessary to introduce action-angle variables for fixed λ. The action variables have already been defined; we need only find the conjugate angles φ_i. The canonical transformation $(p, q) \to (I, \varphi)$ may be (locally) defined by a multivalued generating function $S(I, q; \lambda)$, much as in the one dimensional case (see Appendix 6 and [Ar7]), such that: $pdq = Id\varphi + dS' = -\varphi dI + dS$, where $S' = S - I.\varphi$ is the Legendre transformation of S. The relation $p = \partial S/\partial q$ is satisfied and the angle variables are defined by $\varphi = \partial S/\partial I$. More precisely, it is easy to determine S locally in I, and one can cover $K - \eta$ by a finite number of open balls $B_1(\sigma), \ldots, B_p(\sigma)$, independent of λ and with common radius $\sigma < \eta/2$, such that the local generating function $S_i(I, q, \lambda)$ is defined on $B_i(\sigma) \times T^n$ as follows:

1) We choose a value of q, say q^*, such that the plane $q = q^*$ intersects every torus $T(I, \lambda)$ for $I \in B_i(\sigma)$. As this point of intersection is generally not unique (see Figure 9.1), we choose one of the points of intersection which describes a continuous curve in phase space for $I \in B_i(\sigma)$; this is possible for sufficiently small σ.

2) Then we define the function $S(I,q)$ on this domain by $S(I,q) = \int_{q^*}^{q} p\, dq$, where the integral is calculated along a path on the torus $T(I,\lambda)$. It depends only on the homotopy type of the path and the multivaluedness of this function is precisely measured by the integrals $I_i = 1/2\pi \oint_{\gamma_i} p\, dq$, and the angles may be defined by $\varphi_i = \partial S(I,q)/\partial I_i$. These variables are also multivalued, and the variation of φ_j along the basis loop γ_i is equal to $\partial[2\pi I_i]/\partial I_j = 2\pi\delta_{ij}$, so that the nomenclature "angles" is justified.

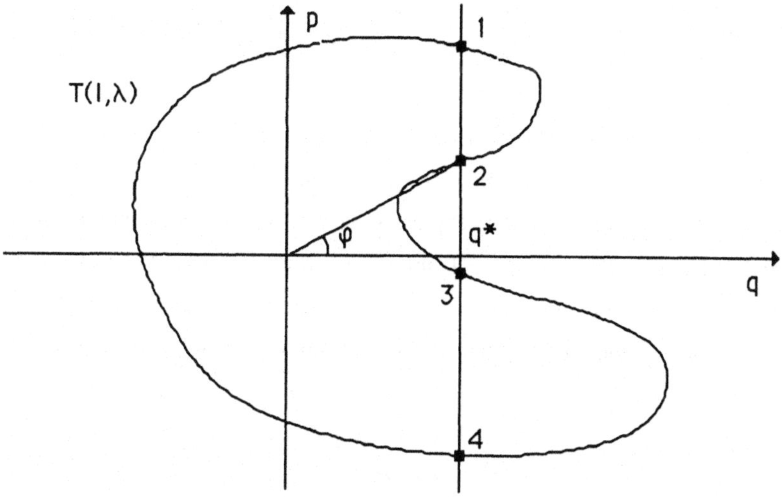

Figure 9.1: The plane $q = q^*$ may intersect the torus $T(I, \lambda)$ in several points. These points correspond to distinct values of the momenta p_i or alternatively of the angles φ_j.

We also introduce the balls $B_1(2\sigma), \ldots, B_p(2\sigma)$ with the same centers, so that, as long as its variation in action is less than σ, any trajectory with initial condition in $B(\sigma) \times T^n$ remains in the domain $B(2\sigma) \times T^n$ on which the local generating function is still defined (provided σ was chosen sufficiently small). To prove the theorem, we may now restrict our attention to the set of initial conditions $[B(\sigma) \cap (K-\eta)] \times T^n$, where $B(\sigma) = B_i(\sigma)$ for a particular i with associated local generating function $S(I, q, \lambda)$. The reader will perhaps excuse these cumbersome formalities that are unfortunately necessary to the statement of the local theorem.

On the ball $B(2\sigma)$ and for given λ the Hamiltonian H is expressed in action-angle variables and therefore replaced by the Hamiltonian:

(1) $$K_0(I, \lambda) = H\,(p[I, \varphi, \lambda], q[I, \varphi, \lambda], \lambda).$$

We have just defined a canonical transformation depending on the parameter λ: $(I, \varphi) = T(p, q, \lambda)$. If λ now varies ($\lambda = \varepsilon t$), the Hamiltonian $H(p, q, \varepsilon t)$ is replaced as usual in the variables (I, φ) by the time-dependent Hamiltonian $K(I, \varphi, \varepsilon t)$ such that:

(2) $$p dq - H(p, q, \varepsilon t)dt = I d\varphi - K(I, \varphi, \varepsilon t)dt + d(S(I, q, \varepsilon t) - I\varphi),$$

or:

(3) $$K(I, \varphi, \varepsilon t) = H(p, q, \varepsilon t) + \partial[S(I, q, \varepsilon t)]/\partial t,$$

and we may write $K(I, \varphi, \varepsilon t)$ in the form:

(4) $$K(I, \varphi, \varepsilon t) = K_0(I, \varepsilon t) + \varepsilon K_1(I, \varphi, \varepsilon t),$$

where $K_1(I, \varphi, \varepsilon t)$ is the derivative $\partial S(I, q, \lambda)/\partial\lambda$ evaluated at the point $(I, q(I, \varphi, \varepsilon t), \lambda = \varepsilon t)$. K is thus expressed as the sum of the integrable, slow-time dependent Hamiltonian K_0 and the Hamiltonian perturbation εK_1 . We may give a more explicit form of the perturbation term by introducing the function $\mathcal{S}(I, \varphi, \lambda) = S(q[I, \varphi, \lambda], I, \lambda)$ and writing (with an implicit summation over i):

(5) $$\begin{aligned}\partial\mathcal{S}(I, \varphi, \lambda)/\partial\lambda &= (\partial S/\partial\lambda)(q, I, \lambda) + \\ &+ [(\partial S/\partial q_i)(q, I, \lambda)][(\partial q_i/\partial\lambda)(I, \varphi, \lambda)] \\ &= (\partial S/\partial\lambda)(q(I, \varphi, \lambda), I, \lambda) + [p_i\partial q_i/\partial\lambda](I, \varphi, \lambda).\end{aligned}$$

The derivative $\partial S(I, q, \lambda)/\partial\lambda$ at the point $(I, q(I, \varphi, \varepsilon t), \lambda = \varepsilon t)$ is then expressed by:

(6) $$\partial S(I, q, \lambda)/\partial\lambda = (\partial\mathcal{S}(I, \varphi, \lambda)/\partial\lambda)(I, \varphi, \varepsilon t) - (p_i\partial q_i/\partial\lambda)(I, \varphi, \varepsilon t).$$

Before applying the averaging theorem, it is perhaps worthwhile to point out the differences separating this case from the problem with a single degree of freedom. We recall that, following Arnold (see Section 8.6), it was then possible to add an auxiliary action variable making the unperturbed Hamiltonian autonomous. A version of the KAM theorem could then be applied, yielding adiabatic invariance of the action for all time. It is tempting to try the same thing here, but the procedure fails for apparently technical reasons (it may be useful to actually do it); in fact even if the first part of the procedure were feasible, one would not apply the KAM theorem, since the invariant tori do not partition the energy surface in more than two dimensions, but rather Nekhoroshev's theorem. This would furnish a much stronger statement than the one we prove here, but that statement would be simply *false*, which explains, a posteriori, why one cannot reduce the problem to the perturbation of an autonomous Hamiltonian, and why we give a *non-Hamiltonian* proof of the (optimal) theorem using Neistadt's averaging result.

To do this, we write the equations of motion:

(7) $$\begin{aligned}dI/dt &= -\varepsilon\partial K_1(I, \varphi, \varepsilon t)/\partial\varphi; \\ d\varphi/dt &= \partial K_0(I, \varepsilon t)/\partial I + \varepsilon\partial K_1(I, \varphi, \varepsilon t)/\partial I.\end{aligned}$$

We are then led to consider the system:

$$\begin{aligned} &dI/dt = -\varepsilon\partial K_1(I, \varphi, \lambda)/\partial\varphi \\ (8) \qquad &d\varphi/dt = \partial K_0(I, \lambda)/\partial I + \varepsilon\partial K_1(I, \varphi, \lambda)/\partial I \\ &d\lambda/dt = \varepsilon \end{aligned}$$

with the constraint $\lambda(0) = 0$.

Let us ignore this constraint for the moment and apply the averaging theorem from Chapter 6 to this non-Hamiltonian system with slow variables I and λ, for trajectories with initial conditions in $W = B(\sigma) \times [-1, 0] \times T^n$. The associated averaged system is simply:

$$(9) \qquad dJ/dt = 0; \quad d\lambda/dt = \varepsilon.$$

This system satisfies the hypotheses required by Theorem 2 of chapter 6 with the approximation $\rho(\varepsilon)/2$ on the time interval $[0, 2/\varepsilon]$. There thus exists a partition $W = W' \cup W''$ such that:

- The measure of W'' is of the order $\sqrt{\varepsilon}/\rho(\varepsilon)$.
- For every initial condition in W' the variation in action $\| I(t) - I(0) \|$ remains less than $\rho(\varepsilon)/2$ on the time interval $[0, 2/\varepsilon]$.

We note that, throughout the time interval $[0, 2/\varepsilon]$, the trajectories with small variation in action remain in the domain $B(2\sigma) \times [-1, 2] \times T^n$ on which the system is defined. It remains for us to deduce the adiabatic theorem from this result, by showing that the constraint on λ is not essential.

Let $(I_0, \varphi_0) = (I(0), \varphi(0))$ be an initial condition in $B(\sigma) \times T^n$ such that the drift in action on the time interval $[0, 1/\varepsilon]$ is greater than $\rho(\varepsilon)$; in other words, (I_0, φ_0) is in V''. Then for the initial condition $(I_0, 0, \varphi_0)$ in $B(\sigma) \times [0, 1] \times T^n$, the separation between the exact trajectory (I, λ) and the averaged trajectory (J, λ) surpasses $\rho(\varepsilon)$ on the time interval $[0, 1/\varepsilon]$.

On the time interval $[-1/\varepsilon, 2/\varepsilon]$, let us now consider the trajectory $(I(t), \lambda(t), \varphi(t))$ of system (8) with initial condition $(I_0, 0, \varphi_0)$. Every point $(I(-t^*), \lambda(-t^*), \varphi(-t^*))$ of that portion of the trajectory described during the time interval $[-1/\varepsilon, 0]$, or the portion described on the interval $[-t', 0]$ if the trajectory leaves $B(2\sigma) \times [-1, 0] \times T^n$ at time $-t' > -1/\varepsilon$, may be chosen as the initial condition for an exact trajectory such that the separation in I between this trajectory and the corresponding averaged trajectory surpasses $\rho(\varepsilon)/2$ at some point in time before $1/\varepsilon$. This follows readily from the fact that the maximal variation of the action on the time interval $[-t^*, 1/\varepsilon]$ must be greater than or equal to its maximal variation on the time interval $[0, 1/\varepsilon]$ (see Figure 9.2).

Consequently, we can associate to every initial condition (I_0, φ_0) in $V'' \cap [B(\sigma)\times T^n]$, that is, a "bad" initial condition for the adiabatic theorem with maximal action variation $\rho(\varepsilon)$, a set of points $D(I_0, \varphi_0)$ in W'', namely the set $\{ (I(t), -\varepsilon t, \varphi(t)),$ $t \in [-t', 0] \}$ where t' depends on (I_0, φ_0), takes values less than or equal to $1/\varepsilon$, and admits a strictly positive uniform lower bound c/ε.

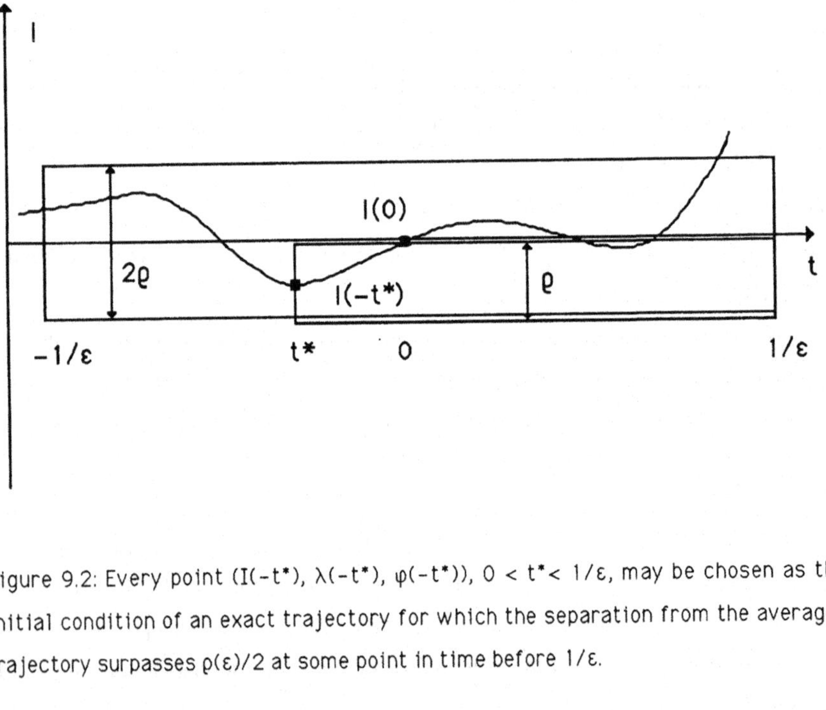

Figure 9.2: Every point $(I(-t^*), \lambda(-t^*), \varphi(-t^*))$, $0 < t^* < 1/\varepsilon$, may be chosen as the initial condition of an exact trajectory for which the separation from the averaged trajectory surpasses $\rho(\varepsilon)/2$ at some point in time before $1/\varepsilon$.

If the adiabatic theorem did not hold, the measure of $V'' \cap [B(\sigma)\times T^n]$ would be greater than $c\sqrt{\varepsilon}/\rho(\varepsilon)$ for all real positive c. By a simple argument, already used in Chapter 4 to estimate the measure of the initial conditions to be excluded, we deduce that the measure of the union of the sets $D(I_0, \varphi_0)$ for (I_0, φ_0) in $V'' \cap [B(\sigma)\times T^n]$, and thus the measure of W'', would then be greater than $c\sqrt{\varepsilon}/\rho(\varepsilon)$, which contradicts the averaging theorem used here (for an approximation of $\rho(\varepsilon)/2$ on the time interval $2/\varepsilon$).

The measure of $V'' \cap [B(\sigma)\times T^n]$ is thus less than $c\sqrt{\varepsilon}/\rho(\varepsilon)$, and the measure of V'' is of the order $\sqrt{\varepsilon}/\rho(\varepsilon)$, which establishes the adiabatic theorem. ∎

9.3 The behavior of the angle variables

For such integrable Hamiltonians, one may also study the adiabatic behavior of the angle variables in the case where the parameter on which the Hamiltonian depends is a periodic vector function of the slow time εt. We give here a brief summary of the recent work by J.H. Hannay ([Han]) and M. Berry ([Ber4]) and refer to the original papers for interesting applications of the formula derived below (see (7)); in Chapter 10 we will briefly examine the quantum conterpart of these results.

We begin with a family of Hamiltonians with at least two parameters, denoted by the *vector* λ, which are periodic functions of the slow time εt, with period 1. We assume again that for each fixed value of λ, the Hamiltonian $H(p, q, \lambda)$ is integrable. As before, we may change to action-angle variables via a time dependent variable transformation and so obtain the Hamiltonian $K_0(I, \lambda(\varepsilon t)) + \varepsilon K_1(I, \varphi, \lambda(\varepsilon t))$.

Let $T(I(0), \lambda(0))$ be an invariant torus for the Hamiltonian K_0 and ω denote the frequency vector of K_0. The interesting quantity to calculate in the adiabatic limit is the shift:

$$\Delta\varphi(I(0)) = \lim_{\varepsilon\to 0} [\varphi(1/\varepsilon) - \varphi(0) - \int_0^{1/\varepsilon} \omega\,(I(t), \lambda(\varepsilon t))\, dt]$$

of the angle variables during one period for an initial condition on $T(I(0), \lambda(0))$, assuming the above adiabatic limit is meaningful for that initial condition.

As we saw previously, the Hamiltonian perturbation may be expressed as $[\partial S(I, q, \lambda)/\partial\lambda_j].[d\lambda_j(\varepsilon t)/dt]$ (implicit summation is assumed), where S is the local generating function of the canonical transformation, and, according to expression (6) of the previous section, this perturbation may also be written in the form:

$$\partial \mathcal{S}/\partial\lambda_j . d\lambda_j/dt \; - \; p_k \partial q_k/\partial\lambda_j . d\lambda_j/dt,$$

where $\mathcal{S}(I, \varphi, \lambda) = S(q[I, \varphi, \lambda], I, \lambda)$ and the functions are calculated at the point $(I, \varphi, \lambda(\varepsilon t))$.

The evolution of the angle variables φ_i is given by Hamilton's equations:

(1) $\quad d\varphi_i/dt = \omega_i(I, \lambda) + \varepsilon\partial K_1/\partial I_i(I, \varphi, \lambda); \quad \omega_i(I, \lambda) =_{def} \partial K_0(I, \lambda)/\partial I_i\,.$

We thus have:

(2) $$\varphi_i(1/\varepsilon) = \varphi_i(0) + \int_0^{1/\varepsilon} \omega_i(I(u), \lambda(\varepsilon u))\, du + \int_0^{1/\varepsilon} \varepsilon\partial K_1/\partial I_i\, (I(u), \varphi(u), \lambda(\varepsilon u))\, du$$

and we are concerned with evaluating the second term. To start with, we may return to the adiabatic invariance of the action. Let $\rho(\varepsilon)$ be a function greater than $c\sqrt{\varepsilon}$ and tending to 0 with ε. We will only consider initial conditions such that the action is conserved to within $\rho(\varepsilon)$ (the set V' introduced above), in which case we may take the action to be constant in the integral to be calculated, since the error committed goes to 0 with ε, uniformly on V'. It remains to evaluate:

$$(3) \qquad \lim_{\varepsilon \to 0} \int_0^{1/\varepsilon} \varepsilon \partial K_1/\partial I_i \, (I(0), \varphi(u), \lambda(\varepsilon u))\, du$$

or:

$$(4) \qquad \lim_{\varepsilon \to 0} \int_0^{1/\varepsilon} \partial/\partial I_i \, [\partial S/\partial \lambda_j . d\lambda_j/dt - p_k . \partial q_k/\partial \lambda_j . d\lambda_j/dt]\, dt,$$

the integrand being estimated at the point $(I(0), \varphi(t), \lambda(\varepsilon t))$. We may pass from integration over time to integration over the closed curve C in parameter space which is described by the vector parameter λ during the time interval $[0, 1/\varepsilon]$. Upon interchanging the order of integration in parameter space and differentiation with respect to action, we obtain the integral:

$$\partial/\partial I_i \, [\int_C \{dS - p_k dq_k\}(I(\lambda), \varphi(\lambda), \lambda)],$$

where the derivative is calculated at the point $I(0)$ and where dS and dq_k are differential forms *in parameter space*.

In the limit $\varepsilon \to 0$, it is natural to average over the fast variables φ (cf. [Han]). This presumes that when $\varepsilon \to 0$, the trajectory $(I(t), \varphi(t))$ with initial condition $(I'(0), \varphi(0))$, where $I'(0)$ is near $I(0)$, tends to a trajectory $I = I'(0)$ which ergodically fills the torus, a point that appears difficult to demonstrate rigorously. Assuming it does hold, in order to study the motion in the limit $\varepsilon \to 0$, we may replace the differential form $dS - p_k dq_k$ by its *average on the torus* $T(I)$ defined as:

$$(5) \qquad \langle dS - p_k dq_k \rangle (I, \lambda) =_{def} (1/2\pi)^n \int_{T(I)} \{dS - p_k dq_k\}(I, \varphi, \lambda)\, d\varphi.$$

The adiabatic shift of the angle variables is then given by:

$$(6) \qquad \Delta\varphi_i(I(0), C) = \partial/\partial I_i \, [\int_\Gamma \langle dS - p_k dq_k \rangle (I, \lambda)\,],$$

where the derivative is evaluated at the point $I(0)$.

The first term vanishes upon integration since $\langle dS \rangle$ is an exact differential, and we thus obtain:

$$(7) \qquad \Delta\varphi_i\,(I(0), C) = -\,\partial/\partial I_i \, [\int_C \langle p_k dq_k \rangle (I, \lambda)] = -\,\partial/\partial I_i \, [\int_S d^2 w\,]$$

where the derivative is again at $I = I(0)$, S is an arbitrary oriented surface in parameter space with boundary C, and $d^2 w$ is the 2-form *in parameter space:*

$$d^2 w(I, \lambda) =_{def} \langle dp_k \wedge dq_k \rangle (I, \lambda)$$

obtained by averaging the 2-form $dp_k \wedge dq_k(I, \varphi, \lambda)$ over the torus $T(I)$.

In Chapter10, we will see how formula (7) relates to an analogous formula in quantum mechanics.

9.4 The ergodic adiabatic theorem

Here we prove the second result mentioned at the outset concerning one-parameter families of Hamiltonians with n degrees of freedom whose flows are ergodic on the energy surface for every fixed value of the parameter. Then the "volume enclosed by the energy surface" is an adiabatic invariant as will be shown below and no other invariant exists. We remind the reader that what follows is a somewhat clarified version of Kasuga's proof ([Kas]).

Consider a one-parameter family of Hamiltonians $H(p, q, \lambda)$, where again the parameter $\lambda \in [0, 1]$ is simply the slow time $\lambda = \varepsilon t$. For convenience we also set $\tau = 1/\varepsilon$, the length of the time interval on which the system will be considered. We assume that:

- H is a C^2 function of the variables p_i , q_i and λ;
- For every fixed λ in $[0, 1]$, the energy sufaces of the Hamiltonian $H(p, q, \lambda)$ that is, the surfaces in phase space given by $H(p, q, \lambda) = E$, are compact;
- For each fixed λ in $[0, 1]$, there exists an open interval $(E(\lambda), E'(\lambda))$ such that for each E in this interval the flow of the Hamiltonian $H(p, q, \lambda)$ possesses no fixed point on the energy suface $H(p, q, \lambda) = E$;
- For almost every pair (λ, E) in $\bigcup_{\lambda \in [0, 1]} \{\lambda\} \times (E(\lambda), E'(\lambda))$, the flow of $H(p, q, \lambda)$ is ergodic on the energy suface $H(p, q, \lambda) = E$.

On $[0, 1] \times \mathbb{R}^{2n}$ (the product of the parameter domain and phase space), we now define the function $I(\lambda, \tilde{p}, \tilde{q})$ as the measure of the volume in phase space enclosed by the energy surface of $H(p, q, \lambda)$ on which $(\tilde{p}, \tilde{q})$ is located:

(1) $\qquad I(\lambda, \tilde{p}, \tilde{q}) =_{def} \mathrm{mes}\{ (p, q) \text{ such that } H(p, q, \lambda) \leq H(\tilde{p}, \tilde{q}, \lambda) \}.$

This function depends only on λ and on the energy $E = H(\tilde{p}, \tilde{q}, \lambda)$, and we will sometimes denote it by $I(\lambda, E)$. We note that although knowledge of $H(p, q, \lambda)$ is enough to determine I, this of course does not make it a functional of the Hamiltonian.

For fixed $\lambda \in [0, 1]$, the domain $D(\lambda, I, I')$ in $\mathbb{R}^{2n}$ is defined as the set of points between the energy surfaces of $H(p, q, \lambda)$ corresponding to E and E' such that $I(\lambda, E) = I$

and $I(\lambda, E') = I'$:

$$D(\lambda, I, I') = \{ (p, q) \in \mathbb{R}^{2n} \text{ such that } I < I(\lambda, p, q) < I' \}.$$

We next define $I(\lambda) = I(\lambda, E(\lambda))$ and $I'(\lambda) = I(\lambda, E'(\lambda))$. We assume that $\cap_{\lambda\in[0,1]}(I(\lambda), I'(\lambda))$ is nonempty and we set:

$$[I^*, I'^*] = \cap_{\lambda\in[0,1]}[I(\lambda), I'(\lambda)].$$

For every initial condition (p_0, q_0) in $D(0, I^*, I'^*)$, and every value $\tau = 1/\varepsilon > 0$ of the adiabaticity parameter, we define the maximal variation of I as:

$$(2) \qquad \Delta(p_0, q_0, \tau) =_{def} \operatorname{Sup}_{\lambda\in[0,1]} | I(\lambda, p_\tau(\lambda), q_\tau(\lambda)) - I(0, p_0, q_0) |$$

where $(p_\tau(\lambda), q_\tau(\lambda))$ denotes the solution of Hamilton's equations for the Hamiltonian $H(p, q, \lambda = \varepsilon t = t/\tau)$, expressed as a function of the slow time λ:

$$(3) \qquad dp_\tau/d\lambda = -\tau\partial H/\partial q; \quad dq_\tau/d\lambda = \tau\partial H/\partial p,$$

provided this solution may be extended to $\lambda = 1$. Otherwise we set $\Delta = \infty$.

Finally, for every real positive δ and every value of the adiabaticity parameter τ, we define $L(\tau, \delta)$ as the set of initial conditions (p_0, q_0) in $D(0, I^*, I'^*)$ such that the variation of I on the corresponding trajectory does not exceed δ: $\Delta(p_0, q_0, \tau) < \delta$. The adiabatic theorem may now be stated.

Theorem:

For every $\delta > 0$, and every segment $[I_1, I_2]$ contained in (I^*, I'^*), the measure of the set of intial conditions in $D(0, I_1, I_2)$ for which the variation of I is greater than δ tends to 0 in the adiabatic limit $\tau \to \infty$ (or $\varepsilon \to 0$):

$$\lim_{\tau\to\infty} \operatorname{mes} [D(0, I_1, I_2) - L(\tau, \delta)] = 0.$$

Throughout the proof, we will work in the domain:

$$D = \cup_{\lambda\in[0,1]} \{\lambda\} \times D(\lambda, I^*, I'^*).$$

To begin the proof, for each λ in $[0, 1]$ and each I in (I^*, I'^*), we introduce a surface measure $\nu_{\lambda,I}$ on the energy surface $\Sigma(\lambda, I)$ of the Hamiltonian $H(p, q, \lambda)$ (determined by $H(p, q, \lambda) = E$, where E satisfies $I(\lambda, E) = I$), such that:

$$(3) \qquad \int_{D(\lambda,I^*,I'^*)} f(p, q)\, dpdq = \int_{I^*}^{I'^*} dI \left[\int_{\Sigma(\lambda,J)} f(p, q)\, d\nu_{s,I}\right]$$

for continuous functions with support in $D(\lambda, I^*, I'^*)$. It is easy to see that this measure is unique and is given by:

$$(4) \qquad d\nu_{\lambda,I} = [\| \nabla H(p, q, \lambda) \| (dI(\lambda, E)/dE)]^{-1}\, d\sigma_{\lambda,I},$$

where $d\sigma_{\lambda,I}$ designates the natural surface measure on $\Sigma(\lambda, I)$ and the derivative dI/dE is evaluated at the point $(\lambda, E(\Sigma(\lambda, I)))$. This measure is normalized:

$$(5) \qquad \int_{\Sigma(\lambda,I)} d\nu_{\lambda,I} = 1.$$

In fact the volume in phase space between two neighboring energy surfaces E and $E + \delta E$ varies inversely with the gradient of the Hamiltonian:

$$(6) \qquad dI/dE = \int_{\Sigma(\lambda,I)} d\sigma_{\lambda,I} / \| \nabla H \|.$$

This implies (5) and explains the meaning of (4); in symbols:

$$dpdq = dEd\sigma/\| \nabla H \| = dId\sigma/[\| \nabla H \|.(dI/dE)].$$

We now state

Lemma 1:

$$\int_{\Sigma(\lambda,I)} [\partial I/\partial\lambda](\lambda, E)\, d\nu_{\lambda,I} = 0.$$

To see this, let $G(I)$ be a C^1 function whose derivative has compact support in $[I^*, I'^*]$. Set $F(p, q, \lambda) = G(I(\lambda, p, q))$. Then:

$$(7) \qquad \int_{D(\lambda,I^*,I'^*)} F\, dpdq = \int_{D(\lambda,I^*,I'^*)} F\, dI(\lambda, p, q) d\nu_{\lambda,I}$$
$$= \int_{I^*}^{I'^*} G(I) \{ \int_{\Sigma(\lambda,I)} d\nu_{\lambda,I} \} dI = \int_{I^*}^{I'^*} G(I)\, dI.$$

The value of the integral is thus independent of λ. Consequently:

$$\int_{D(\lambda,I^*,I'^*)} \partial F/\partial\lambda\, dpdq = 0,$$

or:

$$(8) \qquad \int_{D(\lambda,I^*,I'^*)} dG/dI.\partial I/\partial\lambda\, dpdq = 0,$$

and by using the decomposition of Liouville's measure:

$$(9) \qquad \int_{I^*}^{I'^*} dG/dI \{ \int_{\Sigma(\lambda,I)} \partial I/\partial\lambda(\lambda, E)\, d\nu_{\lambda,I} \} dI = 0,$$

from which the lemma follows, since G was arbitrary. ■

For each fixed λ in $[0, 1]$, we denote by Φ_λ the flow of the Hamiltonian $H(p, q, \lambda)$ on the space $\mathbb{R}^{2n}$, and by Φ the product of these flows; that is, the flow on $[0, 1] \times \mathbb{R}^{2n}$ whose restriction to $\{\lambda\} \times \mathbb{R}^{2n}$ coincides with Φ_λ. The following lemma is immediate, and reexpresses the ergodicity on the energy surfaces.

Lemma 2:

Every complex valued function $f(\lambda, p, q)$ in $L^2(D, d\lambda dpdq)$ which is invariant under the flow Φ is orthogonal to $\partial I/\partial\lambda$:

$$\int_D f^*\, \partial I/\partial\lambda\, d\lambda dpdq = 0$$

where f^* denotes the complex conjugate of f.

In fact, for almost every λ, the restriction of f to $\{\lambda\} \times D(\lambda, I^*, I'^*)$ is invariant under the flow Φ_λ and, because of the ergodicity of this flow, is a function of only the volume $I(\lambda, p, q)$. There thus exists a function g on $[0, 1] \times [I^*, I'^*]$ such that $f(p, q, \lambda) = g[\lambda, I(\lambda, p, q)]$. Consequently, by Lemma 1 it follows that:

$$(10) \qquad \int_D f^* \, \partial I/\partial\lambda \, d\lambda dpdq = \int_0^1 d\lambda \, [\int_{D(\lambda,I^*,I'^*)} f^* \, \partial I/\partial\lambda \, dpdq]$$
$$= \int_0^1 d\lambda \, [\int_{D(\lambda,I^*,I'^*)} g^*(\lambda, I)) \, \partial I/\partial\lambda \, dpdq] =$$
$$= \int_0^1 d\lambda \, [\int_{I^*}^{I'^*} g^*(\lambda, I) \, \{\int_{\Sigma(\lambda,I)} \partial I/\partial\lambda \, d\nu_{\lambda,I}\} \, dI] = 0. \quad \blacksquare$$

Following the Hilbert space approach of classical mechanics (see Appendix 2), we see that the flow Φ which preserves the measure $d\lambda dpdq$ induces, on $L^2(D, d\lambda dpdq)$, a one parameter group of unitary operators $U(t) = \exp[tL_H]$ parametrized by t.

The generator L_H (Poisson bracket with H) is first defined on the space $C_0{}^1(D^0)$ composed of C^1 functions with compact support in the interior of D:

$$(11) \qquad L_H[f] = \{ H, f \} = \sum_{1 \leq i \leq n} (\partial H/\partial p_i \partial f/\partial q_i - \partial H/\partial q_i \partial f/\partial p_i).$$

We then extend it to the space $L^2(D, d\lambda dpdq)$ by introducing the self-adjoint extension, again denoted by L_H.

The evolution of a function $f(\lambda, p, q)$ in $L^2(D, d\lambda dpdq)$ is then governed by the equation:

$$(12) \qquad df(\lambda, p, q)/d\lambda = \partial f(\lambda, p, q)/\partial\lambda + \tau\{H, f\} = \partial f(\lambda, p, q)/\partial\lambda + \tau L_H [f].$$

The kernel of L_H is the space of functions invariant under the flow Φ. We may thus reformulate Lemma 1 as follows: The function $\partial I/\partial\lambda$ is orthogonal to the kernel of L_H.

In addition, the image under L_H of the space $C_0{}^1(D^0)$ is dense in the orthogonal complement of $\mathrm{Ker}(L_H)$, which provides the proof of

<u>Lemma 3</u>:

For each $\eta > 0$, there exists a function $f_\eta(\lambda, p, q)$ in $C_0{}^1(D^0)$ such that:

$$(13) \qquad \| \{ H, f_\eta \} - \partial I/\partial\lambda \|_{L^2(D)} < \eta/2 \, [\mathrm{mes}(D)]^{1/2}$$

with the product measure $d\lambda dpdq$.

The second member has been written so as to facilitate subsequent calculations. It is important to note that this lemma is the analogue of solving the linearized equation in Chapter 6. When the system is ergodic, there are no more tori, and one may no longer make use of the Fourier series that were so essential in the previous case. We are thus led to use functional analysis to find an approximate solution, which of course yields a less precise result. Applying Schwarz's lemma, we find:

$$(14) \qquad \int_D |\{H, f_\eta\} - \partial I/\partial s| \, d\lambda dp dq < \eta/2.$$

We will now prove a lemma whose statement is very close to that of the theorem, and from which the theorem will follow by using a "bootstrap" argument. We first set $M = \mathrm{Sup}_D \, |\partial I/\partial\lambda(\lambda, p, q)|$. Now let $[I_1, I_2]$ be a segment contained in (I^*, I'^*) and let a, b be real numbers in $[0, 1]$ such that:

$$0 < b - a < \mathrm{Inf}\,([I_1 - I^*]/2M, [I'^* - I_2]/2M).$$

Each trajectory whose initial condition at time $\lambda = a$ lies in $D(a, I_1, I_2)$ is then defined on the time interval $[a, b]$ and is contained in D. We designate by $L(a, b, \tau, \delta)$ the set of initial conditions (p, q) in $D(a, I^*, I'^*)$ such that, on the corresponding trajectory $(p_\tau(\lambda), q_\tau(\lambda))$, the variation

$$\Delta(a, b, p, q, \tau) = \mathrm{Sup}_{\lambda\in[a,b]} \, |I(\lambda, p_\tau(\lambda), q_\tau(\lambda)) - I(a, p, q)|$$

of I on the time interval $[a, b]$ is greater than δ. We will show that, in the adiabatic limit $\tau \to \infty$, almost all initial conditions in $D(a, I_1, I_2)$ do not belong to $L(a, b, \tau, \delta)$:

<u>Lemma 4</u>:

For all real $\delta > 0$: $\lim_{\tau\to\infty} \mathrm{mes}[D(a, I_1, I_2) - L(a, b, \tau, \delta)] = 0$.

Let $\eta > 0$, let f_η be a function in $C_0^1(D^0)$ satisfying (14), and set $g_\eta = \{H, f_\eta\} - \partial I/\partial\lambda$. The evolution of $I(\lambda, E)$ is given by:

$$(15) \qquad dI(\lambda, E)/ds = \partial I(\lambda, E)/\partial\lambda + \tau L_H[I].$$

Since for each fixed λ, I depends only on the value of the Hamiltonian $H_\lambda(p, q) = H(p, q, \lambda)$, the second term vanishes and $dI/d\lambda = \partial I/\partial\lambda$. For all λ in $[a, b]$ we thus have:

$$(16) \qquad |I(\lambda) - I(a)| = \left|\int_a^\lambda \partial I/\partial\lambda \, d\lambda\right| \leq \int_a^\lambda |g_\eta| \, d\lambda + \left|\int_a^\lambda \{H, f_\lambda\} \, d\lambda\right|.$$

It follows that the variation $\Delta(a, b, p, q, \tau)$ of I on the time interval $[a, b]$ along a trajectory with initial condition (p, q) in $D(a, I_1, I_2)$ is bounded by:

$$\int_a^b |g_\lambda| \, d\lambda + \mathrm{Sup}_{\lambda\in[a,b]} \left|\int_a^\lambda \{H, f_\eta\} \, d\lambda\right|.$$

In addition, the measure of $L(a, b, \tau, \delta) \cap D(a, I_1, I_2)$ is bounded by the integral: $1/\delta \int \Delta(a, b, p, q, \tau) \, dpdq$, where the domain of integration is $D(a, I_1, I_2)$. It follows that:

$$(17)\qquad mes(L(a,b,\tau,\delta)\cap D(a,I_1,I_2)) \leqslant 1/\delta\int dp_0 dq_0 \int_a^b |g_\eta(\lambda,p_\tau(\lambda),q_\tau(\lambda))|\, d\lambda$$
$$+\ 1/\delta \int dp_0\, dq_0\, Sup_{\lambda\in[a,b]}\, |\int_a^\lambda \{H(\lambda,p_\tau,q_\tau),\ f_\eta(\lambda,p_\tau,q_\tau)\}\, d\lambda\, |$$

where the integration over the initial conditions $(p_0\ ,\ q_0)$ is limited to the domain $D(a, I_1\ , I_2)$.

We may interchange the integrals in the first term and (since the flow preserves measure), replace the integration over the initial conditions by an integration in the coordinates $(p_\tau(\lambda),\ q_\tau(\lambda))$. By (14), it then follows that this first term is bounded by $\eta/2\delta$.

As for the second term, since $\tau\{H, f_\eta\} = df_\eta/d\lambda - \partial f_\eta/\partial\lambda$, we have:

$$(18)\qquad |\int_a^\lambda \{H, f_\lambda\}\, d\lambda\, | < 1/\tau \int_a^b |\, \partial f_\eta/\partial\lambda\, |\, d\lambda + 1/\tau\, |\, f_\eta(\lambda) - f_\lambda(a)\, |,$$

which shows that $|\int_a^\lambda\{H, f_\lambda\}\, d\lambda\, |$ is bounded by $c_2(\eta)/\tau\delta$, where c_2 depends on η since f_η does. However, the singular behavior of f_η as η approaches 0 prevents from having information about the behavior of $c_2(\eta)$. The second term of the right hand side of (17) is thus bounded by $c_3(\eta)/\tau$, and it follows that the measure of $L(a, b, \tau, \delta) \cap D(a, I_1\ , I_2)$ is bounded by $(\eta + c_3(\eta)/\tau)/\delta$.

For each fixed value of δ and for all $\zeta > 0$, we may choose $\eta = \delta\zeta/2$. For $\tau > c_3(\eta)/\eta$, the measure of this set of initial conditions is then less than ζ, which completes the proof of Lemma 4 . ▪

It remains only to deduce the adiabatic theorem from this result via a classical "bootstrap" argument. According to Lemma 4 we know that, for each segment $[I_1\ , I_2]$ contained in $(I^*,\ I'^*)$, for all $\delta > 0$, and for fixed Λ' less than

$$\min\ \{[I_1 - I^*\ ]/2M,\ [\ I'^* - I_2\ ]/2M\},$$

the adiabatic limit ($\tau \to \infty$) of $mes[D(0, I_1\ , I_2) - L(0, \sigma', \tau, \delta)]$ is zero. Let Λ be the upper bound of the Λ' less than 1 such that this limit is zero. We will show by contradiction that Λ equals 1, which will prove the theorem. We suppose that for a certain segment $[I_1\ , I_2]$ the upper bound Λ is strictly less than 1. For all $\lambda' < \Lambda$,

$$\lim_{\tau\to\infty}\ mes[D(0, I_1\ , I_2) - L(0, \lambda', \tau, \delta/2)] = 0,$$

and this holds for all δ. Let δ_0 be less than $\min\{(I_1 - I^*\ ),\ (I'^* - I_2)\}$. For λ' sufficiently near Λ, we can find λ'' in $(\Lambda,\ 1]$ such that, for all $\delta > 0$:

$$\lim_{\tau\to\infty}\ mes\ [D(\lambda', I_1-\delta_0\ , I_2+\delta_0\ ) - L(\lambda', \lambda'', \tau, \delta/2)] = 0,$$

by Lemma 4.

Since for every δ less than $2\delta_0$, the trajectories associated with the initial conditions in $D(0, I_1, I_2) \cap L(0, \lambda', \tau, \delta/2)$ reach points in $D(\lambda', I_1 - \delta_0, I_2 + \delta_0)$ at time λ', and may be further extended to time λ'', we deduce from the above that, for all $\delta < 2\delta_0$:

$$\lim_{\tau\to\infty} \mathrm{mes}[\, D(0, I_1, I_2) - L(0, \lambda'', \tau, \delta)\,] = 0.$$

Since $L(0, \lambda'', \tau, \delta)$ is contained in $L(0, \lambda'', \tau, \delta'')$ for $\delta' \leqslant \delta''$, the adiabatic limit of $\mathrm{mes}\,[D(0, I_1, I_2) - L(0, \lambda'', \tau, \delta)]$ vanishes for all $\delta > 0$; but since λ'' is strictly greater than Λ, we have a contradiction. Λ is thus equal to 1, which completes the proof of the adiabatic theorem. ▪

It is easy to see that the volume I is essentially the only adiabatic invariant of the system; in other words, all other invariants are functions of I. For fixed λ, any adiabatic invariant $\varphi(\lambda, p, q)$ of the system may only depend on the value of the Hamiltonian $H_\lambda(p, q)$, since for almost every pair (λ, E), the invariant must be almost everywhere constant on the energy surface $H_\lambda(p, q) = E$, by reason of the ergodicity of the flow. Every adiabatic invariant is thus a function of λ and E alone, or equivalently, of λ and I alone (we repeat that E and I are interdependent, but that E is a function of I *and* λ). We will now show that φ may not depend on λ. We have:

$$(19) \qquad d\varphi/d\lambda = \partial\varphi/\partial\lambda + \partial\varphi/\partial I_\lambda(\partial I_\lambda/\partial\lambda + \tau L_H\,[I_\lambda])$$
$$= \partial\varphi/\partial\lambda + \partial\varphi/\partial I_\lambda . \partial I_\lambda/\partial\lambda.$$

Consequently, for any segment $[a, b]$ contained in $[0, 1]$ and for every λ in $[a, b]$:

$$(20) \qquad \varphi(\lambda, I_\lambda) - \varphi(a, I_\lambda) = \int_a^\lambda \partial\varphi/\partial\lambda\, d\lambda + \int_a^\lambda (\partial\varphi/\partial I_\lambda)(\partial I_\lambda/\partial\lambda)\, d\lambda.$$

Since $\partial I/\partial\lambda = \{H, f_\eta\} + g_\eta = 1/\tau\,[df_\eta/d\lambda - \partial f_\eta/\partial\lambda] + g_\lambda$, it follows that:

$$(21) \qquad \varphi(\lambda, I_\lambda) - \varphi(a, I_a) = \int_a^\lambda \partial\varphi/\partial\lambda\, d\lambda + \int_a^\lambda \partial\varphi/\partial I_\lambda\, g_\eta\, d\lambda -$$
$$- 1/\tau \int_a^\lambda \partial\varphi/\partial I_\lambda\, \partial f_\eta/\partial\lambda\, d\lambda + 1/\tau \int_a^\lambda \partial\varphi/\partial I_\lambda\, df_\eta/d\lambda\, d\lambda.$$

The second term in this sum is bounded in norm by $c\eta$ and the third term is bounded by $c/\tau \mid f_\eta(\lambda) - f_\eta(a) \mid$. Integrating the last term by parts yields:

$$1/\tau\,[\partial\varphi/\partial I\, f_\eta(\lambda) - \partial\varphi/\partial I\, f_\eta(a)] - 1/\tau \int_a^\lambda f_\eta\,[d(\partial\varphi/\partial I)/d\lambda]\, d\lambda.$$

The integrated term is bounded in norm by $c/\tau\, |f_\eta(\lambda) - f_\eta(a)|$ and the integral in the second term is independent of τ. Thus, with the exception of the first one, all terms on the right hand side of (21) tend to zero in the adiabatic limit, which in turn implies:

$$(22) \qquad \lim_{\tau\to\infty} [\varphi(\lambda, I_\lambda) - \varphi(a, I_a) - \int_a^\lambda \partial\varphi/\partial\lambda\, d\lambda\,] = 0.$$

Since I is an adiabatic invariant, φ may only be an adiabatic invariant provided that $\partial\varphi(\lambda, I)/\partial\lambda = 0$, in other words only if φ is a function of the volume I alone.

In concluding this section, we return to the physical significance of the theorem by showing, in a succinct way, without considering questions of indiscernability and volume in configuration space, that for a large class of Hamiltonian systems with ergodic flows on the energy surface, in the thermodynamic limit, entropy is conserved during an (infinitely slow) adiabatic transformation.

For Hamiltonian systems which are ergodic on the energy surface, it is natural to work with the microcanonical ensemble ([Ku]), and to take:

- The (normalized) surface measure $d\nu_{\lambda,\varepsilon}$ as the microcanonical distribution function,
- The measure of the energy surface $\int_{\Sigma(s,E)} d\sigma/\| \nabla H \|$ as the partition function,
- The (classical) entropy per degree of freedom

$$S(E, n) = k/n \log \int_{\Sigma(\lambda,E)} d\sigma/\| \nabla H \|$$

as the thermodynamic potential, where n is the number of degrees of freedom and k is Boltzmann's constant.

We introduce the function:

$$S(E, \Delta E, n) = k/n \log [I(E, n) - I(E-\Delta E, n)],$$

where I is the Liouville measure of the volume enclosed by the energy surface. We have:

$$S(E, \Delta E, n) = k/n \log[(dI/dE)\Delta E],$$

where the derivative has been estimated for some value E' of the energy between $E - \Delta E$ and E. We saw earlier that: $dI/dE = \int_{\Sigma(n,E)} d\sigma/\| \nabla H \|$, and it follows that:

$$S(E, \Delta E, n) = S(E', n) + k/n \log \Delta E.$$

We now pass to the thermodynamic limit $E \to \infty$, $n \to \infty$, $\Delta E \to 0$, with a constant energy per degree of freedom $E/n = \varepsilon$, and with $\Delta E = \varepsilon/n^{\alpha}$, $\alpha > 0$. In this limit, $S(E', n)$ tends toward $S(E, n)$ and

$$k/n \log \Delta E = k/n \log \varepsilon - \alpha k/n \log n$$

tends to 0. Thus $S(E, \Delta E, n)$ tends toward $S(E, n)$ in the thermodynamic limit.

Next, we define the function $S'(E, n) = k.1/n \log I(E, n)$. For so-called *normal* thermodynamical systems, $I(E, n)$ behaves in the thermodynamical limit like $e^{n\varphi(\varepsilon)}$, where the function $\varphi(\varepsilon)$ has positive first derivative and negative second derivative, thus assuring the positivity of temperature and specific heat (in fact, $I(E, n)$ behaves like the volume of the sphere $S^{2n-1}(E)$). The derivative dI/dE behaves in turn like $\varphi'(\varepsilon)e^{n\varphi(\varepsilon)}$. To establish this behavior of the system mathematically it suffices, for example, to demand that the gradient of the Hamiltonian be bounded, and that the energy surface be convex with curvature bounded from above and below. We then have:

$$\lim_{Th} [S(E, \Delta E, n) - S'(E, n)] = \lim_{Th} k/n \log [dI/dE.\Delta E/I(E)]$$
$$= \lim_{Th} k/n \log[\varphi'(\varepsilon)\Delta E]$$
$$= \lim_{Th} k/n \log[\varphi'(\varepsilon)\varepsilon/n^{\alpha}] = 0.$$

This establishes the fact that in the thermodynamic limit $E \to \infty$, $n \to \infty$, $E/n = \varepsilon$, the entropy per degree of freedom $S(E, n)$ is expressed as a function of the volume $I(E, n)$ enclosed by the energy surface:

$$S(E, n) = k/n \log I(E, n).$$

We may thus take $I(E)$ as the microcanonical partition function, then define the entropy per degree of freedom by $S(E, n)$, as above. From the adiabatic invariance of volume, we then deduce the adiabatic invariance of the classical entropy, the various statistical ensembles being equivalent in the thermodynamic limit.

CHAPTER 10: THE QUANTUM ADIABATIC THEOREM

10.1 Statement and proof of the theorem

This chapter is devoted to the quantum adiabatic theorem, and to its connection with the classical cases examined in the preceding chapters. In this section, we present the proof of the quantum theorem in a slightly informal way, without emphasizing the regularity hypotheses nor the domains of definition of the operators involved. The first proof of the theorem appears in an article by M. Born and V. Fock ([Bor]) but it is incomplete in several respects, for example, they consider only discrete spectra, which rarely occur in quantum mechanics. Here we follow the much later proof due to T. Kato ([Kat]).

We begin with a family of Hamiltonians $H(\lambda)$ which are self adjoint on the Hilbert space $\mathcal{H}$ for every value of λ. λ may in fact represent *several* real parameters, in which case we take λ in Λ, a nice (say convex) domain in $\mathbb{R}^d$. As for regularity, we need to work with finite dimensional spectral projections of the $H(\lambda)$ that are of class C^2; for this reason we assume that the function $\lambda \to (i + H(\lambda))^{-1}$, whose values are *bounded* operators, is of class C^2.

For the moment, we will consider a Hamiltonian $H(\lambda(\varepsilon t))$ for which the parameter λ slowly describes a path in Λ. We set $H(\lambda(\varepsilon t)) = H(s)$, where $s = \varepsilon t$ is the slow time and, as in the preceding chapter, we take $\tau = 1/\varepsilon$ to be the adiabaticity parameter. We apologize for this notation, which is more or less standard in this context but differs from our previous notation (in the classical context). The evolution of the system for $s \in [0, 1]$ is then governed by the Schrödinger equation:

(1) $$\partial\psi_\tau/\partial s = -i\tau H(s)\,\psi_\tau$$

where $\hbar$ has been suppressed and the solution is parametrically dependent on τ. We assume that for all $s \in [0, 1]$, the Hamiltonian $H(s)$ admits an eigenvalue $E(s)$ of multiplicity m, independent of s, which is an *isolated* point of the spectrum. $\Pi(s)$ will denote the (orthogonal) spectral projection on the associated m dimensional eigenspace $V(s)$. The adiabatic theorem may then be stated as follows:

<u>Theorem</u>:

If the normalized initial state of the system $\psi_\tau(0) = \psi$ (independent of τ, $\|\psi\| = 1$)

belongs to the eigenspace $V(0)$ associated to the eigenvalue $E(0)$, then:

(2) $$\sup_{s\in[0,1]} \| \psi_\tau(s) - \Pi(s)\,\psi_\tau(s) \| = O(1/\tau) \;\; (= O(\varepsilon)).$$

In other words, in the adiabatic limit the state $\psi_\tau(s)$ of the system at time s belongs to the eigenspace $V(s)$ associated with the eigenvalue $E(s)$ and, moreover, its distance from the eigenspace is of order $1/\tau$. Physically, this means that in the adiabatic limit the state of the system "follows" the eigenspace, thus precluding the transition of the system to another energy level. We note that, as is often the case, the linearity of the equations makes the quantum theorem mathematically simpler than the corresponding classical theorems.

For the proof, we may assume that $E(s) = 0$ for all s by replacing the Hamiltonian $H(s)$ by $H(s) - E(s)$. This introduces only one phase, transforming $\psi_\tau(s)$ to:

$$\psi'_\tau(s) = \psi_\tau(s)\exp[i\tau\textstyle\int_0^s E(u)\,du].$$

We next introduce the "adiabatic transformation" which, as we shall see, allows us to follow the system in the limit $\tau \to \infty$. It is given by the family of operators $W(s)$ of rank m defined by:

(3) $$W(0) = \Pi(0), \qquad dW/ds = d\Pi/ds.W(s).$$

It is easy to verify that $W(s)$ is a partial isometry mapping $V(0)$ onto $V(s)$ and $V(0)^\perp$ onto $\{0\}$:

(4) $$W(s).W(s)^* = \Pi(s), \qquad W(s)^*.W(s) = \Pi(0),$$

where W^* denotes the adjoint of W.

For fixed τ, let $\{U_\tau(s),\, s \in [0, 1]\}$ be the one parameter group of unitary evolution operators associated to the Hamiltonian $H(s)$. $U_\tau(s)$ is defined by:

(5) $$U_\tau(0) = 1, \quad dU_\tau(s)/ds = -i\tau H(s)U_\tau(s).$$

If $W(s)$ is to effectively describe the evolution, $U_\tau(s)^*W(s)\,\psi$ must remain approximately constant ($U_\tau(s)^* = U_\tau(s)^{-1}$). We thus calculate:

(6) $$d/ds[U_\tau{}^*W\,\psi] = i\tau U_\tau{}^*HW\,\psi + U_\tau{}^*dW/ds\,\psi.$$

The first term in this expression vanishes, since the image of $W(s)$ is the eigenspace $V(s)$ associated with the eigenvalue 0 of $H(s)$ and, since $W(s)$ satisfies (3):

(7) $$d/ds[U_\tau{}^*W\,\psi] = U_\tau{}^*d\Pi/dsW\,\psi.$$

Now, since Π is a projection, $\Pi.d\Pi/ds.\Pi = 0$ (differentiate $\Pi^2 = \Pi$) and since

$\Pi.W = \Pi$ because $V(s)$ is the image of $W(s)$, we have:

(8) $$d/ds[U_\tau{}^* W\,\psi] = U_\tau(s)^*(1 - \Pi(s)).d\Pi/ds W(s)\,\psi.$$

Because 0 is an isolated point of the spectrum (a positive distance η from the rest of the spectrum), there exists a bounded reduced resolvant $R(s)$ such that:

$$1 - \Pi(s) = H(s)R(s).$$

If we decompose $H(s)$ as a sum of projections via the spectral theorem (see e.g. [Re]):

$$H(s) = \int_{\mathbb{R}} E\, d\Pi_E\,,$$

then we may define $R(s)$ by:

$$R(s) = \int_{|E|\geq\eta} E^{-1}\, d\Pi_E\,,$$

which also shows that $R(s)$ is a bounded operator with norm less than or equal to $1/\eta$. We thus have:

(9) $$d/ds[U_\tau{}^* W\,\psi] = U_\tau{}^*(s)H(s)R(s)d\Pi/ds.W(s)\,\psi = -\, i/\tau\, dU_\tau{}^*/ds.K(s)\,\psi,$$

where $K(s)$ is a bounded operator and even has finite rank m. We may thus write:

(10) $$U_\tau(s)^*W(s)\,\psi - \psi = -\, i/\tau \int_0^s dU_\tau{}^*/d\sigma.K(\sigma)\,\psi\, d\sigma = -\, i/\tau\, [U_\tau(s)^*K(s)\,\psi - K(0)\,\psi] + i/\tau \int_0^s U_\tau{}^*(\sigma).dK/d\sigma\,\psi\, d\sigma$$

by integrating by parts. It follows that:

(11) $$W(s)\,\psi - U_\tau(s)\,\psi = -\, i/\tau\, [K(s)\,\psi - U_\tau(s)K(0)\,\psi] + i/\tau\, U_\tau(s) \times \int_0^s U_\tau{}^*(\sigma)dK/d\sigma\,\psi\, d\sigma$$

and consequently:

(12) $$\sup_{s\in[0,\,1]} \| U_\tau(s)\,\psi - W(s)\,\psi \| \leq c/\tau$$

which proves the theorem ($U_\tau(s)\,\psi = \psi_\tau(s)$), and shows that $W(s)$ provides a good approximation to the true evolution for τ large. ■

In fact it is easy to see that the theorem remains true if the multiplicity of the eigenvalue is constant except for a finite number of values of s in $[0, 1]$ where simple crossings, of finite order, occur between eigenvalues (see [Kat]).

We note that it is also possible to generalize the quantum theorem to the case where the Hamiltonian is no longer autonomous but T-periodic in time (for fixed values of the parameters) by using the time T map (monodromy matrix, Floquet exponents) to reduce the system to a normal form. For details and physical applications, we refer the reader to [LoG1-4].

10.2 The analogy between classical and quantum theorems

In this section, we present a mathematical analogy based on Koopman's viewpoint (cf. Appendix 2) for classical systems; we will discuss the semiclassical limit ($h \to 0$) later.

For classical systems we will adopt the notation of Chapter 9 and we begin with a restatement in a functional form of the theorem in the ergodic case, following [LoP].

Let $[I_1, I_2]$ be a segment contained in (I^*, I'^*) and let ψ be a *fixed* real valued function in $L^2([I_1, I_2])$. For every fixed value of s in $[0, 1]$, we define the function $\tilde{\psi}_s$ in $L^2(\mathbb{R}^{2n})$ by:

(1) $\tilde{\psi}_s(p, q) = \psi(I(s, p, q))$.

The evolution of a function f in $L^2(\mathbb{R}^{2n})$ along the trajectories of the flow of the Hamiltonian $H(p,q,s)$ (s fixed) is given by the equation:

(2) $df/ds = -i\tau L_H[f]$,

where L_H is the Liouville operator $L_H[f] = i\{H, f\}$ (the i is added here to make the operator symmetric), and $s = \varepsilon t = t/\tau$ is the slow time. The function $\tilde{\psi}_s$ is constant on the energy surfaces of the Hamiltonian $H(p, q, s)$ for fixed s. It is thus invariant under the flow of this Hamiltonian and belongs to the kernel $N(s)$ of the associated Liouville operator.

The correspondence between the classical and quantum theorems goes as follows. The *linear* Liouville operator plays the same role as the Schrödinger operator in the quantum case. Functions on the phase space $\mathbb{R}^{2n}$ correspond to quantum wave-functions. Finally, functions which are invariant under the Hamiltonian flow (for fixed s), and are thus eigenfunctions of the Liouville operator for the eigenvalue 0, correspond to zero energy quantum eigenstates.

Now, consider the initial state $\tilde{\psi}_0(p, q) = \psi(I(0, p, q))$. If we allow this function to evolve under the flow of the *time dependent* Hamiltonian $H(p, q, s)$ until time s, we obtain the function ψ_s which satisfies:

(3) $\tilde{\psi}_0(p_0, q_0) = \psi_s(p_\tau(s), q_\tau(s))$,

where $(p_\tau(s), q_\tau(s))$ denotes the solution of Hamilton's equations for $H(p, q, s)$ with initial condition (p_0, q_0).

We assume that the function ψ has compact support and is peaked at a particular value I_0 of I. For all s, the support in phase space $\mathbb{R}^{2n}$ of the function $\tilde{\psi}_s$ will be contained in a neighborhood of the energy surface of $H(s)$ which encloses the volume I_0 ; note that for different values of s, the support of the function will be localized in the neighborhood of *different* energy surfaces which however enclose the *same* volume I_0. A priori, the support of the function ψ_s may, in contrast, depend in a completely different way on s. The adiabatic invariance of the volume is equivalent to the L^2 convergence (uniformly in s) of ψ_s to the function $\tilde{\psi}_s$ in the adiabatic limit $\tau \to \infty$. We summarize by rephrasing the:

Adiabatic Theorem:

For any real valued L^2 function ψ with support in (I^*, I'^*), in the adiabatic limit $\tau \to \infty$, the function ψ_s, obtained as the evolution of $\psi(I(0, p, q))$ under the flow of the time dependent Hamiltonian $H(p, q, s)$, converges in L^2 (uniformly in s) to the function $\tilde{\psi}_s(p, q) =_{def} \psi(I(s, p, q))$.

This says that in the adiabatic limit the function ψ_s remains in the kernel of the Liouville operator on the slow time interval $[0, 1]$, which is precisely the content of the *quantum* adiabatic theorem (for the eigenvalue 0).

The spectral theory of the Liouville operator is quite complex, particularly since the eigenvalues are almost never isolated, and for this reason the proof of the quantum theorem does not carry over to the classical case. In particular, the order $1/\tau$ estimate of the remainder obviously cannot be obtained as it can in the quantum case. Nevertheless, to better understand in what way the theorems are "the same", we complete this subsection by showing that the easily derived equation:

(4) $$d\tilde{\psi}_s/ds(p_0, q_0) = d\psi/dI.(\partial I/\partial s)(s, p_0, q_0)$$

is the exact analog of the adiabatic transport equation (see formula 3 of Section 1):

(5) $$d\tilde{\psi}_s/ds = d\Pi(s)/ds\ \tilde{\psi}_s .$$

To do this, we begin by defining the orthogonal projection $\Pi(s)$ of $L^2(\mathbb{R}^{2n})$ on the kernel $N(s)$ of the Liouville operator associated to $H(s)$, i.e., the set of functions invariant under the flow of the Hamiltonian. For a C^∞ function φ with compact support, the operator:

(6) $$\Pi(s)[\varphi](p, q) = \int_{\Sigma(s,p,q)} \varphi\, d\nu_s$$

is given by the average of φ over the energy surface $\Sigma(s, p, q)$ with respect to the normalized measure $d\nu_s$ (see Section 9.4); this operator has norm 1 and may be extended to an orthogonal projction on $L^2(\mathbb{R}^{2n})$. Since the associated function $\tilde{\psi}_s$ belongs to the kernel $N(s)$, we have $\Pi(s)\,\tilde{\psi}_s = \tilde{\psi}_s$ (i.e., $\tilde{\psi}_s$ depends only on the energy), and:

(7) $$d\tilde{\psi}_s/ds = d[\Pi(s)\,\tilde{\psi}_s]/ds = d\Pi(s)/ds\ \tilde{\psi}_s .$$

In fact, the term $\Pi(s)\, d\tilde{\psi}_s/ds$ vanishes, since:

$$(8) \qquad \Pi(s).d\tilde{\psi}_s/ds = \int_{\Sigma(s)} d\tilde{\psi}_s\, d\nu_s = \int_{\Sigma(s)} d\psi/dI.\partial I/\partial s\, d\nu_s$$
$$= d\psi/dI \int_{\Sigma(s)} \partial I/\partial s\, d\nu_s = 0,$$

by Lemma 1 of Section 9.4. We thus see that the function $\tilde{\psi}_s$ satisfies equations (4) and (5), and we will now reduce (5) to (4) by explicitly calculating the operator $d\Pi(s)/ds.\Pi(s)$:

$$(9) \qquad d[\Pi(s)\,\tilde{\psi}_s]/ds(s, p_0, q_0) = \lim_{h\to 0} 1/h\, [\int_{\Sigma(s+h)} \tilde{\psi}_{s+h}\, d\nu_{s+h} -$$
$$- \int_{\Sigma(s)} \tilde{\psi}_s\, d\nu_s]$$

where $\Sigma(s)$ is the level surface of $H(s)$ which contains the point (p_0, q_0), and $\Sigma(s+h)$ is the level surface of $H(s+h)$ containing the same point (see Figure 10.1).

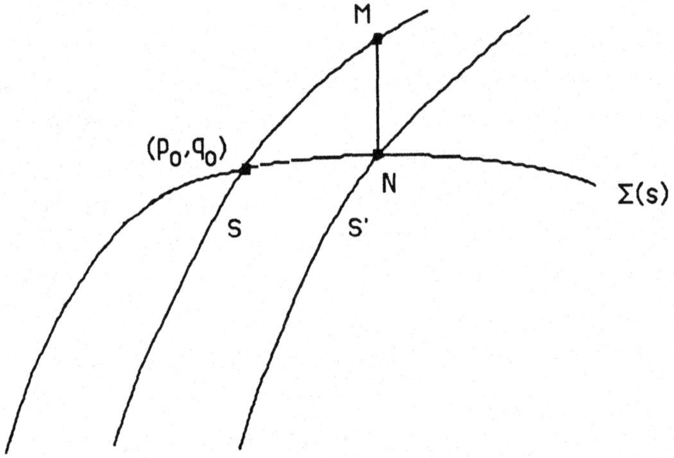

Figure 10.1: S and S' are level surfaces of the Hamiltonian H(s+h) while S(s) is a level surface of H(s).

We decompose the right hand side of (9) by writing:

$$(10) \qquad 1/h\, [\int_{\Sigma(s+h)} \tilde{\psi}_{s+h} d\nu_{s+h} - \int_{\Sigma(s)} \tilde{\psi}_s\, d\nu_s] = 1/h\, [\int_{\Sigma(s)} (\tilde{\psi}_{s+h} - \tilde{\psi}_s)\, d\nu_s] +$$
$$+ 1/h\, [\int_{\Sigma(s+h)} \tilde{\psi}_s\, d\nu_{s+h} - \int_{\Sigma(s)} \tilde{\psi}_s\, d\nu_s] + \text{h.o.t.}.$$

The first term, which is the difference in the integrals on the same energy surface of

$H(s)$, vanishes in the limit $h \to 0$. To see this, we write:

$$(11) \quad 1/h\,[\int_{\Sigma(s)} (\tilde{\psi}_{s+h} - \tilde{\psi}_s)\, d\nu_s\,] = 1/h\,[\int_{\Sigma(s)} \psi(I(s+h, p, q)) - \psi(I(s, p, q))\, d\nu_s]$$

$$= 1/h\,[\int_{\Sigma(s)} d\psi/dI . \partial I/\partial s\, d\nu_s] + O(h)$$

$$= d\psi/dI . \int_{\Sigma(s)} \partial I/\partial s\, d\nu_s + O(h) = O(h),$$

again by Lemma 1 of Subsection 9.4.

To evaluate the second term in (10), for small h we parametrize the points M of the surface $\Sigma(s+h)$ as a function of the points N of $\Sigma(s)$, using the vector n normal to $\Sigma(s)$ and oriented according to $\nabla H(s)$: $M = N + f(N)$.

This parametrization transforms the integral over $\Sigma(s+h)$ into an integral over $\Sigma(s)$. The density of the measure ν_{s+h} is $1/\| \nabla I(M) \|$ on $\Sigma(s+h)$:

$$d\nu_{s+h} = dM/\| \nabla I(M) \|$$

(dM is the surface measure). Consequently, the expression whose limit we seek becomes:

$$(12) \quad 1/h \int_{\Sigma(s)} [\psi(s, N+f(N)n(N)) /\| \nabla I(N+f(N)n(N)) \|] \times$$

$$\times\ (dN + d[f(N)n(N)]) - 1/h \int_{\Sigma(s)} [\psi(s, N)/\| \nabla I(N) \|] dN.$$

The function $f(N)$ is easy to evaluate, for at the point M:

$$(13) \quad H(s+h, M) = H(s, p_0, q_0) + h\partial H/\partial s(s, p_0, q_0) + O(h^2)$$

$$= H(s+h, N) + f(N)n(N)\nabla H(s, N) + O(h^2)$$

$$= H(s, p_0, q_0) + h\partial H/\partial s(s, N) + f(N)n(N)\nabla H(s,N) + O(h^2)$$

by definition of f. This implies:

$$(14) \quad f(N) = h[\partial H/\partial s(s, p_0, q_0) - \partial H/\partial s(s, N)]/\| \nabla H(N) \| + O(h^2)$$

$$= h[\partial I/\partial s(s, p_0, q_0) - \partial I/\partial s(s, N)]/\| \nabla I(N) \| + O(h^2).$$

The expression to be estimated (12) may again be broken into two terms, one corresponding to the variation of the function ψ, the other to the variation of the measure. The latter is zero to first order in h, since the measures have been normalized. It therefore remains to evaluate the term:

$$\int_{\Sigma(s)} [\psi(s, N+f(N)n(N)) - \psi(s, N)]/\| \nabla I(N) \|\, dN,$$

which to first order is:

$$d\psi/dI . \int_{\Sigma(s)} [\partial I/\partial s(s, p_0, q_0) - \partial I/\partial s(s, N)]\, d\nu_s .$$

The second term in this integral vanishes and we therefore arrive to the result:

$$(15) \qquad \lim_{h\to 0} 1/h \,[\int_{\Sigma(s+h)} \tilde{\psi}_{s+h}\, d\nu_{s+h} - \int_{\Sigma(s)} \tilde{\psi}_s\, d\nu_s] = $$
$$= d\psi/dI.\partial I/\partial s(s, p_0, q_0),$$

and this explicit calculation of $d\Pi/ds.\Pi$ shows that equation (5) is nothing but another way of expressing (4).

The same sort of remarks apply to the adiabatic theorem in the integrable case (cf. Section 9.2) by replacing volume with the classical action. This is natural, since the flow is no longer ergodic on almost every energy surface, but is instead ergodic on almost all invariant tori defined by the action variables. To every function ψ in $L^2(\mathbb{R}^n)$, we associate a family of functions $\tilde{\psi}_s$ ($s \in [0, 1]$) defined by $\tilde{\psi}_s(p, q) = \psi(I_s(p, q))$, where the function $I_s(p, q)$ associates to the point (p, q) in $\mathbb{R}^{2n}$ the values of the n action variables on the invariant torus of $H(s)$ passing through that point. For fixed s, the function $\tilde{\psi}_s$ is invariant under the flow of $H(s)$. If we allow the function $\tilde{\psi}_0(p, q)$ to evolve under the flow of the time dependent Hamiltonian $H(p, q, s)$, we obtain the function $\psi_s(p, q)$ again defined by (3). The theorem is stated in the same way as before:

Adiabatic Theorem:

For any function ψ in $L^2(\mathbb{R}^n)$, in the adiabatic limit $\tau \to \infty$ the function ψ_s, obtained as the evolution of the function $\tilde{\psi}_0$ under the flow of the time dependent Hamiltonian, converges in L^2 (uniformly in s) to the function $\tilde{\psi}_s$, invariant under the flow of the Hamiltonian $H(s)$.

We thus see one facet of the unity of the various adiabatic theorems: Quantum, classical integrable, and classical ergodic. The function $\tilde{\psi}_s$ is invariant under the flow of the Hamiltonian $H(s)$ and therefore belongs to the kernel of the Liouville operator associated to this integrable Hamiltonian.

For an integrable Hamiltonian $H(I)$, the associated Liouville operator may be expressed (in action-angle variables):

$$(16) \qquad L_H[f] = i\,\{H, f\} = i\partial H/\partial I.\partial f/\partial\varphi = i\omega(I).\partial f/\partial\varphi,$$

which leads once again to the small divisor problem. If we consider the functions on a torus (rather than $\mathbb{R}^{2n}$), two cases arise, according as to whether the torus is resonant or not:

- If the torus is nonresonant ($\omega(I).k \neq 0$ for every nonzero vector in $\mathbb{Z}^n$, which is the case for almost all values of the actions if the frequencies of the Hamiltonian are nondegenerate), then the eigenspace associated to 0 reduces to the space of constant functions on the torus: The Hamiltonian flow is ergodic on the torus.

- If the torus is resonant, there exists a maximal submodule M of $\mathbb{Z}^n$ such that $\omega(I).k = 0$ for all vectors k in M. The eigenfunctions associated to 0 are thus all the functions with resonant Fourier series: $f(\varphi) = \sum_{k\in M} f_k e^{ik.\varphi}$. All of this is simply a reformulation, in a spectral framework, of properties we have seen many times.

If we now consider the Hamiltonian flow on $\mathbb{R}^{2n}$, as the frequencies of the Hamiltonian are nondegenerate, we see that the only eigenfunctions associated to the eigenvalue 0 of the Liouville operator are the functions constant on the invariant tori of the Hamiltonian $H(s)$, i.e., functions of the actions only. The classical adiabatic theorem may thus be reformulated as follows:

In the adiabatic limit, the function ψ_s remains in the kernel of the Liouville operator on slow time intervals of order 1.

10.3 Adiabatic behavior of the quantum phase

We consider once again a family of quantum Hamiltonians $H(\lambda)$ depending on the vector parameter $\lambda \in \Lambda \subset \mathbb{R}^d$ (Λ convex, $d \geq 2$). We assume that each Hamiltonian is a self adjoint operator on the Hilbert space $\mathcal{H}$ with an isolated eigenvalue, i.e., for each $\lambda \in \Lambda$, there is an eigenvalue $E(\lambda)$ of $H(\lambda)$ at a positive finite distance from the rest of the spectrum. We saw that it is not retrictive to assume this isolated eigenvalue $E(\lambda)$ is 0. We further assume that $V(\lambda)$, the eigenspace associated to 0, is of multiplicity greater than or equal to $m \geq 1$ for $\lambda \in \Lambda$, and is of multiplicity *equal* to m on $M \subset \Lambda$; the complement of M thus corresponds to parameter values for which a degeneracy exists.

As in the classical case, where the behavior of the *angle* variables has only recently been examined (see Section 9.3), so it is in the present quantum case, where the first study of the *phase* of the wave function may be found in the article by M. Berry ([Ber3]), which we particularly recommend for physical applications (the interpretation of the Aharonov-Bohm effect, for example). We first present the elementary calculation of the phase evolution for a nondegenerate eigenvalue ($m = 1$, which is still the important case), then, by more or less following B. Simon's note ([Sim]), we give the geometric interpretation of this calculation along with its possible generalizations. This interpretation makes use of certain elementary notions from the theory of connections on complex fiber bundles and so furnishes a simple illustration of these techniques.

We thus consider the case $m = 1$ and, as in Section 1, we denote by $\lambda(s)$, $s \in [0, 1]$ a path C in $M \subset \Lambda$ which is assumed to be *closed* ($\lambda(0) = \lambda(1)$). We express dependence on $\lambda(s)$ simply by s, that is, $H(\lambda(s)) =_{def} H(s)$, $W(\lambda(s)) =_{def} W(s)$, etc. .

For the eigenvalue 0, let $\psi = \psi^*_0$ be an eigenstate of $H(0)$ with norm 1, and let $\psi^*_s = W(s)\,\psi$ be the vector transported along C by means of $W(s)$. Since this is a unitary transformation, we have $\psi^*_1 = e^{i\gamma}\psi^*_0$, and it is natural to seek the phase shift $\gamma = \gamma(C)$ introduced by the adiabatic transport W "during one cycle". This evidently has to do with the *geometry* of the transport, and involves more precisely a property of *holonomy*. We note that if the eigenvalue $E(\lambda)$ is not zero, we have, in addition to γ, the usual *dynamical* phase:

(1) $\psi^*_1 = e^{i\gamma} \exp(-i\tau \oint E(s)\,ds)\,\psi^*_0, \quad (\hbar = 1).$

Suppose first of all that along C we are able to find a regular (C^1 in s) representation $\varphi(s)$ of the eigenstate:

$$H(s)\varphi(s) = 0, \quad \varphi(0) = \psi^*_0, \quad \|\varphi(s)\| = 1.$$

For all $s \in [0, 1]$, we then have:

(2) $\psi^*_s = e^{i\gamma(s)}\varphi(s),$

with $\gamma(0) = 0$, and the phase we wish to calculate is none other than $\gamma(1)$. By definition of $W(s)$, we have:

(3) $(\psi^*_s, d\psi^*_s/ds) = 0$

where $(.,.)$ is the scalar product in $\mathcal{H}$, which, using (2), leads to:

(4) $d\gamma/ds = i(\varphi, d\varphi/ds)$

and therefore:

(5) $\gamma = \gamma(1) = i\int_0^1 (\varphi, d\varphi/ds)\,ds.$

This expression is valid for *any* choice of $\varphi(s)$, and it is easy to verify that the integrand is invariant under the (gauge) transformation $\varphi(s) \to \varphi(s)e^{i\alpha(s)}$, where $\alpha(s)$ is a real function. This allows us to relax the regularity condition on the $\varphi(s)$, which in any case may be constructed locally. In fact, the family $\varphi(s)$ may be constructed with jumps at at most one point, in which case formula (5) remains valid by adding up the contributions of the intervals on which $\varphi(s)$ is regular. We will come back to these questions later.

For the moment, we also assume the existence of a surface S which is diffeomorphic to a disk, is *contained in* M, and has C as its boundary. This is roughly equivalent to assuming that the set $\Lambda - M$ where degeneracies occur is of codimension strictly greater than two; for example the crossing of two nondegenerate eigenvalues of a complex Hamiltonian is generically of codimension three. It is then possible to determine a regular family $\varphi(\lambda)$ of normalized eigenstates ($H(\lambda)\varphi(\lambda) = 0$, $\| \varphi(\lambda) \| = 1$), for $\lambda \in S$, which in particular determines $\varphi(s)$ to be regular on the boundary C of S. Using (5) along with Stoke's theorem, we then write:

$$(6) \qquad \gamma = \gamma(1) = i \int_S (D\varphi, D\varphi) = \int_S V,$$

where D denotes the operation of exterior differentiation *in parameter space* Λ. By introducing the coordinates $\lambda_1, \ldots, \lambda_d$ on M, the 2-form $V = i(D\varphi, D\varphi)$ takes the explicit form :

$$(7) \qquad V = -2 \sum_{i<j} \operatorname{Im}(\partial\varphi/\partial\lambda_i \,,\, \partial\varphi/\partial\lambda_j)\, d\lambda_i \wedge d\lambda_j \,.$$

Formulas (5), (6), and (7), valid in the case of a simple eigenvalue and obtained, as we saw, in very elementary fashion, constitute however the essential result of this study. The adiabatic theorem says that the solution $\psi(s)$ of the nonautonomous Schrödinger equation is close to ψ^*_s for large τ. Using this theorem, along with formulas (1), (5), (6), and (7), one obtains a complete description of the behavior of the wave function in the adiabatic limit.

We remark that the preceding development is interesting only when the Hamiltonian is *not* real; otherwise we may always choose real eigenstates for which the phase vanishes. For this reason, one is especially interested in physical systems immersed in a magnetic field or for which spin must be considered (such examples are examined in [Ber3]), and it is certainly more than chance that these sytems have interesting holonomy properties.

As we mentioned above, it is enlightening to "translate" the preceding development into more geometric terms. We now proceed to do this, which also allows us to briefly examine the case of "stably degenerate" systems ($m \geq 1$). This hypothesis of generic degeneracy is more than simply academic, in as much as it is realized for systems with discrete symmetry (reflection with respect to a plane, etc.) for all values of the parameter λ. We leave the reader to devise relevant physical examples and we just remark that interesting physical phenomena then arise, even for real Hamiltonians (see below).

The set $E = \bigcup_{\lambda \in M} \{\lambda, V(\lambda)\}$ may be viewed as a fiber bundle with base M; the fibers are the (complex) planes of dimension m. This bundle has the peculiarity of being embedded in $M \times \mathcal{H}$, and so is equipped with a natural connection. This connection is none other than the one defined by adiabatic transport. More precisely, the operators $W(s) = W(\lambda(s))$ define, along any curve $\lambda(s)$ in M, an operation of parallel transport which determines the connection ∇. By definition:

(8) $\forall\ \psi \in V(\lambda(0)) \quad \nabla_{d/ds}\,[W(\lambda(s))\ \psi] = 0$

where $\nabla_{d/ds}$ represents the covariant derivative in the direction of the curve $\lambda(s)$. From the properties of $W(s)$ (cf. Section 1), the parallel transport preserves norm, showing that the connection is compatible with the natural Hermitian metric on E (the one inherited from $\mathcal{H}$), and also showing that, during parallel transport, the speed of variation in $\mathcal{H}$ of a vector is orthogonal to the fiber, which characterizes the connection as we shall see below.

For $m = 1$, this reduces to equation (3), the real part of which simply expresses the preservation of norm. It is therefore the imaginary part of this identity which characterizes "adiabatic parallel transport". Formulas (5) through (7) may then be interpreted in the following way: By definition, $\nabla\psi^*_s = 0$, which together with (2) implies that:

$$\nabla\varphi = -i d\gamma/ds\ \varphi.$$

This last identity expresses precisely the fact that the connection form ω (here a scalar) is given by $\omega = -\ i d\gamma$. Here $d\gamma$ is the differential form on M which takes the value $d\gamma/ds$ when applied to the vector tangent to the path $s \rightarrow \lambda(s)$; it is defined via a basis $\varphi(\lambda)$ which always exists locally in Λ. This somewhat complicated translation of a simple formula allows the formulas to be rewritten as follows:

(9) $\gamma(C) = i \oint_C\ \omega = i \int_S\ d\omega.$

We also know that for a connection of rank m on a fiber bundle, the curvature form K, which is a matrix 2-form, is given in the general case by $K = d\omega + \omega \wedge \omega$. The second term vanishes in dimension 1, and we see by (9) that the angle $\gamma(C)$ is expressed as the integral of the curvature form over a surface bounded by the given contour; this illustrates the original relation between the curvature and the holonomy properties of the parallel transport. We remark that the existence of S (diffeomorphic to a disk) implies that E restricted to S is trivial, since S is contractible (this being independent of the dimension m). However, this does not prevent the connection from giving rise to nontrivial properties.

We now briefly turn our attention to the general case (m arbitrary), and in particular we examine the case $m = 2$ with real Hamiltonians (which is not considered in [Ber3] nor in [Sim]). We noticed earlier how this case may arise naturally via e.g. the existence of discrete symmetries for all values of λ.

As above, C is a loop parametrized by s; we denote by $e_i(s)$ ($i=1,\dots, m$) an arbitrary orthonormal basis (a moving frame), defined at least locally in s (and even in $\lambda \in M$) and sufficiently regular (C^1). By $\varepsilon_i(s)$ ($i = 1,\dots, m$) we denote the basis transported parallel along C which coincides with $e_i(s)$ for $s = 0$: $\varepsilon_i(s) = W(s)\, \varepsilon_i(0)$, $\varepsilon_i(0) = e_i(0)$. Since parallel transport preserves the metric, we have:

$$(10) \qquad e_i(s) = U(s)\, \varepsilon_i(s)$$

where $U(s)$ is a unitary operator: $U^* = U^{-1}$, $U(0) = 1$. We may write:

$$(11) \qquad \nabla_{d/ds} e_i(s) = dU/ds\, \varepsilon_i(s) = dU/ds.U^*\, e_i(s)$$

which shows that the connection form is given by the matrix:

$$(12) \qquad \omega = dU.U^*$$

which is just the usual transformation formula for moving frames. In the present case, the transport is defined by:

$$(13) \qquad d\varepsilon_i/ds = d\Pi/ds\, \varepsilon_i$$

which makes sense only for a fiber bundle *imbedded* in $M \times \mathcal{H}$ ($\Pi(s)$ is a projection into $\mathcal{H}$). We thus have:

$$(14) \qquad d\varepsilon_i/ds = (d\varepsilon_j/ds, \varepsilon_i)\varepsilon_j + (\varepsilon_j , \varepsilon_i) d\varepsilon_j/ds \quad \text{(implicit summation)}$$
$$= (d\varepsilon_j/ds, \varepsilon_i)\varepsilon_j + d\varepsilon_i/ds.$$

The connection is characterized, as we mentioned before, by $(d\varepsilon_j/ds, \varepsilon_i) = 0$ ($\forall$ i, j); i.e., by the orthogonality in $\mathcal{H}$ of the speed vector with the fiber $V(s)$. It follows that:

$$(15) \qquad (d\varepsilon_i/ds, \varepsilon_j) = 0 = (dU^*/ds\, e_i , \varepsilon_j) + (U^*\, de_i/ds, \varepsilon_j)$$
$$= (e_i , dU/ds\, \varepsilon_j) + (de_i/ds, U\, \varepsilon_j)$$
$$= (e_i , dU/ds.U^*\, e_j) + (de_i/ds, e_j),$$

which shows that the connection form matrix ω is given by:

$$(16) \qquad \omega_{ji} = - \omega_{ij} = (De_i , e_j)$$

where D is still the exterior differentiation operator with respect to λ. The analogue of the phase $\gamma(C)$ for arbitrary m is of course the matrix $U(1)$, and we have:

$$(17) \qquad U(1) = \exp\left(\oint_C \omega(s)\, ds\right) = \exp\left(\int_S D\omega\right)$$

where ω is given by (16) ($(D\omega)_{ij} = - (De_i , De_j)$).

The case where $m = 2$ and the Hamiltonians are real is particularly simple. We may then assume that the (e_i) and (ε_i) are also real, so that U is the orthogonal matrix:

$$(18) \qquad U(s) = \begin{vmatrix} \cos\theta(s) & -\sin\theta(s) \\ \sin\theta(s) & \cos\theta(s) \end{vmatrix}$$

Here the angle θ is the analog of γ and the connection matrix is simply:

$$(19) \qquad \omega = dU.U^* = d\theta . \begin{vmatrix} 0 & -1 \\ 1 & 0 \end{vmatrix}$$

The angle through which the vector has turned in the plane associated with the eigenvalue 0 is then given by:

$$(20) \qquad \theta(C) = \theta(1) = \oint_C \omega_{21} = \oint_C (De_1 , e_2) = \int_S (De_1 , De_2).$$

The similarity between formulas (5) through (7) and formula (20) is not fortuitous; in fact, to a fiber bundle of real *oriented* planes, one may associate a fiber bundle of complex rays, through the complex structure defined by the operator J such that:

$$(21) \qquad J e_1 = e_2, \quad J e_2 = - e_1 .$$

The analogy between the formulas then becomes an identity, and equation (20) is in some sense a generalization of equation (6) to an arbitrary complex structure. The reader can perhaps find an explicit application.

10.4 Classical angles and quantum phase

In the present section, we will show that the differential form on parameter space introduced in Section 9.3 to study the adiabatic behavior of angle variables for an integrable Hamiltonian is the semiclassical limit of the form V studied above (equation (7) of Section 3). This correspondence is presented in the article by M. Berry ([Ber4]), which we follow here.

Our point of departure is still given by a quantum Hamiltonian depending on a vector parameter $\lambda \in \Lambda$ (in $\mathbb{R}^d$, $d \geq 2$), which is obtained by *quantization* of a classical Hamiltonian $H(p, q, \lambda)$ possessing n degrees of freedom and integrable for all λ. We also assume that, for all λ, the eigenvalue $E(\lambda)$ of this Hamiltonian is nondegenerate and isolated in the spectrum. We saw above that the geometrical variation of the phase of an eigenstate associated to this eigenvalue could be expressed in terms of the 2-form V on parameter space:

$$(1) \qquad V = - \operatorname{Im} [D(\varphi, D\varphi)] = - \operatorname{Im} [D (\int_{\mathbb{R}^n} \varphi^*(q, \lambda) D\varphi(q, \lambda)\, dq)],$$

where $\{\varphi = \varphi(q, \lambda) \in L^2(\mathbb{R}^{2n})\}$ is a regular realization of the eigenfunction with norm 1, and where D designates exterior differentiation on parameter space. The wave function $\varphi(q, \lambda)$ is semiclassically associated to an invariant torus $T(I, \lambda)$ of the integrable Hamiltonian with phase space $\mathbb{R}^{2n}$, the actions I_j satisfying the Bohr-Sommerfeld quantization relation: $I_j = (s_j + \sigma_j)\hbar$, where the n-tuple $(s_1, \ldots, s_n)$ characterizes the quantum state under consideration, and where the σ_j are n real constants (each equal to 1/2 in the original rule). For more details concerning the "semiclassical" terminology as used here, we suggest any of the many existing works, for example the book by V.P. Maslov and M.V. Fedoriuk ([Mas]) or the article by M. Berry ([Ber5]).

Before determining the semiclassical limit of the form V, we again rapidly review action-angle variables in many dimensions, to clarify somewhat the remarks made in Section 9.2 (the notation here is slightly different).

The canonical transformation from the original variables of the Hamiltonian $H(p, q, \lambda)$ to action-angle variables takes place by means of a generating function $S(q, I, \lambda)$ such that:

(2) $$p = \partial S/\partial q, \quad \theta = \partial S/\partial I.$$

This generating function is multivalued in general: To fixed values of the variables q correspond, a priori, several points $(q, p^{(\alpha)})$ on the invariant torus $T(I, \lambda)$ of the Hamiltonian $H(p, q, \lambda)$; these points differ in the values of the momenta p (see Figure 10.2).

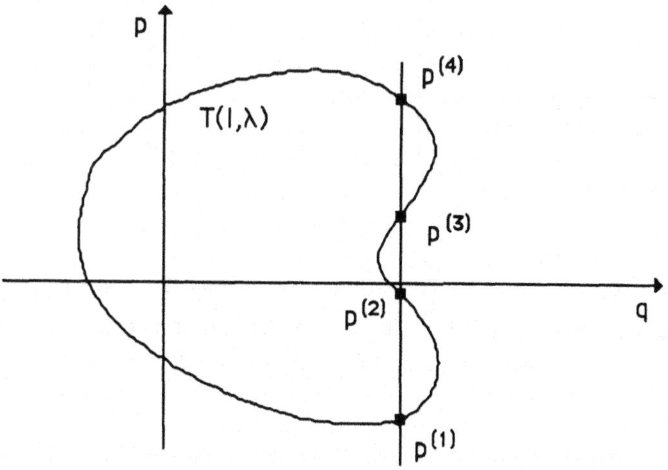

Figure 10.2: The intersection of the torus $T(I, \lambda)$ with a straight line q = Cst generically consists of finitely many points. We denote by $p^{(i)}$ the values of the momenta for the ith point of intersection.

One is thus led to distinguish several branches of the function S, denoted $S^{(\alpha)}(q, I, \lambda)$, such that:

(3) $$p^{(\alpha)} = \partial S^{(\alpha)}/\partial q(q, I, \lambda).$$

The various points on the torus $T(I, \lambda)$ which correspond to the same values of the coordinates q may be distinguished by the angular coordinates at these points, permitting the definition of a *single valued* function $S(\theta, I, \lambda)$ by means of the equation:

(4) $$S(\theta, I, \lambda) = S^{(\alpha)}(q(\theta, I, \lambda), I, \lambda),$$

where the values of the angular coordinates θ determine the branch of the function used.

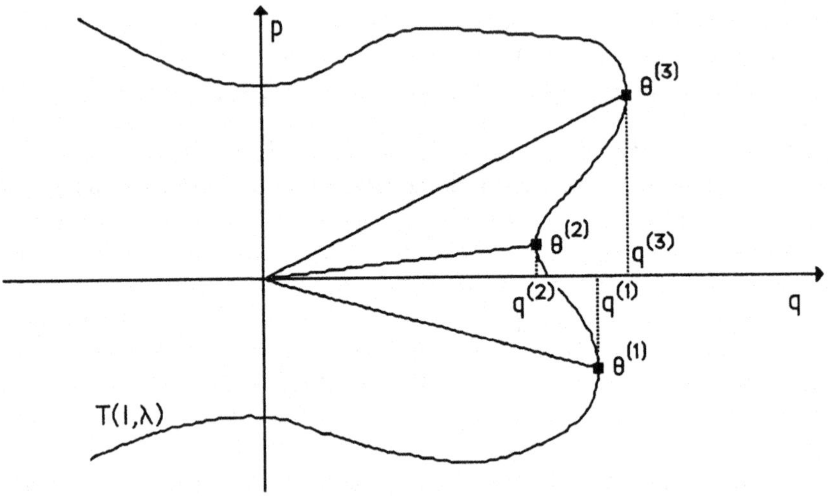

Figure 10.3: On the torus $T(I,\lambda)$ the Jacobian $|\, dq_i/d\theta_j \,|$ vanishes at the points with angular coordinates $\theta^{(1)}$, $\theta^{(2)}$. The function S has several branches, denoted by $S^{(i)}$; the ith branch $S^{(i)}$ is defined for q ranging from $q^{(i-1)}$ to $q^{(i)}$.

The exterior derivative DS of S in *parameter space* is given by the equation:

(5) $$DS = DS^{(\alpha)} + \partial S^{(\alpha)}/\partial q.Dq = DS^{(\alpha)} + p^{(\alpha)}.Dq.$$

The semiclassical limit of the form V is determined with the help of Maslov's important equation connecting the wave function $\varphi(q, \lambda)$ to the classical action, again in the semiclassical limit:

(6) $$\varphi(q, \lambda) = \sum_\alpha a_\alpha(q, I, \lambda) \exp[i/\hbar\, S^{(\alpha)}(q, I, \lambda)].$$

Here the summation occurs over several branches of the function S (the number of branches depends on q for a fixed torus $T(I, \lambda)$), and the complex coefficients a_α, which are real or purely imaginary according to the value of the associated Maslov index, are given by:

(7) $$a_\alpha^{\ 2} = 1/(2\pi)^n.\det(\partial\theta^{(\alpha)}/\partial q) \quad \text{where} \quad \theta^{(\alpha)} = \partial S^{(\alpha)}/\partial I.$$

For more details concerning Maslov's method, we refer the reader to [Mas] and for a vivid picture of quantal-classical interplay to [Ber1].

From the above equation, we deduce:

(8) $$D\varphi(q, \lambda) = \Sigma_\alpha Da_\alpha \exp(i/\hbar\, S^{(\alpha)}) + \Sigma_\alpha i/\hbar\, a_\alpha DS^{(\alpha)} \exp(i/\hbar.S(\alpha)).$$

Only the second term is taken into account in the semiclassical limit; the form V thus has the same semiclassical limit as the expression:

$$-\operatorname{Im} \Sigma_{\alpha,\beta} D \int dq\, [i/\hbar\, a_\alpha a_\beta{}^* DS^{(\alpha)} \exp\{i/\hbar\, (S^{(\alpha)}-S^{(\beta)})\}].$$

The contribution from cross terms ($\alpha \neq \beta$) is zero in the semiclassical limit in view of the nonstationary phase, and V thus has the asymptotic form:

(9) $$-\operatorname{Im} \Sigma_\alpha i/\hbar\, D \int dq\, |\, a_\alpha |^2 DS^{(\alpha)} = -1/\hbar\, \Sigma_\alpha D \int dq\, |\, a_\alpha |^2 DS^{(\alpha)} =$$
$$= -(2\pi)^{-n}\hbar^{-1} D\, [\Sigma_\alpha \int dq\, \det(\partial\theta^{(\alpha)}/\partial q) DS^{(\alpha)}].$$

Introducing the function $S(\theta, I, \lambda)$, this may be rewritten as:

(10) $$-(2\pi)^{-n}\hbar^{-1} D[\int_{T(I,\lambda)} d\theta\, (DS - pDq)] =$$
$$= -(2\pi)^{-n}\hbar^{-1} \int_{T(I,\lambda)} [D^2 S - D(p.Dq)]\, d\theta$$
$$= (2\pi)^{-n}\hbar^{-1} \int_{T(I,\lambda)} D(p.Dq)\, d\theta,$$

where the integration is over the torus $T(I, \lambda)$ corresponding semiclassically to the quantum eigenstate $\varphi(q, \lambda)$, and (10) may be rewritten once more as:

(11) $$(2\pi)^{-n}\hbar^{-1} \Sigma_{1 \leq i \leq n} \int_{T(I,\lambda)} Dp_i \wedge Dq_i\, d\theta.$$

The semiclassical limit of V is thus proportional to the 2-form:

$$W = (2\pi)^{-n} \Sigma_{1 \leq i \leq n} \int_{T(I,\lambda)} Dp_i \wedge Dq_i\, d\theta$$

introduced in Section 9.3 to describe the adiabatic behavior of angle variables for integrable *classical* Hamiltonians:

(12) $$V_{\text{semiclassical}} = W/\hbar.$$

We refer to [Ber4] for explicit calculations illustrating this formula.

10.5 Non-commutativity of adiabatic and semiclasical limits

Based on what we have just seen, it is interesting to note that it is generally not possible to interchange the classical and adiabatic limits for a quantum system depending slowly on time, and that it is necessary to take a detailed account of the way in which "$\hbar \to 0$" (after carefully specifying what this means) and correspondingly of the way in which "$\tau \to \infty$". This is strikingly illustrated by the following example, again due to M. Berry (cf. [Ber2]), to whom we refer for more details.

Consider a particle of mass m moving in a double well potential, for example a quartic potential, the form of which is slowly varying in time. We assume that the right hand well disappears in the limit $t \to -\infty$, and that the left disappears in the limit $t \to +\infty$ (see Figure 10.4).

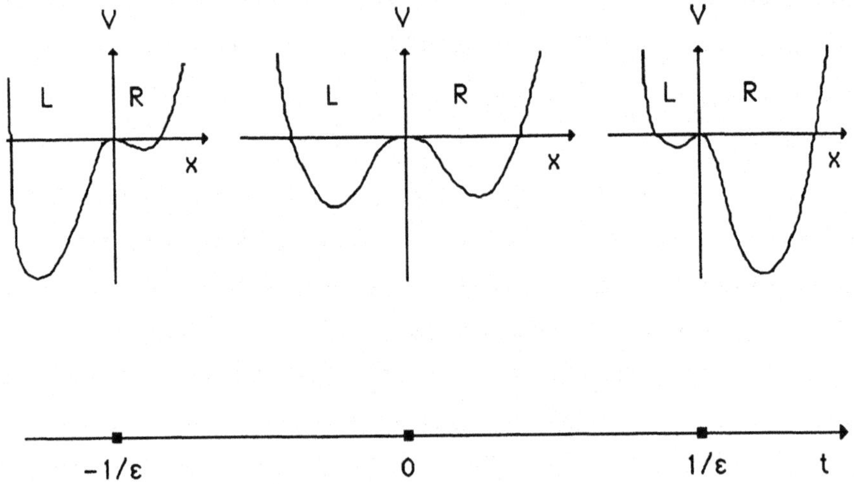

Figure 10.4: The form of the quartic potential slowly varies in time.

Let us take as the initial state at time $-1/\varepsilon$ the nth quantum state localized in the left hand well (outside the well the wave function decreases exponentially fast) with energy $E_{L,n}(-1/\varepsilon)$ less than the local minimum of the potential in the right hand well. We then seek the classical state of the system for large time, assuming the potential is varied adiabatically ($\tau = 1/\varepsilon \to \infty$).

1) If we first pass to the classical limit, we obtain classical motion confined to the left hand well and characterized by the action:

$$(1) \qquad I_L(-1/\varepsilon) = 1/\pi \int_{x'}^{x''} [2m(E_{L,n}(-1/\varepsilon) - V(x, -1/\varepsilon))]^{1/2}\, dx,$$

where x' and x'' are the classical turning points with energy $E_{L,n}(-1/\varepsilon)$. This action is equal to $(n + 1/2)\hbar$ according to the Bohr-Sommerfeld quantization rule.

By *subsequently* applying the (one dimensional) classical adiabatic theorem, we deduce that at time $1/\varepsilon$ the system will be in a classical state still localized in the left hand well and associated, in the adiabatic limit, to the same value of action. The energy $E(t)$ of the system adiabatically follows the potential as a function of time and is given by the equation:

$$(2) \qquad (n + 1/2)\hbar = 1/\pi \int_{x'(t)}^{x''(t)} [2m(E - V(x,t)]^{1/2}\, dx,$$

where $x'(t)$ and $x''(t)$ are the classical turning points in the left hand well of the potential $V(t)$.

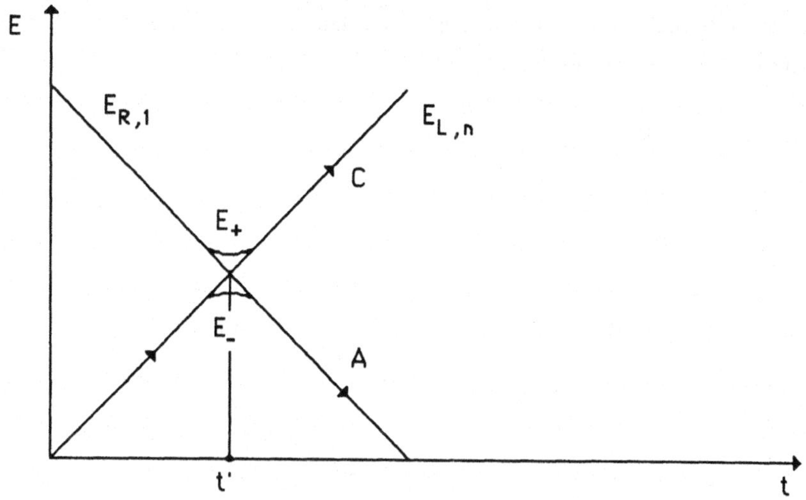

Figure 10.5: If one first passes to the classical limit (C) the system's state at large time is localized in the left hand well. If one interchanges the order of passage to the limit (A) the system's state for large time is the ground state of the right hand well.

2) Let us now interchange the order of passage to the limit. If we first apply the classical adiabatic theorem, we deduce that for all time the quantum state of the system remains, in the adiabatic limit, in an eigenstate associated to the energy $E(t)$. Until a certain time t', the energy $E_{L,n}(t)$ of the nth state in the left hand well remains less than the energy $E_{R,1}(t)$ of the base state of the right hand well, and the system thus remains in the nth quantum state localized in the left hand well (if the potential varies sufficiently slowly

with time, the probability of transition to another quantum state localized on the left is negligible). Neglecting the tunneling effect between the two wells, at time t' there would be a crossing of the eigenvalues $E_{L,n}$ and $E_{R,1}$ (see Figure 10.5); in fact, the tunneling effect prevents the degeneracy of the eigenvalue, and the quantum state of the system is the minimal energy state shared between the two wells. For later times, the system adiabatically follows the variation of the potential and remains in the minimal energy state. For large time, the system therefore passes to the ground state of the right hand well. If we *then* pass to the classical limit, we obtain classical motion confined to the right hand well.

This strikingly illustrates how different outcomes are obtained according to the order in which passage to the limit is realized. In the article cited previously (cf. [Ber2]), the reader will find a discussion of the circumstances under which each of these contradictory results is actually correct.

APPENDIX 1: FOURIER SERIES

1. Analytic functions

Proposition 1:

Let $f(z)$ be a 2π-periodic function of the complex variable z, analytic in the strip $|\mathrm{Im}(z)| < \sigma$ (sometimes called a Bargman strip) and continuous on the closure $|\mathrm{Im}(z)| \leqslant \sigma$, which we shall denote by $D(\sigma)$. Then the norm of the Fourier coefficients

$$(1) \qquad f_k = 1/2\pi \int_0^{2\pi} f(z)e^{-ikz}\, dz$$

(where the integration is over the real axis) is an exponentially decreasing function of $|k|$ ($k \in \mathbb{Z}$):

$$(2) \qquad |f_k| \leqslant \|f\|_{D(\sigma)}\, e^{-\sigma|k|},$$

where $\|f\|_{D(\sigma)}$ is the maximum of $|f|$ on the closed strip $D(\sigma)$.

This elementary result may be shown as follows. The integral of the analytic function $f(z)\exp(-ikz)$ along the rectangular path with sides $[0, \pm(\sigma - \varepsilon)]$, $[\pm(\sigma - \varepsilon), 2\pi \pm i(\sigma - \varepsilon)]$, $[\pm i(\sigma - \varepsilon), 2\pi]$ and $[2\pi, 0]$ is zero. Here ε is small and positive and the sign $\pm$ is taken to be the negative of the sign of k.

Because of the periodicity of $f(z)\exp(-ikz)$, the values of the integral on the vertical sides of the rectangle cancel out. Consequently:

$$(3) \qquad \int_0^{2\pi} f(z)e^{-ikz}\, dz = \int_{-i(\sigma-\varepsilon)}^{2\pi-i(\sigma-\varepsilon)} f(z)e^{-ikz}\, dz.$$

It follows that:

$$(4) \qquad \left|\int_0^{2\pi} f(z)e^{-ikz}\, dz\right| = e^{-|k|(\sigma-\varepsilon)} \left|\int_0^{2\pi} f(z - i(\sigma-\varepsilon))e^{-ikz}\, dz\right|$$
$$\leqslant 2\pi \|f\|_{D(\sigma)}\, e^{-|k|(\sigma - \varepsilon)},$$

from which we deduce formula (2) by taking the limit as ε tends to zero. ∎

It is also easy to derive the analogous result for analytic functions of several complex variables:

Proposition 2:

Let $f(z_1, \ldots, z_n)$ be a function with period 2π in each of the variables z_i, analytic on a domain containing the open domain $|\mathrm{Im}(z_i)| < \sigma$, $i = 1, \ldots, n$, and continuous on its closure $D(\sigma)$. Then the Fourier coefficients:

$$(5) \qquad f_k = 1/(2\pi)^n \int\ldots\int f(z_1, \ldots, z_n) e^{-ik.z} dz_1\ldots dz_n$$

where $k \in \mathbb{Z}^n$ and $k.z = k_1 z_1 + \ldots + k_n z_n$ satisfy the inequality:

$$(6) \qquad \| f_k \| \leqslant \| f \|_{D(\sigma)} e^{-\sigma |k|},$$

where $|k| = |k_1| + \ldots + |k_n|$.

This result is easily obtained by integrating in each variable over a path identical to the one used to prove Proposition 1. ∎

Since the Fourier coefficients decrease exponentially fast one can easily obtain upper bounds for expressions of the form $\sum_{|k|\geqslant N} |k|^r f_k e^{ik.\varphi}$. These bounds are used in the analytic part of the proof of Nekhoroshev's theorem (see Chapter 7).

Proposition 3:

If the 2π-periodic function f is analytic for $|\mathrm{Im}\,\varphi| < \sigma < 1$ and its Fourier coefficients f_k are bounded from above by $Ce^{-\sigma|k|}$ for some constant C (in particular $C = \| f \|_{D(\sigma)}$), then:

- for any β, $0 < \beta < \sigma$, and any integers N and r one has the estimate:

$$\| \sum_{|k|\geqslant N} |k|^r f_k e^{ik.\varphi} \|_{D(\sigma-\beta)} \leqslant C2^{2n+r+2}(n+r-1)/e]^{n+r-1} \times \exp(-\beta N/2)/\beta^{n+r}.$$

- for any β, $0 < \beta < \sigma$, one has:

$$\| f \|_{D(\sigma-\beta)} \leqslant C(4/\beta)^n.$$

We have:

$$(7) \qquad \| \sum_{|k|\geqslant N} |k|^r f_k e^{ik.\varphi} \|_{D(\sigma-\beta)} \leqslant \sum_{|k|\geqslant N} |f_k| |k|^r \| e^{ik.\varphi} \|_{D(\sigma-\beta)}$$
$$\leqslant \sum_{|k|\geqslant N} (Ce^{-\sigma|k|}) |k|^r e^{(\sigma-\beta)|k|}$$
$$\leqslant C \sum_{|k|\geqslant N} |k|^r e^{-\beta|k|}.$$

Since the number of integer-vectors with norm equal to j is less than $2^n(j+1)^{n-1}$, we have:

$$(8) \qquad \| \sum_{|k|\geqslant N} |k|^r f_k e^{ik.\varphi} \|_{D(\sigma-\beta)} \leqslant C2^n \sum_{j\geqslant N} (j+1)^{n+r-1} \exp(-\beta j).$$

From the inequality:

$$[je\beta/2(n+r-1)] \leqslant \exp[j\beta/2(n+r-1)],$$

we deduce:

$$(9) \qquad j^{n+r-1} \leqslant [2(n+r-1)/e\beta]^{n+r-1} \exp[j\beta/2].$$

Therefore:

$$(10) \qquad \| \sum_{|k|\geq N} |k|^r f_k e^{ik.\varphi} \|_{D(\sigma-\beta)} \leq C2^{2n+r-1}[(n+r-1)/e\beta]^{n+r-1} \exp(\beta/2) \times \times \sum_{j\geq N} \exp(-\beta j/2)$$

and, as $1-e^{-\beta/2} \geq \beta/4$, we obtain:

$$(11) \qquad \| \sum_{|k|\geq N} |k|^r f_k e^{ik.\varphi} \|_{D(\sigma-\beta)} \leq C2^{2n+r+2}[(n+r-1)/e]^{n+r-1} \times \times \exp(-\beta N/2)/\beta^{n+r}.$$

The second assertion also is easily proved. We have:

$$(12) \qquad \| f \|_{D(\sigma-\beta)} \leq C\sum e^{-\beta|k|} \leq C(\sum_{l\geq 0} e^{-\beta l})^n \leq C(4/\beta)^n.$$

This result is of course uninteresting when $C = \| f \|_{D(\sigma)}$ but it proves useful in Chapter 7 for other values of C. ∎

To end this section let us recall the Cauchy's inequality which is also used in Chapter 7. Let $f(I, \varphi)$ be an analytic function on the complex domain $K_{\rho,\sigma} = D(K, \rho, \sigma)$ (complex extension of $K \times T^n$, where K is a compact subset of $\mathbb{R}^m$). Then, for any α, $0 < \alpha < \rho$ and β, $0 < \beta < \sigma$:

$$(13) \qquad \| \partial^{|k|+|l|}/\partial I^k \partial\varphi^l \|_{D-(\alpha,\beta)} \leq k!\, l! \alpha^{-|k|}\beta^{-|l|} \| f \|_D$$

where k and l are multi-indices with lengths $|k| = |k_1|+...+|k_m|$ and $|l| = |l_1|+...|l_n|$ and we have used the usual notation ($k! = k_1!...k_m!$).

2. Differentiable functions of a single variable

We shall only recall here a few completely elementary facts (for much more about Fourier series the reader may consult [Whitt] and [Zy]). Let us start with

Proposition 4:

Let f be a class C^{r-1+} periodic function of a single real variable, $r \geq 1$, and let k be a positive integer. Then the Fourier coefficients f_k satisfy:

$$(14) \qquad \| f_k \| = O(|k|^{-r}).$$

In fact, by integrating $r-1$ times by parts, we may express the Fourier coefficient f_k of order $k \neq 0$ in the form:

$$(15) \qquad f_k = (ik)^{-r+1}.1/(2\pi) \int_0^{2\pi} f^{(r-1)}(x)e^{-ikx}\,dx.$$

and by the Riemann-Lebesgue Lemma (valid for functions with bounded variations and a fortiori for Lipschitz continuous functions, see [Whitt]), the integral on the r.h.s. is of size $O(1/k)$ as k goes to infinity. ▪

From this there obviously follows

Proposition 5:

Let $f(x)$ be a class C^{1+} periodic function of one real variable with period 2π. The associated Fourier series then converges absolutely to f.

From Proposition 4 one also derives a bound on the pth order remainder:

$$R_p(x) =_{def} \sum_{|k| \geq p} f_k e^{ikx}.$$

Proposition 6:

The pth order remainder of the Fourier series of a class C^{r-1+} function, $r \geq 1$, satisfies:

$$(16) \qquad R_p(x) = O(p^{1-r}).$$

3. Differentiable functions of several variables

We now consider periodic differentiable functions of n real variables $x_1, \ldots, x_n$ having period 2π in each variable.

Proposition 7:

If the function f is of class C^{n+} in each of its variables, its Fourier series $S(f)$ converges absolutely and the pth order remainder

$$R_p(x_1, \ldots, x_n) = \sum_{|k| \geq N} f_k e^{ik.x}$$

is bounded in norm by c/p. Here again $k \in \mathbb{Z}^n$ and $k.x = k_1x_1 + \ldots + k_nx_n$.

We adopt the norm $|k| = \text{Sup} |k_i|$. Let k be a nonzero integer vector and let i be the index such that $|k| = |k_i|$. By integrating n times with respect to the variable x_i we obtain as in Proposition 4 the bound for the kth order Fourier coefficient:

$$(17) \qquad \| f_k \| \leq c/|k|^{n+1}.$$

From the convergence of the series $\sum_{k \neq 0} |k|^{-n-1}$, we immediately obtain absolute convergence of the Fourier series. The estimate of the remainder follows by considering the tail of the series. ■

We next obtain a much more refined estimate for the remainder of the Fourier series of a function of class C^{2n+1}:

Proposition 8:

Let f be a class C^{2n+1} function of n variables. The pth order remainder $R_p = \sum_{|k|\geq p} f_k e^{ik.x}$ of the Fourier series of f is bounded in norm by $c[\log(p)]^n/p^{2n+1}$.

We make use of several lemmas from the theory of harmonic approximation (approximation of periodic functions by trigonometric polynomials) which are of general interest. For more details on the theory, we suggest the excellent work by J.R. Rice [Ri] which we follow here.

We recall that whenever the Fourier series of a periodic function f converges to f, the series is the best trigonometric approximation to f in L^2, in the sense that it is the trigonometric series which converges most rapidly to f in L^2 norm. This is not true in other norms such as the C^0 norm which we use in the remainder of this appendix (and which gives rise to "Chebyshev approximations").

We first introduce the notion of degree of convergence to a function for a sequence of trigonometric polynomials. We adopt the notation $p = (p_1, \ldots, p_n)$, and we say that the sequence T_p converges to $f(x_1, \ldots, x_n)$ to degree $r(p)$ if $\| f - T_p \| \leq cr(p)$ for some $c > 0$ and if this estimate cannot be improved. Successive approximations of a function are generally obtained as convolutions with a kernel; the important ones for our purposes are the Dirichlet and Jackson kernels.

The Dirichlet kernel is the periodic function:

$$D_q(x) = 1/2\pi \sin\{(q+1/2)x\}/\{2\sin(x/2)\},$$

and its integral is equal to 1. The function $D_q(x) = \Pi_{1\leq i\leq n} D_{q_i}(x_i)$ is called the n-variable Dirichlet kernel. The partial rectangular sum $S_q(x)$, $q = (q_1, \ldots, q_n)$ of the function f, i.e., the sum of Fourier coefficients of order k satisfying the inequalities $|k_1| \leq q_1, \ldots, |k_n| \leq q_n$) is none other than the convolution product of f and the n-variable Dirichlet kernel $D_q(x)$:

$$S_q(x) = D_q(x) * f(x), \qquad (18)$$

where the convolution is defined by $f * g = \int_0^{2\pi} f(y)g(x-y)\,dy$.

The Jackson kernel is the periodic function:

$$J_q(x) = C(q)[\sin(qx/2)/\sin(x/2)]^4,$$

where the normalization constant $C(q)$ is equal to $3/\{(2\pi q)(2q^2+1)\}$ to make the integral of this kernel also equal to one. Again the product $J_q(x) = \prod_{1\leqslant i\leqslant n} J_{q_i}(x_i)$ is called the n–variable Jackson kernel.

We now establish the following

Lemma 1:

Let f be a real valued function of n variables $x_1, \ldots, x_n$, periodic in each variable and of class C^1 in x_1. Suppose U_p is a sequence of polynomials which converges to $\partial f/\partial x_1$ to at least degree $r(p)$: $\| \partial f/\partial x_1 - U_p \| \leqslant cr(p)$. Then there exists a trigonometric approximation T_p of f with degree of convergence at least $r(p)/p_1$.

Let C be the constant part of U_p and let $V_p = U_p - C$ be its variable part. We then have the bound $| C | \leqslant cr(p)$, since:

$$(19) \qquad (2\pi)^n | C | \leqslant \left| \int (\partial f/\partial x_1 - C - V_p)\, dx_1 \ldots dx_n \right| \leqslant (2\pi)^n cr(p).$$

It follows that $\| \partial f/\partial x_1 - V_p \| \leqslant 2cr(p)$. Let W_p be such that $V_p = \partial W_p/\partial x_1$ and $Y_p = f - W_p$. Then:

$$(20) \qquad \| \partial Y_p/\partial x_1 \| \leqslant 2cr(p).$$

We define the trigonometric approximation $Z_q^{\,p}$ $(q \in \mathbb{N})$ of Y_p by the convolution product in the first variable:

$$(21) \qquad Z_q^{\,p}(x) = J_q(x_1) \cdot Y_p(x).$$

We have:

$$(22) \qquad | Z_q^{\,p}(x) - Y_p(x) | \leqslant \int_0^{2\pi} J_q(u)\, | Y_p(x_1-u, x_2, \ldots, x_n) - Y_p(x_1, \ldots, x_n) |\, du$$
$$\leqslant cr(p) \int_0^{2\pi} uJ_q(u)\, du \leqslant cr(p)/q.$$

From this we deduce:

$$(23) \qquad \| Z_{p_1}^{\,p} + U_p - f \| \leqslant cr(p)/p_1,$$

and if we denote the trigonometric approximation $Z_{p_1}^{\,p} + U_p$ of f by T_p, then:

$$(24) \qquad \| T_p - f \| \leqslant cr(p)/p_1.$$

This completes the proof of Lemma 1. ▪

By integrating twice in each variable, we deduce from Lemma 1 the following

Lemma 2:

If the n-variable function f is at least of class C^{2n}, and if there exists a trigonometric approximation of $\partial^{2n} f/\partial x_1^{\,2}\ldots\partial x_n^{\,2}$ of at least degree of convergence $r(p)$, then there exists a trigonometric approximation T_p of f of at least degree of convergence $r(p)/(p_1 \ldots p_n)^2$.

We also estimate the degree of convergence $r(p)$ of trigonometric approximations of Lipschitz functions by means of

Lemma 3:

For any Lipschitz function f, there exists a trigonometric approximation T_p of f of at least degree of convergence $(1/p_1 + \ldots + 1/p_n)$.

We introduce the trigonometric approximation:

(25) $\quad T_p(x) = J_p(x) * f(x)$

where J_p is the n-variable Jackson kernel. Then:

(26) $\quad \| T_p - f \| = \| \int [f(y) - f(x)] J_p(x - y)\, dy_1 \ldots dy_n \|$

and since f is Lipschitz, we have:

(27) $\quad \| T_p - f \| \leqslant c \int \| u \| J_p(u)\, du_1 \ldots du_n$

$$\leqslant c \sum_i \int | u_i | J_{pi}(u_i)\, du_i \leqslant c(1/p_1 + \ldots + 1/p_n),$$

which proves Lemma 3. ▪

From Lemmas 2 and 3, we deduce that for any function f of class C^{2n+1}, there is a trigonometric approximation T_p satisfying:

(28) $\quad \| T_p - f \| \leqslant c(1/p_1 + \ldots + 1/p_n)/(p_1 \ldots p_n)^2.$

We now turn our attention to Fourier series. We have the following:

Lemma 4:

If there exists a trigonometric approximation T_p of the function f of at least degree of convergence $r(p)$, then the partial sums S_p of the Fourier series of f converge to f with at least degree of convergence $r(p)\log(p_1)\ldots\log(p_n)$.

Let $S_q^{\,p}$ be the rectangular partial sum of order $q = (q_1, \ldots, q_n)$ of the Fourier series of

$f - T_p$. It is expressed in terms of the n-variable Dirichlet kernel by:

$$S_q{}^p(x) = (f - T_p)(x) \cdot D_q(u).$$

We therefore have the bound:

$$(29) \qquad \| S_q{}^p \| \leqslant c \| f - T_p \| \prod_i \int D_{qi}(u_i)\, du_i \leqslant c \| f - T_p \| \log(q_1)\ldots\log(q_n).$$

Since for $q_i \geqslant p_i$, $S_q{}^p = S_q - T_p$, we have:

$$(30) \qquad \| S_p - f \| \leqslant \| S_p - T_p \| + \| T_p - f \|$$

$$\leqslant \| S^p{}_p \| + \| T_p - f \|$$

$$\leqslant cr(p)[1 + c \log(p_1)\ldots\log(p_n)]$$

which proves the lemma. ▪

We easily establish Proposition 8 from this lemma. We have:

$$(31) \qquad \| f - S_p \| \leqslant c \log(p_1)\ldots\log(p_n)(1/p_1 + \ldots + 1/p_n)/(p_1\ldots p_n)^2.$$

From this we deduce, for a cubic partial sum C_p of order p, i.e., the sum of the Fourier coefficients of order k satisfying the inequalities $| k_1 | \leqslant p, \ldots, | k_n | \leqslant p$:

$$(32) \qquad \| f - C_p \| \leqslant cn[\log(p)]^n / p^{2n+1}.$$

Since the remainder R_p of the Fourier series of f is simply $f - C_{p-1}$, we obtain the desired bound:

$$(33) \qquad \| R_p \| \leqslant c[\log(p)]^n / p^{2n+1}. \quad ▪$$

APPENDIX 2: ERGODICITY

In this appendix we present a brief review of the elementary definitions and results of the theory of ergodicity. For (much) more information, with an emphasis on the connection between ergodic theory and classical systems, the interested reader may consult, for example, the book by V.Arnold and A. Avez ([Ar8]).

Consider a flow T_t which preserves the probability measure μ on the space X ($\mu(X) = 1$). Because of its fundamental implications in statistical physics, the following question originally stimulated the development of ergodic theory:

Given a function φ on the space X, under what conditions will the time average

$$<\varphi>_T(x^*) = 1/T \int_0^T \varphi(T_t(x^*))\, dt$$

(of the function φ over the trajectory $T_t(x^*)$ with initial condition x^*) converge to the space average $\varphi^* = \int_X f\, d\mu$ of φ as T approaches infinity? Here, one can think of T_t as the time t map of a system of ODE's; the same kind of questions arise in the discrete framework for the iteration of a given transformation. Since the present book is concerned with differential equations however, we shall mainly state the definitions and results for flows.

By taking φ to be the characteristic function χ_A of a set A in X which is invariant under the flow T_t, one immediately shows that a necessary condition for answering this question in the affirmative is that all subsets invariant under the flow have measure 0 or 1. Flows with this property are said to be *ergodic*. Birkhoff's ergodic theorem says that this property is also sufficient:

Birkhoff's Ergodic Theorem:

For any function φ in $L^1(X, \mu)$, and for almost all x, the temporal average

$$<\varphi>_T = 1/T \int_0^T \varphi \circ T_s(x)\, ds$$

converges, when $T \to \infty$, to a function $M(\varphi)$ in $L^1(X, \mu)$ which is invariant under the flow and which has the same space average as φ.

In particular, if the flow is ergodic, the temporal average $<\varphi>_T(x)$ of φ over the trajectory with initial condition x converges in $L^1(X, \mu)$, for almost all x, to the spatial average φ^* of φ.

It is then interesting to formulate an analogous question in the space $L^2(X, d\mu)$ of square integrable functions, with its Hilbert space structure, and to change perspective slightly, by considering, not trajectories of the flow on X, but rather the evolution of functions in $L^2(X, d\mu)$, a change which is equivalent to passing from Schrödinger's to Heisenberg's representation in quantum mechanics. In this way one associates to the flow T_t a one parameter group of evolution operators $\{ U_t , t \in \mathbb{R} \}$ on $L^2(X, \mu)$ defined by:

(1) $\qquad U_t [f](x) = f[T_t(x)].$

One easily verifies that measure preservation under the flow is equivalent to unitarity of the operators U_t . This point of view is largely due to Koopman, and it has the advantage of reducing many aspects of the problem to the study of *linear* operators on a Hilbert space (see also [Re], Chapter VII).

In this framework, the first question which arises is the following: Under what conditions does the time average

$$<\varphi>_T (x) = 1/T \int_0^T U_s[f]\, ds(x)$$

converge in L^2 norm to the constant function $\varphi^* = \int_X f\, d\mu$? This question is answered by Von Neumann's ergodic theorem, which may be stated as follows:

<u>Von Neumann's Ergodic Theorem</u>:

Let V be the vector subspace of $L^2(X, \mu)$ consisting of functions invariant under the flow T_t ; that is, V is the eigenspace associated to the eigenvalue 1 of the operator U_1 . Let P be the orthogonal projection onto V. Then as $T \to \infty$, the time average of an arbitrary function φ in $L^2(X, \mu)$ converges in the L^2 sense to the function $P\varphi$:

(2) $\qquad \lim_{T\to\infty} \| 1/T \int_0^T U_s[\varphi]\, ds - P\varphi \| = 0.$

In fact, the time average $\lim_{T\to\infty} <\varphi>_T$ and the space average φ^* coincide (in L^2 norm) if and only if $P\varphi = \int_X \varphi\, d\mu$. Since the space average is simply the scalar product of φ and the constant function 1, the necessary and sufficient condition for equality of the averages is that the multiplicity of the eigenvalue 1 of the operator U_1 be precisely 1; in other words, the only L^2 functions invariant under the flow are the constant functions. This property may thus be viewed as an alternate definition of ergodicity.

We now proceed to prove Von Neumann's theorem by first restricting our attention to the discrete time case. In passing to the continuous case, it suffices to consider the operator U in the proof below as successive applications of the flow T_t with $t = 1$. The statement is as follows:

Von Neumann's Ergodic Theorem (discrete form):

Let $\mathcal{H}$ be a Hilbert space, U a unitary operator (the evolution operator), V the eigenspace corresponding to the eigenvalue 1 of the operator U, and P the orthogonal projection onto this eigenspace. For each vector φ in $\mathcal{H}$, the time average

$$<\varphi>_N = 1/N \sum_{0 \leqslant n < N} U^n(\varphi)$$

converges in norm to the projection $P\varphi$.

Since U is unitary, $\mathcal{H}$ may be decomposed as :

(3) $\mathcal{H} = \mathrm{Ker}(I-U^*) \oplus \mathrm{clos}(\mathrm{Im}(I-U))$,

where clos(A) denotes the closure of the set A. If φ belongs to $\mathrm{Ker}(I - U^*)$, then $U\varphi = \varphi$, $P\varphi = \varphi$, and the theorem holds.

Suppose now that φ belongs to the image $\mathrm{Im}(I - U)$ of $I - U$, and set $\varphi = \psi - U\psi$. Since

(4) $1/N \sum_{0 \leqslant n < N} U^n(\varphi) = (\psi - U^N \psi)/N$,

we have the bound:

(5) $| <\varphi>_N | \leqslant 2 \| \psi \|/N$,

so that tends to zero as N approaches infinity. Since $P\varphi = 0$, the theorem holds for φ in $\mathrm{Im}(I - U)$.

This latter result extends to the case where $\varphi \in \mathrm{clos}(\mathrm{Im}(I - U))$, which proves the theorem. ▪

It is interesting how disarmingly simple the above proof is, compared to the somewhat tricky proof of Birkhoff's theorem, thus illustrating the power of Hilbert space techniques.

We now give a third definition of ergodicity, which may be compared with the definitions of strong and weak mixing given below. In fact, using Von Neumann's theorem, it is easy to show that T_t is ergodic if and only if for two arbitrary measurable sets A and B, as T approaches infinity the time average $1/T \int_0^T \mu[T_s A \cap B]\, ds$ converges to the product $\mu(A).\mu(B)$ of the measures of A and B; we thus have convergence of $\mu[T_t A \cap B]$ to $\mu(A).\mu(B)$ in the Césaro sense.

There is a hierarchy of statistical properties stronger than ergodicity: weak mixing (which implies ergodicity), mixing (which implies weak mixing), the properties of

Bernouilli, Lebesgue spectra, K and C systems, and so on. Here we limit ourselves to recalling the definitions of mixing properties which further illustrate the relation between dynamical and spectral properties.

We saw previously that preservation of measure under the flow is equivalent to the unitarity of the evolution operators U_t and that ergodicity is equivalent to 1 being an eigenvalue of multiplicity 1 of the evolution operators U_t. We now give similar definitions of the mixing properties; proof of their equivalence is a simple exercise.

Weak Mixing:

1) The flow T_t is weakly mixing if for two arbitrary measurable sets A and B:

$$\lim_{T\to\infty} 1/T \int_0^T |\mu(T_t(A)\cap B) - \mu(A).\mu(B)|\, dt = 0. \tag{6}$$

2) The flow T_t is weakly mixing if and only if 1 is the only eigenvalue of U_1 and it occurs with multiplicity 1.

Mixing:

1) The flow T_t is mixing if and only if for any two measurable sets A and B:

$$\lim_{t\to\infty} \mu(T_t(A) \cap B) = \mu(A).\mu(B). \tag{7}$$

2) The flow T_t is mixing if and only if it is weakly mixing and the spectrum of the restriction of U_1 to the orthogonal complement of the eigenspace associated to the eigenvalue 1 is purely absolutely continuous.

To end this very brief discussion of ergodicity, it may be useful to simply state the underlying classification problem. Let (X, μ, T) and (X', μ', T') be two dynamical systems; here T and T' are measure preserving applications from X (X') to itself. When are these systems equivalent, in the sense that there exists an almost everywhere bijective transformation $S: X \to X'$ which is measure preserving and such that $T = S^{-1}T'S$? This problem is not completely solved, but spectral theory furnishes invariants defined by means of the associated unitary operators on $L^2(X, \mu)$ and $L^2(X', \mu')$. The Kolmogorov-Sinaï entropy is also an important, *nonspectral* invariant, but any detailed discussion of it would of course take us too far afield.

APPENDIX 3: RESONANCE

1. Phases, frequencies and nondegeneracy conditions

In the present work, we are essentially interested in systems of the form:

$$dI/dt = \varepsilon f(I, \varphi, \varepsilon), \quad d\varphi/dt = \omega(I) + \varepsilon g(I, \varphi, \varepsilon), \quad (I, \varphi) \in \mathbb{R}^m \times T^n.$$

and particular instances of it, especially the Hamiltonian case, with governing Hamiltonian function:

$$H(I, \varphi) = H_0(I) + \varepsilon H_1(I, \varphi) \quad (m = n, \omega = \nabla H_0).$$

In general, one studies strictly non isochronous systems in which the unperturbed frequency vector $\omega(I)$ satisfies a nondegeneracy condition guaranteeing that the total measure of the resonances is zero and that the variation of the slow variables I in an arbitrary direction actually causes the frequencies to vary. To this effect, one uses either Kolmogorov's or Arnold's nondegeneracy condition, the latter being also called the "isoenergetic nondegeneracy condition" in the Hamiltonian case.

A system satisfies *Kolmogorov's condition* if and only if the map $I \to \omega(I)$ has maximal rank (equal to n) at every point. This implies $m \geqslant n$ and in the Hamiltonian case it amounts to requiring that the Hessian $\det(\nabla^2 H_0)$ be nonzero.

A system satisfies *Arnold's condition* if and only if the map, which associates to each point $I \in \mathbb{R}^n$ the point in $n - 1$ dimensional projective space defined by the frequency vector $\omega(I)$, has maximal rank, equal to $n - 1$; this implies that $m \geqslant n - 1$. In other words, assuming, with no loss of generality, that $\omega_n(I) \neq 0$, one demands that the map

$$I \to (\omega_1(I)/\omega_n(I), \dots, \omega_{n-1}(I)/\omega_n(I))$$

be of maximal rank. In practice, this can be checked by noting that it is equivalent to the map $(\lambda, I) \to \lambda\omega(I)$ having rank n at the point $(1, I)$, where λ is a real homogeneity parameter. In the Hamiltonian case, because the Hamiltonian is a first integral, one wants this condition to be satisfied when the variables are restricted to vary on the unperturbed energy surface (which is ε-close to the perturbed one). This amounts to requiring that the map $(\lambda, I) \to (\lambda\omega(I), H_0(I))$ from $\mathbb{R}^{n+1}$ to $\mathbb{R}^{n+1}$ be of maximal rank at $(1, I)$, i.e., have nonvanishing Jacobian determinant. This is called Arnold's determinant and is also equal to the Hessian of the map $(\lambda, I) \to \lambda H_0(I)$ evaluated at $(1, I)$.

In this appendix we recall elementary geometric properties of resonances in the case where one of these two conditions is satisfied.

2. Resonances; elementary definitions

A point $I \in \mathbb{R}^m$ is said to be *resonant* if there exists a nonzero vector k in $\mathbb{Z}^n$ such that the commensurability relation $\omega(I).k = 0$ between components of the frequency vector ω is satisfied at the point I. The set of resonant points associated to a fixed vector k is generically a hypersurface of dimension $m - 1$ in the base space. Such an equation defines a simple resonance (i.e., of dimension 1, with a single vector k).

The same resonance is of course associated with two vectors k and k' in $\mathbb{Z}^n$ such that one is an integral multiple of the other; the *order* of a simple resonance is the norm of the smallest vector in $\mathbb{Z}^n$ defining the resonance.

3. Resonant modules

One next generalizes the notions formulated above to resonances of arbitrary dimension.

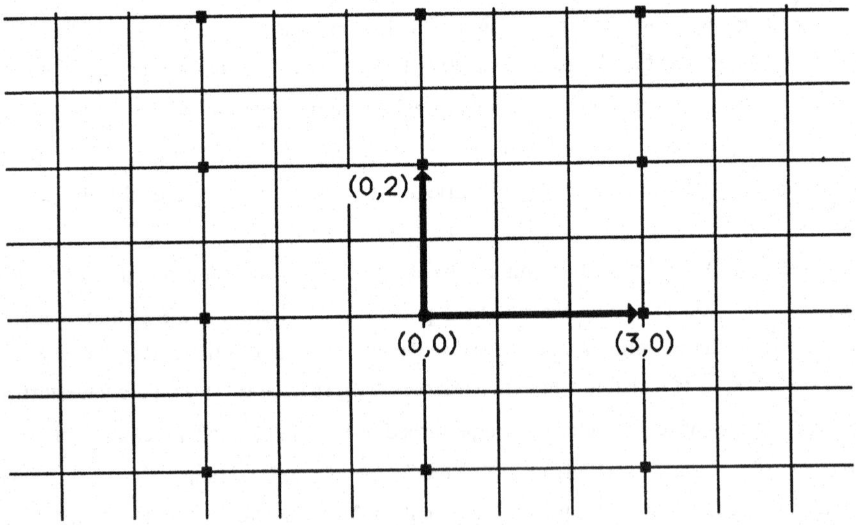

Figure A3.1: Schematic representation of a resonant module: Grid points are vectors of $\mathbb{Z}^2$, black dots correspond to those vectors which belong to the resonant module, the two bold arrows are the basis vectors of the module.

The equations defining the resonances are linear in k; for fixed I, the equation $\omega(I).k = 0$ is satisfied by vectors k belonging to a discrete subgroup, or module, of $\mathbb{Z}^n$, called the *maximal resonant module* associated to the point I. In this context the terms module, basis and dimension may be used interchangeably with group, generators and order; we will use the former terminology in this work. The submodules of the maximal resonant module will also be called maximal) resonant modules (see Figure A3.1).

To every submodule M of $\mathbb{Z}^n$ may be associated a resonance (perhaps empty) defined as the set of points I of the base satisfying $\omega(I).k = 0$ for each vector k in M.

The maximal module M defining a resonance has a basis $(k_1, ..., k_r)$ of vectors in $\mathbb{Z}^n$; r, which is the dimension of the free module M, will also be called the *dimension* of the resonance. Generically the corresponding surface in $\mathbb{R}^m$, defined by the r equations $\omega(I).k_i = 0$, has codimension r for $r \leqslant m$. We may consider the set of nonresonant points as the resonance associated with the module $\{0\}$.

4. Geometry of the resonances

The resonances are naturally nested or stratified; every resonance of dimension r is included in a resonance of dimension $r - 1$.

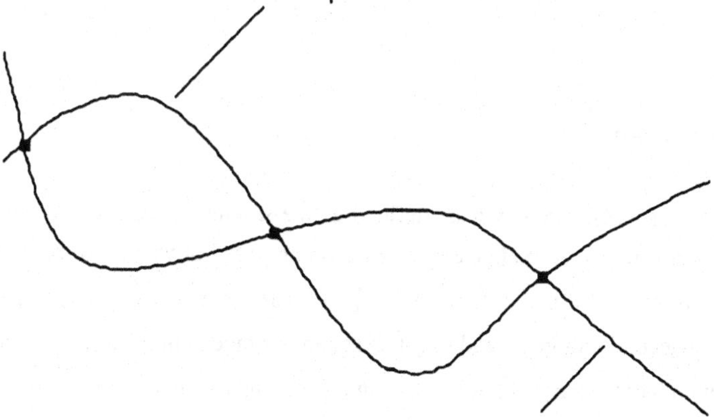

Figure A3.2: The resonance are naturally nested: The intersection of the two resonances of dimension one $\omega(I).k_1 = 0$ and $\omega(I).k_2 = 0$ is a resonance of dimension two.

The geometry of the resonances on the base $\mathbb{R}^m$ depends on the number of phase variables n. For single phase systems ($n = 1$), the resonances occur at the zeros of the associated frequency.

In two-phase systems ($n = 2$) with nondegenerate frequencies, double resonances can only arise at points where both frequencies vanish simultaneously. The resonance condition is of the form

$$k_1\omega_1(I) + k_2\omega_2(I) = 0, \; k=(k_1, k_2) \in \mathbb{Z}^2 - \{(0, 0)\},$$

and if - say - $\omega_2(I) \neq 0$ at a particular point I, this is locally equivalent to $\omega_1/\omega_2(I) = -k_2/k_1$. In other words, the resonances are determined by the fact that a certain function ((ω_1/ω_2) or (ω_2/ω_1)) takes on rational values. Therefore, since this function is single-valued, resonant hypersurfaces can intersect *only* at the points where $\omega(I) = 0$. It is largely this separation of the resonances which explains the simplicity of two-phase systems.

In nondegenerate systems with more than two phases the resonances are no longer disjoint, and their intersections usually form a dense set in phase space, as illustrated by the following example in $\mathbb{R}^2 \times T^3$:

$$dI_1/dt = 0; \quad dI_2/dt = 0; \quad d\varphi_1/dt = I_1; \quad d\varphi_2/dt = I_2; \quad d\varphi_3/dt = 1.$$

Here the resonance surfaces are the straight lines given by the equations $k_1 I_1 + k_2 I_2 + k_3 = 0$ with integer k_i's.

In such systems, it is possible for trajectories to wander long distances passing from resonance to resonance, a phenomenon called Arnold diffusion in the Hamiltonian case.

5. Order of resonance

The *order* of a resonance will be defined as the smallest integer p such that the associated maximal resonant module admits a basis composed of vectors with norm less than p. One may set $|k| = \sup_i |k_i|$ or $|k| = \sum_i |k_i|$ as a norm on $\mathbb{Z}^n$. The higher the order of a resonance, the more spread out its resonant module is in $\mathbb{Z}^n$. A p-module will be a module with basis vectors of norm less than p (this slightly unorthordox terminology is useful when there is no risk of confusion).

6. The ultraviolet cutoff

The resonances form a dense and often complicated subset of the base space. However, the effect of high-order resonances is generally small, and in the proofs one uses this fact by taking into account only a finite number (depending on ε) of low-order resonances. More precisely, one restricts attention to resonances whose associated maximal module is an N-module, where the integer $N(\varepsilon)$ is suited to the particular problem under consideration. The use of such an *ultraviolet cutoff*, first employed by Arnold in his proof of the KAM theorem, is of course linked to the decay of the Fourier spectrum of the pertubation terms, and thus to the regularity in phase of these functions, the optimal case being that of analytic perturbations, with exponentially decaying spectra (see Appendix 1).

7. Resonant zones

The solution in φ of linear equations of the form $\omega(I).\partial S(I, \varphi)/\partial\varphi = h(I, \varphi)$, which arise naturally in perturbation theory (see Chapters 3 through 8), is generally not possible because of the presence of small divisors $\omega(I).k$ in the φ-Fourier expansions. This presents problems not only on the resonances strictly speaking, where the divisors vanish, but also in the neighborhood of resonances, where the divisors are "too small.". One is thus led to define a resonant zone of variable size around each resonance taken into account, so that the linear equation may be solved on the complement of these zones. Since the most dangerous resonances are the low-order ones, it is often desirable to define zones whose sizes decrease with increasing order (see, for example, Chapters 4 and 7).

The resonant zones will be defined by equations of the form:

$$| \omega(I).k_i | \leqslant \nu \quad (i = 1,.., r),$$

where the k_i form a basis for the *maximal* module defining the resonance. Notice that although k and nk, $n \in \mathbb{Z}$, generate the same resonance, the resonant zones $| \omega(I).k | \leqslant \nu$ and $| \omega(I).nk | \leqslant \nu$ are different.

8. Primitive modules and resonant Fourier series

We are interested in the following simple algebraic question encountered for example in Chapters 4 and 7:

Is it possible to extract from a Fourier series $\sum_k c_k e^{ik.\varphi}$ the "resonant subseries" corresponding to a given module M of dimension r ?

The answer is yes, if the submodule is maximal. In this case we shall see that it suffices to define new angle variables $\varphi' = (\varphi'_1, \ldots, \varphi'_n)$ which are appropriate linear combinations of the old angle variables, in such a way that the resonant terms do not depend on the first r variables. The nonresonant terms may then be eliminated by averaging over these variables. Before describing this variable change in precise terms, we first recall some simple facts from linear algebra.

Let M be a module of dimension r, $(k_1, \ldots, k_r)$ a basis for M, and A the $r \times n$ matrix with integer entries in which the ith row is the vector k_i. The following theorem concerning the reduction of matrices with integer entries may then be proved:

<u>Theorem</u>:

The matrix A is similar to an $n \times r$ matrix D with integer entries of the form:

$$\begin{bmatrix} d_1 & 0 & 0 & 0 & 0 & 0 \\ 0 & d_2 & 0 & 0 & 0 & 0 \\ 0 & & d_i & 0 & 0 & 0 \\ 0 & & & d_r & 0 & 0 \end{bmatrix}$$

where d_i divides d_j for $i \leqslant j$.

In other words, there exists an invertible matrix P of order r and an invertible matrix Q of order n, each with integer entries and unit determinant, such that $A = PDQ$.

This theorem is proved by explicitly constructing the matrices P and Q using the following elementary matrices which describe simple manipulations on the rows and columns of a given matrix:

1) $T_{ij}(p) = 1 + pE_{ij}$ (E_{ij} is the matrix whose only nonzero entry is a 1 located at the intersection of the ith row and the jth column). $T_{ij}(p)^{-1} = T_{ij}(-p)$.

2) $S_i = 1 - 2E_{ii}$, which is involutive. ▪

From the invariance of these quantities we see that $d_i = \Delta_i / \Delta_{i-1}$, where Δ_i is the greatest common divisor of the relative minors of order i of the matrix A. The d_i's are the invariants of the module and this is said to be *primitive* if all its invariants are equal to 1.

Let us now consider a resonance and an associated resonant module M. Let $\omega(I) = \Omega$ be the vector of frequencies with components $(\omega_1(I), ..., \omega_n(I))$ at the point I of the resonance. Since the module M is resonant, $A\Omega = 0$, which may be written $PDQ\Omega = 0$. Because P is invertible, it follows that $DQ\Omega = 0$. Since D is composed of a diagonal block of order r and a zero block of order $(n-r) \times r$, the latter condition $DQ\Omega = 0$ is equivalent to requiring that Ω satisfy $\Pi Q\Omega = 0$, where Π is the matrix identical to D except for its diagonal block, which is the identity matrix of order r.

This means that every resonant module of dimension r may be embedded in a primitive module of dimension r. Consequently, the maximal module associated to the resonance is necessarily primitive, a fact which is not quite intuitively evident at first sight. The maximal module admits a basis whose coordinates in the standard basis for $\mathbb{Z}^n$ are given by the rows of the matrix ΠQ, and the preceding discussion in fact yields an effective method for constructing the maximal module. The replacement of a module by a primitive module is the mutidimensional generalization of the replacement, for simple resonances, of the resonance condition $\omega.nk = 0$ by the condition $\omega.k = 0$, where k is a vector with relatively prime coordinates.

We note that the set of resonant frequencies associated to a given primitive module is a subspace of $\mathbb{R}^n$ of dimension $n - r$ which admits a basis consisting of the images under Q^{-1} of the last $n - r$ basis vectors in $\mathbb{R}^n$.

We are now in a position to answer the initial question concerning resonant Fourier series. We use the change of variables $\varphi_i = \sum_j R_{ij}\, \varphi'_j$ defined by an nth order invertible matrix R with integer entries, and we seek to eliminate all terms in the integral

$$(2\pi)^{-r} \int \sum c_k \exp(ik.R\varphi')\, d\varphi' = (2\pi)^{-r} \int \sum c_k \exp\,(i[{}^tRk].\varphi')\, d\varphi'$$

except for the resonant terms associated to the maximal module M (here ${}^tR$ denotes the transpose of R).

Thus, for k in the resonant module M, ${}^tRk.\varphi'$ must not depend on $\varphi'_1, ..., \varphi'_r$. This entails that the first r components of ${}^tR\,k$ are zero, or in other words that $\Pi\, {}^tRk = 0$ for k in M. This leads in turn to the choice $R = {}^tQ$, where the matrix Q is defined above. It is now easy to verify that with this definition of the new angle variables φ' ($\varphi' = {}^tQ^{-1}\varphi$); averaging over φ' eliminates precisely the nonresonant terms.

APPENDIX 4: DIOPHANTINE APPROXIMATIONS

In perturbation theory, the transition from formal calculations of series to deeper analytic results requires a quantitative control of the small divisors that inevitably show up in one form or another, as is briefly explained below for the case of normal forms (cf. Appendix 5). These divisors assume different and yet similar forms, according to the problems considered. For example:

i) In the problem of the conjugation of an analytic vectorfield to its linear part in the neighborhood of the origin (Siegel's theorem; cf. Appendix 5), one comes across expressions of the form $\lambda_s - (m.\lambda)$ with $\lambda = (\lambda_1, \ldots, \lambda_n) \in \mathbb{C}^n$, $m = (m_1, \ldots, m_n) \in \mathbb{N}^n$.

ii) If the vectorfield is periodic in t (with period 2π), one is required to work with the combinations $\lambda_s - (m.\lambda) - ik$, where the notation is as before and $k \in \mathbb{Z}$ (cf. [Ar8], §26).

iii) For the study of integrable Hamiltonian perturbations in Hamiltonian mechanics, one finds the form (cf. Chapter 7 and Appendix 8): $\omega.k$, $\omega = (\omega_1, \ldots, \omega_n) \in \mathbb{R}^n$, $k = (k_1, \ldots, k_n) \in \mathbb{Z}^n$, where $\omega(I)$ is the frequency vector of the unperturbed Hamiltonian.

In the first two problems, since the m_i are positive, a simple lower bound on the small divisors is possible on an entire region of $\mathbb{C}^n$ where λ takes its values (the so-called Poincaré domain). It was C.L. Siegel ([Siel-4]) who first had the idea of combining Diophantine estimates and quadratic convergence in Newton's method (cf. Appendix 8) in order to control the small divisors when λ belongs to a certain subset (of full measure) of the complement of Poincaré's domain (this is since called Siegel's domain). In Arnold's book [Ar8], one can find a proof of Siegel's theorem which states that an analytic locally invertible map is biholomorphically conjugate to its linear part in the neighborhood of a fixed point if the vector λ composed of the eigenvalues of the map tangent at the fixed point satisfies a certain Diophantine condition.

In this appendix we state (and partly prove) the approximation theorem for small divisors of the form iii), which are the ones most often encountered in mechanics. For an introduction to the arithmetic theory of Diophantine approximation, we refer the reader to the book by J.W.S. Cassels [Cas].

For $\tau, \gamma > 0$, we write:

(1) $$\Omega(\tau, \gamma) =_{def} \{ \omega \in \mathbb{R}^n, \forall k \in \mathbb{Z}^n - \{0\}, |\omega.k| \geq \gamma |k|^{-\tau} \}$$

where $|k| = \sup_i |k_i|$ and we set $\Omega(\tau) =_{def} \bigcup_{\gamma>0} \Omega(\tau, \gamma)$.

Theorem:

There are three cases to distinguish, according to the values of τ:

i) $0 < \tau < n - 1$; then $\Omega(\tau) = \emptyset$.

ii) $\tau = n - 1$; $\Omega(\tau)$ has Lebesgue measure zero and Hausdorf measure n.

iii) $\tau > n - 1$; $\mathbb{R}^n - \Omega(\tau)$ has zero measure, and more precisely,

$$\text{mes}\,\{(\mathbb{R}^n - \Omega(\tau, \gamma)) \cap B_R\} \leqslant C(\tau)\gamma R^{n-1},$$

where B_R denotes the ball in $\mathbb{R}^n$ of radius R.

Remarks:

- Assertion iii), proved below, is the simplest, and moreover the only result to be used in perturbation theory up to now; analogous assertions are proved in the same way, mutatis mutandis, for different kinds of small divisors.
- We refer the curious reader to Cassel's book for proofs of i) and part of ii) (i.e., $\text{mes}\,\Omega(\tau) = 0$, a result due to A. Kinchine). The result on the Hausdorf measure is due, in this form, to W.M. Schmidt ([Sc1-2]).

Assertion i) is a direct consequence of Dirichlet's theorem, which we recall below (cf. [Cas], Chapter 1).

Theorem (Dirichlet):

For all $\theta = (\theta_1, \ldots, \theta_m) \in \mathbb{R}^m$, the inequality

$$(2) \qquad \| p.\theta \| = \| p_1\theta_1 + \ldots + p_m\theta_m \| < |p|^{-m},$$

where $p \in \mathbb{Z}^m$, has infinitely many solutions.

Here, $\| . \|$ denotes the positive fractional part of a real number; in other words:

$$\forall\, \theta \in \mathbb{R},\ \| \theta \| =_{def} \inf_{q\in\mathbb{Z}} |\theta - q|.$$

This theorem, which we have not stated in full generality, is itself a direct consequence of a simple geometrical lemma on linear forms due to H. Minkovski.

Returning to ω and k, we prove i) assuming Dirichlet theorem. One can assume without loss of generality that $\omega_n \neq 0$. We thus set $\theta_i = \omega_i/\omega_n$ $(i = 1, \ldots, n-1)$ and apply the theorem for $m = n - 1$, which states that the inequality:

$$(3) \qquad |\omega.k| < |\omega_n| (\sup_{1\leqslant i\leqslant n-1} |k_i|)^{-(n-1)},$$

admits an infinity of solutions. We remark that for these solutions,

$$(4) \qquad |\omega_n k_n| \leqslant 1 + \left|\sum_{1\leqslant i\leqslant n-1} k_i\omega_i\right| \leqslant 1 + (\sup_{1\leqslant i\leqslant n-1} |k_i|)(\sum_{1\leqslant i\leqslant n-1} |\omega_i|)$$
$$\leqslant (\sup_{1\leqslant i\leqslant n-1} |k_i|)(\sum_{1\leqslant i\leqslant n} |\omega_i|).$$

It then follows from (3) that:

$$(5) \qquad |\omega.k| \leqslant |k|^{-(n-1)}(\sum_{1\leqslant i\leqslant n} |\omega_i|)/|\omega_n|^{n-2} = C(\omega)|k|^{-(n-1)}.$$

Assertion i) of the theorem now follows immediately from the fact that this inequality has solutions for arbitrarily large $|k|$.

We will not prove the assertions ii) of the theorem, by far the most difficult part (cf. [Cas] and [Sc1-2]).

iii) The proof here is elementary. For each $k \in \mathbb{Z}^n - \{0\}$, the equation $\omega.k = 0$ defines a plane in $\mathbb{R}^n$ passing through 0 (these are the resonant planes in mechanics; see Appendix 3). The measure in $\mathbb{R}^n$ of the set (the resonant zone):

$$\{\omega : |\omega.k| \leqslant \gamma |k|^{-\tau}\} \cap B_R \quad (\gamma \text{ and } k \text{ fixed})$$

is equal to $c\gamma R^{n-1}|k|^{-\tau-1}$, and thus:

$$(6) \qquad \text{mes}\,\{(\mathbb{R}^n - \Omega(\tau, \gamma)) \cap B_R\} \leqslant c\sum_{k\in\mathbb{Z}^n-\{0\}} \gamma R^{n-1}|k|^{-\tau-1}$$
$$\leqslant C(\tau)\gamma R^{n-1}.$$

and the sum is convergent if and only if $\tau > n - 1$ (the convergence is equivalent to that of $\int_{r>0} r^{n-\tau-2}\,dr$). ∎

APPENDIX 5: NORMAL FORMS

In order to study a system of equations that is a small perturbation of a system one knows how to integrate, one idea (the oldest) consists of trying to expand the solutions of the perturbed system in terms of the small parameter by working in the neighborhood of a solution of the original system. As is well known, this technique often results in *secularities*, that is, corrective terms which grow (usually algebraically) with time. The discussion of the *asymptotic* nature of series thus derived goes back essentially to Poincaré, but is not our main topic here. As a very simple and yet representative example, we consider the case of two linear oscillators with nearly equal frequencies:

$$d^2x/dt^2 + \omega^2 x = 0 \quad \text{and} \quad d^2x/dt^2 + (\omega+\varepsilon)^2 x = 0.$$

By viewing the second equation as a perturbation of the first, we are led to a solution of the form

$$\sin[(\omega+\varepsilon)t] = \sin(\omega t) \sum_{n\geq 0} (\varepsilon t)^{2n}/(2n)! + \\ + \cos(\omega t) \sum_{n>0} (-1)^{n+1} (\varepsilon t)^{2n-1}/(2n-1)!$$

We make a point of the fact that, though apparently trivial, this example appears in disguised form in much more complex situations, and it underscores the fact that in general it is much more difficult to calculate "phases" than "amplitudes," so that "elliptic" theory is often more sensitive to perturbations than "hyperbolic" theory (which includes "damping").

Numerous techniques, including averaging, have been devised to repare or eliminate secularities. The underlying idea in most cases is simple, and relies on the fact that a near identity coordinate change in the phase space of the original system can transform it to a simpler form (perhaps even to the simplest possible form). Since a coordinate change is time independent (by definition; we restrict our attention for the moment to autonomous systems), it cannot give rise to secularities. We also remark that this point of view belongs to the classifying trend of mathematics which tends to group objects in equivalence classes; here equivalence refers to the group of admissible coordinate changes; see below.

The first systematic theory of *normal forms* or forms to which one wishes to transform nearby systems, may be found in Poicaré's thesis. We very briefly and informally examine Poincaré's work here, stressing only some of the concepts we find important and refering the interested reader to chapter 5 of Arnold's book [Ar8], which we also follow in part (this chapter provides a good introduction to the theory of *non-Hamiltonian* normal forms; for the Hamiltonian theory, we suggest Appendix 8 and the references therein, especially [Ben4]). The reader will also find in [Brj], together with many interesting results (some of them new) a history of normal form theory for singular vector field. We shall not give further references in this appendix, since normal forms under various guises are pervasive in modern perturbation theories.

In the case examined by Poincaré, one is interested first in the conjugation of an analytic vectorfield to its linear part at a nondegenerate critical point. This problem is embodied in equation:

$$(1) \qquad dx/dt = Ax + v(x)$$

where $x \in \mathbb{C}^n$, $A \in GL_n(\mathbb{C})$ and $v(x)$ is analytic in a neighborhood of the origin. We should point out that we are concerned here with the "continuous version" of the theory (for vectorfields), rather than with the parallel "discrete version" (for mappings). In general, the results and proofs are easily transposed from one version to the other.

Though it is not essential to do so, for convenience we assume that A is diagonalizable with eigenvalues $\lambda_1, \dots, \lambda_n$, and we work with an eigenbasis $\{e_1, \dots, e_n\}$. We denote the vector with components λ_i by $\lambda \in \mathbb{C}^n$, while the "perturbation" $v(x)$ of order $r \geqslant 2$ (the small parameter being distance from the origin) appears in the form:

$$(2) \qquad v(x) = \sum_{1 \leqslant s \leqslant n} \left(\sum_{|m| \geqslant r} a_m x^m \right) e_s ,$$

where $x^m =_{def} x_1^{m_1} \dots x_n^{m_n}$, $|m| =_{def} m_1 + \dots + m_n$ and the series converges for sufficiently small $\|x\|$. We will write $v = v_r + v^*$ where v_r is homogeneous of degree r.

Let us now try to eliminate v_r, the lowest order term in the perturbation, by the coordinate change:

$$(3) \qquad y = x + f(x),$$

where $f(x)$ is a priori holomorphic of order r; in fact, we will see that f can be taken to be a homogeneous vector polynomial of degree r. We have:

$$(4) \qquad dy/dt = dx/dt + \partial f/\partial x . dx/dt$$
$$= Ax + v_r(x) + v^*(x) + \partial f/\partial x . (Ax + v_r(x) + v^*(x)),$$

and, by inverting (3) in the neighborhood of the origin:

$$(5) \qquad dy/dt = Ay + v_r(y) + (\partial f/\partial y . Ay - Af(y)) + w(y),$$

where $w(y)$ is of order at least $r+1$. Here we proceed formally without taking into account the essential problems of radii of convergence, bounds on the remainders, etc.

f is thus determined by the equation:

$$(6) \qquad L_A(f) =_{def} \partial f/\partial y.Ay - Af(y) = -v_r(y).$$

In order to solve it, we note that the operator L_A is linear, preserves degree, and is diagonal in the basis of the series corresponding to the basis $\{e_s\}$. All of this may be translated into the following result, obtained by straightforward calculation:

$$(7) \qquad L_A(y^m e_s) = [m.\lambda - \lambda_s].y^m e_s$$

with $m.\lambda =_{def} m_1\lambda_1 + ... + m_n\lambda_n$. If the numbers $[m.\lambda - \lambda_s]$ are nonzero for the values of $m \in \mathbb{N}^n$ corresponding to the expansion of v_r , we then choose:

$$(8) \qquad f(x) = \sum_{1\leq s\leq n} (\sum_{|m|=r} [m.\lambda - \lambda_s]^{-1} a_m x^m) e_s$$

and the change of variables (3) then transforms the system (1) to the form:

$$(9) \qquad dy/dt = Ay + w(y),$$

where w is at least order $r + 1$.

It is instructive at this point to comment in a somewhat general way on these calculations, as they contain, in their simplicity, the essential elements of the theory:

1. We begin with a system of a certain class: Continuous, C^p ($p \leq \infty$), real analytic, complex analytic, Hamiltonian, etc. It may be added that in this context, questions of regularity are not only of academic interest.

2. We also have the group of transformations preserving this class: Homeomorphisms, C^p or analytic diffeomorphisms, biholomorphic isomorphisms, canonical transformations, etc.

3. The system is "conjugated" (acted on) by elements of this group of transformations in an attempt to transform it to its simplest possible form: Linear, integrable Hamiltonian, etc.

4. In general the terms of the perturbation are graduated in a certain order: By the small parameter ε, by the order of the monomials, by the order of the harmonics in the Fourier series. Some of these various graduations are more or less equivalent; for example, graduating by degree, as we already mentioned, becomes graduation by the small parameter "distance from the origin" via the transformation $x \rightarrow \varepsilon x$.

5. In proceeding order by order, one must solve at each step the *linearized equation* of conjugation, here (6). We have met such situations many times in the course of this book, in the non-Hamiltonian as well as in the Hamiltonian case. This linearized equation describes the infinitesimal action of the unperturbed vectorfield on the (infinite dimensional) Lie algebra of the group of transformations being considered: Lie brackets of vectorfields, Poisson brackets of Hamiltonian vectorfields. At each step, one finds a

transformation which is "tangent" to the sought after normalizing transformation.

6. To solve this linearized equation, one must invert the linearized operator (here L_A) on a finite dimensional space (here the space E_r of homogeneous vector polynomials of degree r), whose eigenvalues are precisely the *small divisors* (here $[\lambda_s - (m.\lambda)]$). As the operator has been put in diagonal form, it suffices to invert the eigenvalues. Note that, since the possibility of solving the equation depends on the cokernel $E_r/L_A(E_r)$ and L_A describes an adjoint action, the linearized equation is sometimes termed "homological".

7. Certain eigenvalues tend to become small or vanish, especially as the order increases (the order here is r; we consider $[\lambda_s - (m.\lambda)]$ for $|m| \leq r$). The operator is thus not injective, or at least the norm of its inverse grows rapidly. It is then impossible to eliminate the terms corresponding to very small denominators (or to zero denominators) and these are called *resonant* denominators. We note that it is often unnecessary to distinguish between exactly resonant and almost resonant terms, if the object is to preserve the uniformity of the expressions obtained.

8. Here we enter the most difficult part of the subject, which involves, according to the particular problem and desired result, combining two options in variable proportion:

i) Not eliminating certain terms corresponding to vanishing or very small eigenvalues; this gives rise to "resonant normal forms". In the case of analytic fields, we arrive in this way after N steps to an equation of the form:

$$(10) \qquad dy/dt = Ay + v_{r_1} + \ldots + v_{r_p} + w(y)$$

where $w(y)$ is at least order $r + N$ and each v_{r_i} is resonant (or almost resonant):

$$(11) \qquad v_{r_i}(x) = \textstyle\sum_{1\leq s\leq n}(\sum_{|m|=r_i} a_m x^m)\, e_s$$

with $|\lambda_s - (m.\lambda)|$ equal to zero (or very small) whenever $a_m \neq 0$. We refer to [Ar8] for further details, as well as to the section in [Gu] devoted to normal forms, which discusses the algebraic calculations connected with these questions. We will briefly return to resonant normal forms in Appendix 8 and they are used extensively in Chapters 4 (non-Hamiltonian framework) and 7 (Hamiltonian case).

ii) The second option consists of "controling" the small divisors which appear when one undertakes to eliminate certain terms in the perturbation. In this case one chooses $\lambda = (\lambda_1, \ldots, \lambda_n) \in \mathbb{C}^n$ (in other words, the linear part A of the field at the singular point) in such a way that there are no resonances, or only a finite number (Poincaré's domain), or so that the small divisors satisfy a Diophantine condition (cf. Appendix 4). As the literature concerning these questions involves a large part of the modern development of perturbation theory, we can only touch on the Hamiltonian case (cf. again Chapter 7 and Appendix 8).

In conclusion, we should stress that from the point of view of normal forms,the theory of averaging in the general case (cf. Chapters 2, 5, and 6) appears quite elementary, as it simply represents a transformation to normal form through *first* order. It is only in the case of systems possessing further specific properties that stronger results are possible with averaging, such as in the case of single-phase averaging (Chapter 3) or for Hamiltonian systems (Chapter 7).

APPENDIX 6: GENERATING FUNCTIONS

In using canonical transformations in mechanics, for example to put a Hamiltonian in simplest form, one employs essentially two tools, *generating functions* and *Lie series*, which will be the topics of Appendices 6 and 7. Our aim here is to achieve some computational facility with generating functions, without entering into a detailed examination of the mathematical basis of this classical technique. For details we refer the reader to any treatise on classical mechanics, for example Goldstein ([Go]) or, for a less detailed treatment, Arnold ([Ar7]).

We begin with a Hamiltonian $H(p, q, t)$, where $(p, q) \in \mathbb{R}^{2n}$. The principle of least action in phase space ensures that Hamilton's canonical equations are equivalent to extremizing the integral:

(1) $\quad \delta \int pdq - Hdt = 0$

in which the limits of integration (q_0, t_0) and (q_1, t_1) are fixed (but $p_0 = p(t_0)$ and $p_1 = p(t_1)$ may vary) and where we have used the classical notation $pdq = \sum p_i\, dq_i$. This may be otherwise expressed as the fact that the vectorfield defining Hamilton's equations is given by the characteristics of the Poincaré-Cartan form $\omega = pdq - Hdt$.

We first look for transformations $(p, q) \to (P, Q)$ which preserve the form of Hamilton's equations. We see that the integrands in (1) must be connected by a relation of the form:

(2) $\quad \lambda\, (pdq - Hdt) = PdQ - Kdt + \Phi$

where $K(P, Q, t)$ is the new Hamiltonian, Φ is a closed differential form ($d\Phi = 0$) and $\lambda \in \mathbb{R} - \{0\}$. If $\Phi = 0$, we find the rather uninteresting changes of scale:

(3) $\quad P_i = \mu p_i\,; \quad Q_i = \nu q_i\,; \quad \lambda = \mu\nu; \quad K = \lambda H.$

We now suppose that $\lambda = \mu = \nu = 1$. Since Φ is closed, it is locally exact, and on any simply connected region of phase space we may therefore write $\Phi = dS$, where S is the *generating function* of the transformation $(p, q) \to (P, Q)$. There then remains:

(4) $\quad pdq - Hdt = PdQ - Kdt + dS.$

Since the transformation is independent of time ($\partial S/\partial t = 0$), we must have:

(5) $\quad K(P, Q) = H(p, q), \quad pdq = PdQ + dS,$

and in particular:

(6) $\quad dp \wedge dq = dP \wedge dQ,$

so that the transformation is symplectic. Returning to (5), we see that $S = S(q, Q)$ and:

(7) $$pdq = PdQ + \partial S/\partial q.dq + \partial S/\partial Q.dQ,$$

and so:

(8) $$p_i = \partial S/\partial q_i, \quad P_i = -\partial S/\partial Q_i.$$

But in this formulation, (q, Q) must form a coordinate system on phase space (at least locally) which is not always the case (think of the identity transformation). Practically speaking, this requires that a certain Jacobian (in this case $\det \partial Q_i/\partial p_j$) be nonzero. For this reason we use generating functions depending on different variables; the various possible forms of the *same* tranformation (when they exist, cf. again the identity) are connected via Legendre transformations, which may act on any pair of variables (P_i, Q_i). We thus obtain 2^n kinds of generating functions of the form:

$$S = S(q, P_{i_1}, \ldots, P_{i_k}, Q_{i_{k+1}}, \ldots, Q_{i_n}).$$

The transformation formulas are easy to obtain; we will treat the case $k = n$ for notational simplicity. Starting with (5) and writing $S_1 = S$, we have:

(9) $$pdq = PdQ + dS_1 = -QdP + d(S_1+PQ) = -QdP + dS_2,$$

where $S_2(q, P) = S_1 + PQ$ is the Legendre transformation of S_1 in the variables Q $(P = -\partial S_1/\partial Q)$. From (9) we have:

(10) $$p = \partial S_2/\partial q, \quad Q = \partial S_2/\partial P.$$

Here it is the variables (q, P) that form a coordinate system. This form of the transformation is well suited to perturbation theory, as (10) shows that qP is the generating function of the identity. In order to treat small perturbations, we will thus use functions of the form $qP + \varepsilon S(q, P)$ which generate canonical transformations ε-close to the identity.

As an example of application, we show how to construct the *action-angle* variables for Hamiltonians with one degree of freedom (see also [Go] or [Ar7] §50 for example); these are needed as a starting point in almost all perturbation techniques and are used extensively in Chapters 7, 8 and 9 where we recall the construction for integrable Hamiltonians with any number of degrees of freedom. In one degree of freedom, any Hamiltonian $H(p, q)$ is (locally) integrable (H provides the necessary conserved quantity) and one then seek a canonical change of variables $(p, q) \rightarrow (I, \varphi)$ to the so-called action-angle variables which fulfill the following two conditions:

- H is a function of I alone: $H(p, q) = h(I)$.

- φ is 2π periodic; more precisely, since I is a constant on any level curve Σ_h (defined by $H(p, q) = h$) of H, we suppose that these curves are closed, at least for a certain range in h to which we restrict ourselves, and that φ is 2π periodic along any of these.

We construct the transformation by means of a generating function of the second type $S(q, I)$ such that:

(11) $$p = \partial S(q, I)/\partial q, \quad \varphi = \partial S(q, I)/\partial I, \quad H(p, q) = h(I).$$

This implies:

(12) $$dS_{(I=cst)} = pdq, \quad S(I, q) = \int_{q_0}^{q} pdq$$

where the integral is taken over the level curve $I = cst$ (or $H = cst$) and the bottom end $q_0(I)$ is a fixed point on this curve which may vary smoothly with I. One has:

(13) $$\Delta S = \oint_{h(I)=h} pdq = A(h)$$

that is, the area enclosed by the level curve $H = h$. $\varphi = \partial S/\partial I$ is a multivalued valued function on Σ_h and, after interchanging derivatives, one finds that $\Delta\varphi = d/dI(\Delta S)$; since this must be 2π by definition, one must have $\Delta S(I) = 2\pi I$ (up to a constant) which implies:

(14) $$I(h) = A(h)/2\pi.$$

It is here assumed that the function $h \to A(h)$ (and thus $h \to I(h)$) is invertible. We thus arrive to the classical definition of the action variable. In the variables (I, φ) I is a constant of the motion and φ evolves with the constant velocity $\omega(I) = \partial h/\partial I = 2\pi(dA/dh)^{-1}$. In particular, the period of the motion along the level curve Σ_h is given by:

(15) $$T(h) = 2\pi/\omega = dA/dh.$$

One word about the arbitrariness in the definition of the action-angle variables: As we have seen, (I, φ) can be changed into $(I + c, \varphi + c'(I))$ where c is a constant, and c' a (smooth) function. We give below the example of the harmonic oscillator, which is used in Chapter 8. In this case, the transformation formulas can be found in a quite straightforward way (using (14) for example) but we present the complete calculation for illustrative purposes. We have:

(16) $$H = 1/2\,(p^2 + \omega^2 q^2) = h, \quad p = \pm\,(2h - \omega^2 q^2)^{1/2},$$

$$I = 1/2\pi \oint pdq = 2/\pi \int_0^{q_m} (2h - \omega^2 q^2)^{1/2}\, dq, \quad q_m = (2h)^{1/2}/\omega$$

that is, $I = h/\omega$ as expected, which is the area of the ellipse $H = h$. h is now given by $h(I) = \omega I$.

The other transformation formula can be spelled out as:

(17) $p = p(q, I) = (2\omega I - \omega^2 q^2)^{1/2}$

$S = S(q, I) = \int_0^q (2\omega I - \omega^2 q^2)^{1/2}\, dq$

$\varphi = \partial S/\partial I = \omega \int_0^q (2\omega I - \omega^2 q^2)^{-1/2}\, dq$

from which one finds:

(18) $p = (2I)^{1/2} \cos\varphi, \quad q = (2I)^{1/2}/\omega \sin\varphi$

which as we noticed could have been obtained much more easily. For further use, we note that this transformation can also be effected through a generating function of the first kind, namely $S_1(q, \varphi) = 1/2\, \omega q^2 \operatorname{cotg}\varphi$.

Before concluding, let us briefly examine the nonautonomous case, which presents no particular difficulty. (4) is still valid according to the principle of least action, or the characteristics of the Poincaré-Cartan form, but H, K and S now depend on t. Expanding, we find:

(19) $pdq - Hdt = PdQ - Kdt + \partial S/\partial q.dq + \partial S/\partial Q.dQ + \partial S/\partial t.dt,$

or:

(20) $p = \partial S/\partial q, \quad P = -\,\partial S/\partial Q, \quad K(P, Q, t) = H(p, Q, t) + \partial S/\partial t(q, Q, t).$

As before we use Legendre transformations, with t playing the role of a parameter, to obtain for example with $S_2 = S_2(q, P, t)$:

(21) $p = \partial S_2/\partial q, \quad Q = \partial S_2/\partial P, \quad K = H + \partial S_2/\partial t.$

We end with two remarks concerning the advantages and disadvantages of the generating function formalism.

Remark 1: The greatest advantage is that one obtains exact formulas for noninfinitesimal transformations. Outside of perturbation theory - in other words when there is no small parameter - generating functions, in connection with Hamilton-Jacobi theory, provide the most explicit means of writing canonical transformations.

Remark 2: However, this method is not as explicit as might first appear, since the formulas (for example (20)) mix new and old variables. It is thus necessary to invert to obtain (P, Q) as a function of (p, q), which is usually impossible in explicit form. Such problems involving *mixed variables* often arise in practice. In particular, even for transformations close to the identity it is not always clear how to determine domains in (p, q) and (P, Q) which correspond bijectively or even such that the second is the image of the first. On this point we may refer to the "technical lemmas" at the end of Arnold's article [Ar3]. From this point of view Lie series are certainly more manageable (cf. Appendices 7 and 8), but again their effective use is confined to the construction of near identity transformations.

We also take this opportunity to notice that [Sie5] and [Gia] offer general surveys of the techniques of Hamiltonian perturbation theory.

APPENDIX 7: LIE SERIES

1. Overview of the method

In this section we first review the method of Lie series in the simplest case, partly following the works [Car] and [Ra], then we examine certain generalizations of this simplest case. The aim of the Lie formalism is to construct near-identity transformations $z \to z'$, not necessarily canonical, on a space $\mathbb{R}^m$. In general, one begins with a dynamical system which is a perturbation of a system one knows how to integrate, and one tries to reduce it to a simpler form by means of one or more changes of variables (cf. the method of normal forms, Appendix 5). The principal advantage of the Lie series formalism resides in the fact that it is generally easier to obtain expressions in the new variables for functions originally given in terms of the old variables. In view of the applications of the method in Chapter 7, we end this appendix with a more specialized discussion of the formalism in the Hamiltonian case.

We denote the change of variables by $z' = z + \varepsilon h(z)$, where h is a diffeomorphism from the domain $D \subset \mathbb{R}^m$ $(z \in D)$ to the domain $D' \subset \mathbb{R}^m$ $(z' \in D')$. The basic idea of the Lie method is to consider the inverse diffeomorphism $(1 + \varepsilon h)^{-1}$ as the application at time ε of the flow T_ε associated to a differential system:

(1) $\quad dz/d\tau = w(z),$

where the notation τ is used to avoid any confusion between this auxiliary time τ and the "physical" time t arising in the dynamical system under study. We will later examine the case where the vectorfield w also depends on τ, and even on t. In the present section we simplify matters by taking w autonomous (with respect to τ) and independent of t. Therefore, on D':

(2) $\quad z = (1 + \varepsilon h)^{-1}(z') = T_\varepsilon(z').$

In anticipation of more complex cases, we note here that, since the vectorfield w is autonomous, the map T_ε may be considered as the application at time 1 of the field εw, and the inverse map $T_\varepsilon^{-1} = (1 + \varepsilon h)$ may be viewed as the application at time ε (resp. 1) of the field $-w$ (resp. $-\varepsilon w$).

We now consider a function $f(z)$ on the domain D. On D', its expression in terms of the new variables is given by:

(3) $T_\varepsilon[f](z') = f(z[z']) = f \circ T_\varepsilon(z')$.

It was in order to obtain this simple transformation formula for functions that the old variables were defined (via T_ε) in terms of the new ones, rather than conversely (we shall often drop the subscript ε). Of course, the other possibility may be put to use just as well and in fact is used by J.Cary (transcribed in [Lic]) and other authors. We warn the reader that these two conventions, although equivalent, may cause some confusion in the more complex cases to be examined below; in particular our formulas are equivalent to those in [Car] and [Lic], provided T is replaced by T^{-1}.

In a slightly more mathematical vein, we would like to add one more cautionary remark. The method of Lie series deals essentially with functions, and transformations thereof; therefore in the applications, the variables can almost always be considered as dummy, and manipulations are performed in a perfectly automatic way. However when one tries to keep track of what is really happening, several confusions are possible which can be essentially traced back to the following:

- Two points of view are possible when one wants to perform a change of variables, usually termed "active" and "passive" (one may think of a "change of basis" in linear algebra): The first one, which we shall adopt, consists in viewing this operation as an "effective" transformation of the space into itself, whereas the second considers it as a reparametrization of the points, that is, as a change of coordinates. In this respect, the generating vector field w for example should be considered as a function of z', and T^{-1} as generated by the vector field $-w(z)$, a function of the variable z.
- In a similar way, one should carefully distinguish, for a given function, between its functional form , in which the variable involved plays no role, and the value it assumes once the argument is given.

Again, these two problems, which parallel similar questions in tensor algebra and differential geometry, may be the source of many confusions, sign errors etc. Fortunately, as already mentionned, the method and the formulas it produces can be used blindly in the applications.

Returning to the T operator, we first note that, when the function f is real analytic,$T_\varepsilon [f](z')$ may be expressed as a power series in ε, called the Lie series. If we define the function φ by:

(4) $\varphi(z', \tau) = f \circ T_\tau(z')$,

where T_τ is the flow associated with the vectorfield w, we then have:

(5) $$f \circ T_\varepsilon(z') = \varphi(z', \varepsilon) = \sum_{n=0}^{\infty} \varepsilon^n/n! \, [\partial^n \varphi(z', \tau)/\partial\tau^n]_{\tau=0},$$

provided ε is less than the radius of convergence of the series. Let D be the operator representing differentiation with respect to the vector tangent to the trajectory $T_\tau(z')$:

(6) $$D[g](T_\tau(z')) = \sum_i [w_i \partial g/\partial z_i](T_\tau(z')).$$

We have:

(7) $$\partial\varphi/\partial\tau \, (z', \tau) = D[f](T_\tau(z')),$$

and consequently:

(8) $$(Tf)(z') = f \circ T_\varepsilon(z') = \sum_{n=0}^{\infty} \varepsilon^n/n!.D^n[f](z') = (\exp \varepsilon D)[f](z').$$

We note that one obtains an *implicit* expression for the new coordinates by choosing $f = z_i$, or more accurately, $f = \pi_i$ the ith coordinate projection; all this reduces to $z = Tz'$, and at the same time it serves to justify this notation. This expression may then be inverted order by order, which in this case simply amounts to replacing ε by $-\varepsilon$ in the series.

Let us examine the particular case of Hamiltonian systems. To every near-identity *canonical* transformation $z = (p, q) \to z' = (p', q')$, one associates a Hamiltonian $\chi(z')$ such that the inverse change of variables $z' \to z$ coincides with the application at time ε of the flow T_ε generated by the vectorfield defined by the Hamiltonian χ. An analytic function f transforms as before according to equation (3), and the Lie series of $T_\varepsilon[f]$ may be expressed as a function of the Poisson bracket with the Hamiltonian χ:

(9) $$L_\chi[g] = \{\chi, g\} = \sum_i [\partial g/\partial q_i \, \partial\chi/\partial p_i - \partial g/\partial p_i \, \partial\chi/\partial q_i],$$

since the derivative operator D here is the Liouville operator L_χ. (8) then reads:

(10) $$(Tf)(z') = f \circ T_\varepsilon(p', q') = \sum_{n=0}^{\infty} \varepsilon^n/n!.[L_\chi]^n \, [f](p', q').$$

In particular, if $f = H$ is the Hamiltonian for the t-evolution of the system, equation (10) gives the new Hamiltonian directly in terms of the new variables:

(11) $$H'(z') =_{def} (TH)(z') = H(Tz') = H(z).$$

That the old and new hamiltonians should be thus related (that is via T) can be proven as follows; for any function $f(z)$, the following holds:

(12) $$\begin{aligned} \partial f/\partial t \, (z) &= \{H, f\}(z) = \{TH, Tf\}(z') \\ &= \partial/\partial t \, [Tf](z') = \{H', Tf\}(z') \end{aligned}$$

where the last equality of the first line holds because T is canonical and therefore preserves Poisson brackets, whereas the equality of the second line *defines* the new hamiltonian H'. Identification confirms that $H' = TH$.

We shall now give a precise lemma which serves to estimate the remainders of Lie series in the Hamiltonian case (the general case being similar). This simple lemma, taken from [Ben5-6], is used extensively in Chapters 7 and 8 as a useful tool to prepare iteration lemmas, the common ingredient of most perturbation theories. We use (I, φ) instead of (p, q) as the common canonical variables to comply with the notations of these two chapters. Here it can also be seen how Lie series methods partly avoid the use of mixed variables and thus allow for a simple determination of the domain on which a given function χ defines an invertible canonical transformation, this being more complicated when working with generating functions, as we noticed at the end of Appendix 6.

If K is a compact convex subset of $\mathbb{R}^n$, we set as usual (cf. Chapter 7):

$$K_{\rho,\sigma} = D(K, \rho, \sigma) = D = \{ (I, \varphi) \in \mathbb{C}^{2n}, \operatorname{Re} I \in K, \| \operatorname{Im} I \| \leqslant \rho,$$
$$\operatorname{Re} \varphi \in T^n, \| \operatorname{Im} \varphi \| \leqslant \sigma \},$$

where ρ and σ are positive. For $\delta < \rho$ and $\xi < \sigma$, we set $D - (\delta, \xi) = D(K-\delta, \rho-\delta, \sigma-\xi)$, where $K - \delta$ is as usual the set of centers of balls of radius δ contained in K. We will work in the space $A(D)$ of functions continuous on $D(K, \rho, \sigma)$ and analytic on the interior of this domain, equipped with the sup norm: $\| f \|_D = \operatorname{Sup} \{| f(x) |, x \in D\}$.

<u>Lemma 1</u>:

Let $\chi \in A(D)$ and $\delta, \xi > 0$, $\delta < \rho$ and $\xi < \sigma$. We assume that:

(13) $\qquad \| \partial\chi/\partial I \|_{D-(\delta/2,\xi/2)} \leqslant \xi/2, \quad \| \partial\chi/\partial\varphi \|_{D-(\delta/2,\xi/2)} \leqslant \delta/2.$

Then, the time 1 map of the flow with Hamiltonian χ, denoted by $T = (T_I, T_\varphi)$ is defined on $D' = D(K', \rho-\delta, \sigma-\xi)$ such that:

(14) $\qquad K - \delta \subset K' \subset K - \delta/2, \quad K - \delta \subset T_I(K') \subset K - \delta/2.$

Moreover, T satisfies:

(15) $\qquad \| T_I - 1 \| \leqslant \delta/2, \quad \| T_\varphi - 1 \| \leqslant \xi/2.$

Finally, it defines an operator, still denoted by $T: A(D) \to A(D')$ by $Tf =_{def} f \circ T$ and such that, for $f \in A(D)$:

(16) $\qquad \| Tf \|_{D'} \leqslant \| f \|_D, \quad \| Tf - f \|_{D'} \leqslant \| \{\chi, f\} \|_D$

$\qquad\qquad \| Tf - f - \{\chi, f\} \|_{D'} \leqslant 1/2 \, \| \{\chi, \{\chi, f\}\} \|_D .$

<u>Proof</u>:

The estimates (15) follow easily from (13), and the first inequality of (16) is also immediate. The remaining two are proved using first and second order Taylor formulas:

$$(17) \qquad (Tf - f - \{\chi, f\})(I, \varphi) = 1/2\, d^2/dt^2\, [f \circ T_{t'}](I, \varphi)$$
$$= 1/2\, \{\chi, \{\chi, f\}\}.T_{t'}(I, \varphi), \quad t' \in (0, 1),$$

where T_t is the flow of the Hamiltonian χ ($T_1 = T$). It remains to define D' and to demonstrate the inclusions (14). We set:

$$D' = \{ (I, \varphi) \in D - (\delta/2, \xi/2),\ T_t(I, \varphi) \in D - (\delta/2, \xi/2) \ \text{for } t \in [0, 1] \}.$$

By (11), $D' \supset D - (\delta, \xi)$ and:

$$T(D') = \{ (I, \varphi) \in D - (\delta/2, \xi/2),\ T_{-t}(I, \varphi) \in D - (\delta/2, \xi/2) \ \text{for } t \in [0, 1] \}$$

shows that $T(D') \supset D - (\delta, \xi)$, which proves the inclusions (14) by suitable choice of K'. ■

To make use of this lemma, we must produce bounds for the right hand sides of the estimates (16). This is done most easily as follows: Assuming that the functions f and χ are analytic on an open set G in the complex plane which properly contains D (for example such that $G - (\alpha, \beta) \supset D$), the relevant derivatives are then controlled by means of Cauchy's formula:

$$\| \partial^{|k|+|l|} g / \partial I^k \partial \varphi^l \|_{G-(\alpha,\beta)} \leqslant k!\,l!\,\alpha^{-|k|} \beta^{-|l|} \| g \|_G ,$$

valid for an analytic function g on G and for multiindices $k = (k_1, \dots, k_m)$, $l = (l_1, \dots, l_m)$, ($|k| = k_1 + .. + k_n$, $|l| = l_1 + .. + l_n$). We refer to Chapter 7 for more details, where such estimates are combined, in a typical way, with the classical estimates on the remainder of the Fourier series of analytic functions, as recalled in Appendix 1. We now come to the more elaborate cases of the use of Lie series.

2. The case of several successive changes of variables

It would be natural at this point to study variable changes defined by equations of type (1) in which the field w depends on τ and t. However, we delay this generalization to the next section in order to first study an instructive special case. This special case corresponds to successive iterations, in increasing order, of the previously defined variable changes. We will show that such a succession of transformations may be brought about in at least three ways, the third of which uses as generator a vectorfield

$w = w(z, \tau)$, polynomial or real analytic in τ. This latter case is important, as it arises when one seeks to simultaneously effect several perturbation steps for autonomous systems (w is still assumed independent of t).

Returning to the general (non-Hamiltonian) case, we consider the product of p successive near-identity transformations $T_1, \ldots, T_p$ such that $\| T_k - 1 \|$ is of order ε^k. All series will be convergent for small enough ε and for smooth enough vectorfields, and we will pass to the limit $p = \infty$, for which we may need to assume $T_k = 1$ for large k.
One may proceed in three ways:
1) By successively applying transformations, with $T = .. \circ T_k \circ .. \circ T_1$, employing at each step the Lie method detailed in the preceding section.
2) By considering T^{-1} as the application at time ε of the flow associated to the *autonomous* system:

$$dz/d\tau = w'_1(z) + \varepsilon w'_2(z) + \ldots + \varepsilon^{k-1} w'_k(z) + \ldots$$

3) By applying the Lie method to $T = .. \circ T_k \circ \ldots \circ T_1$, considering T^{-1} as the application at time ε of the flow associated to the *nonautonomous* system:

$$dz/d\tau = w''_1(z) + \tau w''_2(z) + \ldots + \tau^{k-1} w''_k(z) + \ldots .$$

1) In the first method, one associates to the kth transformation $z_{k-1} = T_k(z_k)$ the vectorfield w_k such that the inverse transformation T_k^{-1} coincides with the application at time ε of $\varepsilon^{k-1} w_k$. T_k is then given by the formula $T_k[f](z) = (\exp \varepsilon^k D_k)[f](z)$, where D_k is the derivative operator associated to w_k. It then follows that:

$$(18) \qquad T[f](z') = T_p \circ \ldots \circ T_1[f](z') = (\textstyle\prod_k (\exp \varepsilon^k D_k))[f](z').$$

Writing the series in explicit form and regrouping terms of the same order, we find:

$$(19) \qquad T[f](z') = \textstyle\sum_{n=0}^{\infty} \varepsilon^n T_n f(z') = \sum_{n=0}^{\infty} \varepsilon^n f_n(z'); \quad f_0 = f,$$

where:

$$(20) \qquad T_n f = f_n = \textstyle\sum (D_j)^{k_j} \circ \ldots \circ (D_1)^{k_1}[f],$$

the sum being taken over all integer k-tuples such that:

$$k_1 + 2k_2 + \ldots + jk_j = n, \quad k_i > 0 \text{ for all } i \text{ (in particular } j < n).$$

We note that for large n, the number of terms in the sum for f_n is equivalent to $\exp[\pi(2n/3)^{1/2}](4n\sqrt{3})^{-1}$.

2) In the second method, T^{-1} is the application at time ε of the flow associated with the autonomous vectorfield $w'_1(z) + \varepsilon w'_2(z) + \ldots + \varepsilon^{k-1} w'_k(z) + \ldots$. If D'_i is the derivative

operator associated with w'_i, we obtain the transformation formula:

$$(21)\qquad T[f](z') = \sum_{n=0}^{\infty} \varepsilon^n/n!.(D'_1 + \ldots + \varepsilon^{k-1}D'_k + \ldots)^n[f](z')$$
$$= \sum_{n=0}^{\infty} 1/n!.(\varepsilon D'_1 + \ldots + \varepsilon^k D'_k)^n[f](z').$$

Consequently, since a priori D'_i and D'_j do not commute for $i \neq j$, $T[f](z')$ is expressed as a Lie series of the form (2) with coefficients (for $n \geq 1$):

$$(22)\qquad T_n f = f_n = \sum 1/j!.D'_{k_j} \circ \ldots \circ D'_{k_1}[f],$$

where the summation is restricted to integer j-tuples satisfying:

$$k_1 + \ldots + k_j = n,\quad k_i > 0 \text{ for all } i \text{ (in particular } j \leq n).$$

This sum comprises 2^{n-1} terms, a considerably greater number than that obtained by the first method, at least for large n.

3) Finally, in the third method T^{-1} is the application at time ε of the flow associated with the nonautonomous vectorfield $w''(z, \tau) = w''_1(z) + \tau w''_2(z) + \ldots + \tau^{k-1}w''_k(z) + \ldots$. The Lie series of $T[f](z')$ has the coefficients:

$$(23)\qquad T_n f = f_n = 1/n!.(d^n f/d\tau^n)_{\tau=0},$$

where the derivative is calculated along a trajectory of the field $w''(z, \tau)$ with initial condition z'. Since the vectorfield is not autonomous, the associated derivative operator $D''(\tau)$ depends explicitly on τ, and in expressions involving higher order derivatives, the temporal derivatives must be taken into account. Thus, to second order we have:

$$(24)\qquad d^2f/d\tau^2 = d/d\tau(w''(z, \tau)df/d\tau) = w''d^2f/d\tau^2 + \partial w''/\partial\tau.df/d\tau.$$

By recursion the following formula is easily established:

$$(25)\qquad d^n f/d\tau^n = \sum_{1 \leq k \leq n} (n-1)!/[(n-k)!(k-1)!](\partial^{k-1}D''/\partial\tau^{k-1}) \times$$
$$\times [d^{n-k} f/d\tau^{n-k}].$$

Decomposing the operator D as:

$$(26)\qquad D'' = D''_1 + \tau D''_2 + \ldots + \tau^{k-1}D''_k + \ldots,$$

we find the recursion formula for the coefficients in the Lie series of $T[f]$:

$$(27)\qquad T_n f = f_n = 1/n \sum_{1 \leq k \leq n} D''_k[f_{n-k}].$$

This may be written out explicitly as:

$$(28)\qquad T_n f = f_n = \sum [k_1(k_1+k_2).+\ldots+(k_1+\ldots+k_n)]^{-1} D_{k_j} \circ \ldots \circ D_{k_1}[f],$$

where the sum is again restricted to j-tuples such that:

$$k_1 + \ldots + k_j = n,\quad k_i > 0 \text{ for all } i \text{ (in particular } j \leq n).$$

This sum therefore comprises the same number of terms (2^{n-1}) as in the preceding case.

We see that the first method allows calculation of the f_n with the least number of terms, and this may be useful in practice (but only from $n = 4$ on), especially since the Lie series method can be implemented rather easily on computers, using formal languages (this is particularly effective for Hamiltonian systems with polynomial Hamiltonians). The second method is not widely used, and the third, oft-used method has the advantage of generalizing to the t-dependent case, as we shall see in the next section. Of course, to define the *same* transformation by means of all three methods, it will suffice to identify the fields w, w', and w'' by identifying, order by order, the terms in the Lie series given respectively by formulas (20), (22), and (28).

All of the above formulas may be applied without difficulty to the particular case of Hamiltonian systems and canonical transformations, in which case they permit calculation of the new Hamiltonian in terms of the new variables.

In the first method, the vectorfield $\varepsilon^{k-1} w_k$ arises from a Hamiltonian $\varepsilon^{k-1} \chi_k$, and the nth coefficient of the Lie series is expressed as in (20) with the D_k replaced by the operators $L_k =_{def} L_{\chi_k}$. We may proceed in the same way for the second and third methods, where the Lie series are defined respectively by (22) and (28), with D'_k replaced by $L_{\chi'_k}$ and D''_k by $L_{\chi''_k}$ as required.

3. The general case

We now examine, in the Hamiltonian case, a general transformation $T(\tau, t)$ induced by the time τ of the Hamiltonian field $\chi = \chi(z', \tau, t)$. Although we restrict our consideration to the Hamiltonian case, especially in view of the application to the adiabatic series in Chapter 8, the non-Hamiltonian case can be treated in exactly the same way. Here, the original coordinates $z = (p, q)$ and the new coordinates $z' = (p', q')$ belong to $\mathbb{R}^{2n}$. t is the time of the system under consideration with Hamiltonian (for the t-evolution) $H = H(z, t)$. We will see that the formalism allows for considerable symmetry in the treatment of the variables τ and t, and we concentrate on finding the expression of the new Hamiltonian $H' = H'(z', t)$ in terms of the new variables; one can compare the formulas below with those obtained via the generating function formalism (see Appendix 6). Finally, we point out that interesting examples of Lie series calculations may be found in the article by J. Carry ([Car]) as well as in the book [Lic]; we also refer to Chapter 8 (as well as the above references) for the definition and use of adiabatic series.

By definition, we have:

(29) $\partial z/\partial \tau = \{\chi , z\} = L_\chi(z), \quad z(0) = z'.$

Since $z = Tz'$, we may rewrite this in operator form (analogous to the evolution of observables in quantum mechanics):

(30) $\partial T/\partial \tau = L_\chi T, \quad \partial T^{-1}/\partial \tau = - T^{-1}L_\chi ,$

the second equation being an immediate consequence of the first.

The exact meaning of these equations should in fact be given in terms of transformations of functions, rather than points. For a function $f(z)$, one can write:

(31) $\partial f(z)/\partial \tau = \partial/\partial \tau\ (Tf)(z') = \partial/\partial \tau\ (f \circ T)(z') = (Df \circ L_\chi T)(z')$

where Df is the differential application of f. Formula (30) is a useful (and sometimes dangerous) abreviation for this, corresponding to the case when f is the identity, or rather one of the coordinate projections.

If in addition $z = z(t)$ undergoes a Hamiltonian evolution in t, one has:

(32) $\partial/\partial t\ f(z) = \{H, f\}(z) = \{TH, Tf\}(z')$

$= \partial/\partial t\ (Tf)(z') = d/dt\ (Tf)(z') - (\partial T/\partial t\ f)(z'),$

because the canonical transformation T is t-dependent. Now, $d/dt\ (Tf)(z') = \{H', Tf\}(z')$ by definition of the new Hamiltonian H'. In addition, since T is canonical for any value of t, $\partial T/\partial t$ is also canonical, and one can write an equation analogous to (30) for it, say $\partial T/\partial t = L_R T = \{R, T\}$, where the value of the Hamiltonian R will be calculated below. (32) is now rewritten as:

(33) $\{TH, Tf\}(z') = \{H', Tf\}(z') - \{R , Tf\}(z')$

that is:

(34) $H'(z', t) = (TH)(z', t) - R(z', t)$

in which one notices that R is naturally a function of z', just as χ is. It only remains to compute the value of R. We again write:

(35) $\partial T/\partial t = L_R T, \quad \partial T/\partial \tau = L_\chi T,$

and we equate the mixed derivatives of T obtained by means of these two formulas, which gives:

(36) $L_{\partial R/\partial \tau} = L_{\partial \chi/\partial t} + L_{\{\chi, R\}},$

and we may in fact require that:

(37) $\partial R/\partial \tau = \partial \chi/\partial t + \{\chi, R\},$

since this is true up to a function independent of z' which does not alter the bracket L_R, the only important quantity here. (37) is a transport equation for R, and we find, using

(30) and the fact that $R_{\tau=0} = 0$:

(38) $$R(z', t) = T(\tau, t) \int_0^\tau T^{-1}(\theta, t).\partial\chi/\partial t\ (z', \theta, t)\ d\theta$$

in which some cumbersome parentheses have been omitted. Thus we get the important formula:

(39) $$H'(z', t) = (TH)(z', t) + T(\tau, t)\int_0^\tau T^{-1}(\theta, t).\partial\chi/\partial t\ (z', \theta, t)\ d\theta.$$

We may now return to the usual case of perturbation theory, where all the quantities are assumed expandable (and expanded) in a small parameter ε. We set:

$$H = H(z, \varepsilon, t) = \sum_{n=0}^{\infty} \varepsilon^n H_n(z, t)$$

(40) $$H' = H'(z', \varepsilon, t) = \sum_{n=0}^{\infty} \varepsilon^n H'_n(z', t)$$

$$\chi = \chi(z', \tau, t) = \sum_{n=0}^{\infty} \tau^n \chi_{n+1}(z', t),$$

and we consider the transformation $T = T(\varepsilon, t)$ induced by χ at time $\tau = \varepsilon$. As in the preceding section, for a function $f(z, t)$ we set:

(41) $$(Tf)(z', t) = \sum \varepsilon^n T_n f = \sum \varepsilon^n f_n .$$

The formulas analogous to (27) and (28) in Section 2 remain valid, as the introduction of t as a parameter does not alter them. Explicitly:

(42) $$T_n f = f_n = 1/n \sum_{1\leq k\leq n} L_k(T_{n-k}[f]);\quad f_0 = f$$

$$T_n f = f_n = \sum [k_1(k_1+k_2)+\ldots+(k_1+\ldots+k_n)]^{-1} L_{k_j}\circ\ldots\circ L_{k_1}[f]$$

with $k_1 + \ldots + k_j = n$, $k_i > 0$ for all i ($L_k =_{def} L_{\chi_k}$).

We also note that T is unitary, so that $T^{-1} = T^*$ and the series for T^{-1} may be derived from the series for T by changing the order and the sign of the factors L_{χ_k} (since these operators are anti-selfadjoint). All of this may also be easily deduced from equations (30). The first formulas read:

$$T_0 = 1$$

$$T_1 = L_1$$

(43) $$T_2 = 1/2\, L_2 + 1/2\, (L_1)^2$$

$$T_3 = 1/3\, L_3 + 1/6\, L_1L_2 + 1/3\, L_2L_1 + 1/6\, (L_1)^3.$$

Finally, to obtain the new Hamiltonian H', it suffices to identify the series in (39), or better still its derivative with respect to τ (at $\tau = \varepsilon$):

(44) $$\partial\chi/\partial t + \{\chi, H'\} = \partial H'/\partial\varepsilon - T\,\partial H/\partial\varepsilon.$$

This leads to a system of equations first described by Deprit. As usual in perturbation theory, this is solved order-by-order *simultaneously* for H'_n and χ_n . At n^{th} order

($n \geq 1$), it is easily checked that the equation takes the form:

(45) $$\partial\chi_n/\partial t + \{\chi_n , H_0\} = n(H'_n - H_n) - F_n$$

where F_n depends only on terms already determined. Explicitly:

(46) $$F_n = \sum_{1 \leq k \leq n} (L_{n-k}H'_k + kT_{n-k}H_k).$$

This is once again a transport equation for χ_n , and H'_n is chosen so as to make H' as simple as possible while at the same time avoiding small divisors; i.e., one eliminates secularities (the average of F_n) together perhaps with some nearly resonant terms. The reader may consult [Ca] and [Lic] for some simple examples, as well as Chapter 8 for the case of the harmonic oscillator with slowly varying frequency, in which case the ordering is somewhat different, due to this slow (adiabatic) variation. We end this appendix by spelling out the first three of equations (45) (at 0th order, $H'_0 = H_0$):

$$\partial\chi_1/\partial t + \{\chi_1 , H_0\} = H'_1 - H_1$$

(47) $$\partial\chi_2/\partial t + \{\chi_2 , H_0\} = 2(H'_2-H_2) - L_1(H'_1+H_1)$$

$$\partial\chi_3/\partial t + \{\chi_3 , H_0\} = 3(H'_3-H_3) - L_1(H'_2+2H_2)$$
$$- L_2(H'_1+1/2\, H_1) - 1/2\, (L_1)^2 H_1 .$$

Note that one only needs to know χ_{n-1} in order to compute H'_n .

APPENDIX 8: HAMILTONIAN NORMAL FORMS

In Appendix 5 we briefly stressed the importance of normal forms in modern perturbation theory, particularly in the theory of Hamiltonian perturbations. We repeat that it is usually the difficulties that arise in transforming to normal form which best display the features of a problem, and that these difficulties show up in the solution of the linearized conjugation equation (the homological equation) in the form of small divisors (see Appendix 5).

The proofs of Siegel's, Kolmogorov's, and Nekhoroshev's theorems thus follow the same general outline, the latter's proof being the most classical, at least in its analytical part. In all cases the transformation to normal form takes place recursively, using different iteration lemmas according to the desired result (cf. Chapter 7). As a matter of fact this appendix will repeat part of the material of the latter chapter, but in a less formal way which we hope may provide a broader understanding of the important phenomena.

We remark that Lie series methods (cf. Appendix 7), as well as generating function methods, can in principle be used to effect many canonical transformations in a single step. Although these methods - especially the Lie series - may be more practical in applications, they do not give better results than those obtained by recurrence (nevertheless, for perturbations having only finitely many harmonics, cf.[Gio]).

Below we give a brief exposition of the various iteration methods and the corresponding lemmas. Three cases, each equally relevant in the non-Hamiltonian case, are to be distinguished:

1 Simple (integrable) normal forms

2 Resonant normal forms

3 Quadratic convergence (Newton, Siegel, Kolmogorov).

We urge the reader to consult the article [Ben4], which we follow in part, for a quick but well-organized review of the various results in the theory of Hamiltonian perturbations; the authors view these results as being determined by the possibilities which are left open by the Poincaré-Fermi theorem on the nonexistence of analytic integrals for generic perturbations. We also apologize for adopting here a notation (close to that in [Ben4]) which differs slightly from that used in the main part of the text, as it is better suited to the purpose of this appendix.

In all cases, one studies perturbations of an integrable Hamiltonian given in terms of action-angle variables:

$$(1) \qquad H(I, \varphi) = h(I) + \varepsilon f(I, \varphi); \quad I \in K \subset \mathbb{R}^n, \quad \varphi \in T^n,$$

where K is compact - and even say convex - and h and f have a certain regularity; one commonly used hypothesis, particularly in the preceding chapters, is the following: Let H be analytic (and continuous at the boundary) on a complex extension D of $K \times T^n$:

$$(2) \qquad K_{\rho,\sigma} = D(K, \rho, \sigma) = D = \{ (I, \varphi) \in \mathbb{C}^{2n}, I \in K, \| \mathrm{Im}\, I \| \leqslant \rho, \mathrm{Re}\, \varphi \in T^n, \| \mathrm{Im}\, \varphi \| \leqslant \sigma \},$$

the norm $\| . \|$ being defined as $\| v \| =_{def} \sup_k | v_k |$, and for functions, $\| f \|_A =_{def} \sup_{x \in A} \| f(x) \|$; ρ and σ are strictly positive constants.

Using Cauchy's formula, it is possible to control the various functions' derivatives with respect to I; their Fourier coefficients can be controlled as well using the exponential decay of the series, which also follows from Cauchy's formula in the φ variables (cf. Appendix 1).

Remark:

More generally, in (1) it is possible to write $\varepsilon f(I, \varphi, \varepsilon)$: the small parameter ε controls the strength of the perturbation and often has a physical meaning. Nevertheless, as we shall see, for example in Newton's method, it is often necessary to consider the perturbation as a whole, so that its norm itself becomes the small parameter (one writes $f(I, \varphi)$, with $\| f \| = \varepsilon$). In any case, the modifications necessary to treat the perturbation order by order are clear enough and we omit them in this brief résumé.

We now seek to reduce the size of the perturbation; for this purpose we make use of the lemma which is proved in Section 1 of Appendix 7 to effect a canonical transformation by means of the Lie series generated by the auxiliary Hamiltonian $\chi(I, \varphi)$ defined on D:

$$(3) \qquad \exp(\varepsilon L_\chi) H = h + \varepsilon f + \varepsilon\{\chi, h\} + \varepsilon^2\{\chi, f\} + O(\varepsilon^2)$$

and we have the estimate (cf. Appendix 7):

$$(4) \qquad \| \exp(\varepsilon L_\chi) H - H - \varepsilon\{\chi, H\} \| \leqslant \varepsilon^2/2 \, \| \{\chi, \{\chi, H\}\} \|.$$

The term of order ε can thus be written: $f + \{\chi, h\} = f + \omega(I).\partial\chi/\partial\varphi$, where $\omega = \nabla h$ and the dot indicates the scalar product. It remains to eliminate *part* of this term by solving the *linearized equation* (see below). The three previously listed cases may now be examined.

1. Simple normal forms

This is the simplest case, distinguished here for historical and pedagogical reasons, since it is in fact included in the resonant case, as we shall see. We write f in terms of its Fourier series:

$$(5) \qquad f(I, \varphi) = \sum_{k\in\mathbb{Z}^n} f_k(I)e^{ik\varphi} = f_0(I) + \tilde{f}(I, \varphi)$$

and seek to eliminate f; the reader may easily verify that the average or *secular part* f_0 cannot be eliminated. We are thus led to choose χ as a solution to the linear equation:

$$(6) \qquad \tilde{f} + \omega(I).\partial\chi/\partial\varphi = f_0$$

and so, up to an unimportant function of I alone:

$$(7) \qquad \chi(I, \varphi) = \sum_{k\neq 0} \chi_k(I)e^{ik\varphi}; \quad \chi_k(I) = if_k(I)(\omega(I).k)^{-1}.$$

The small divisors make their first appearance in (7), and must be controlled if convergence of the series defining χ is to be obtained.

<u>Remarks</u>:

1. It is important to notice that obstacles arise at the very first step in the method, which in general cannot be made to work on an open subset of action space (cf. below and the version of Poincaré's theorem in [Ben1]). In any case, after this first step, the perturbation takes the form of a convergent series, and thus is apparently more complicated than in (1) (but see the remark following (1)).

2. It is also worthwhile to notice that the convergence of the series for χ does of course not guarantee the convergence of the series for $\exp(\varepsilon L_\chi)$ (see Appendix 7) which is to be studied separately. However, this does not prevent $\exp(\varepsilon L_\chi)$ from existing as the time ε of the flow generated by χ, and satisfying (3) and (4). The problem of the radius of convergence of the series for $\exp(\varepsilon L_\chi)$, where χ solves the linearized equation (6), is treated in [Gio] for the special case where the initial perturbation contains a finite number of harmonics. It is curious to note that this problem, of considerable practical importance, has to our knowledge been studied very little by astronomers, even in the classical (and equivalent) framework of transformations defined by generating functions.

The first question is: By using (7) to solve the equation with small divisors (6), in what cases can all the oscillating terms be eliminated to first order from the perturbation? It may be said (with hindsight!) that these cases are essentially as follows:

<u>i) All but finitely many f_k vanish</u>:

That is, the first order term of the initial perturbation comprises only finitely many harmonics (here the perturbation may depend on ε). Moreover, for these remaining harmonics we must have $\omega(I).k \neq 0$. Two subcases now appear:

(a) $\omega = (\omega_1, \ldots, \omega_n)$ is a constant vector; in other words, the unperturbed Hamiltonian describes a set of n noninteracting harmonic oscillators with incommensurable frequencies ω_i ($\omega.k \neq 0$ if $k \neq 0$). This is no doubt the simplest possible case and was first examined by Birkhoff ([Bi]), who studied the motion in the neighborhood of a nondegenerate elliptic fixed point, represented by a Hamiltonian of the form:

$$(8) \qquad H(x, y) = \sum_j \omega_j /2 \, (x_j^2 + y_j^2) + \sum_{r>2} \tilde{f}_r(x, y),$$

where f_r is a polynomial of degree r. The small parameter here is the distance from the origin (via the scaling $(x, y) \to (\varepsilon x, \varepsilon y)$). By transforming to action-angle variables for the harmonic oscillator: $I_i = 1/2 \, (x_i^2 + y_i^2)$, φ_i being the angle in the plane (x_i, y_i), then dividing by ε, we obtain the form:

$$(9) \qquad H(I, \varphi) = \omega.I + \sum_r f_r(I, \varphi),$$

where f_r is now a trigonometric polynomial.

We may thus eliminate the perturbation to first order in the neighborhood of the origin by solving (6) (the sum in (7) is finite). We thus obtain, in accordance with (3):

$$(10) \qquad H^{(1)} = \exp(\varepsilon L_\chi) H = \omega.I + \varepsilon\langle f_1\rangle(I) + \varepsilon^2(f_2(I, \varphi) + \{\chi, f_1\} + 1/2 \, \{\chi, \omega.\partial\chi/\partial\varphi\}) + O(\varepsilon^3),$$

where $\langle f\rangle_1 = f_{1,0}$ is the average of f_1. Since f_1, f_2 and χ posess only a finite number of harmonics, the operation may be reiterated to eliminate terms of order 2, and so on. After r such steps we obtain the Hamiltonian:

$$(11) \qquad H^{(r)} = \omega.I + \sum_{i=1}^{r} \varepsilon^i h_i(I) + \varepsilon^{r+1} f^{(r+1)}(I, \varphi).$$

The construction oulined above is due to G. D. Birkhoff, and the sum appearing in (11) is usually called the *Birkhoff series*. It is, of course, an asymptotic series, and in general the norm of $f^{(r+1)}$ grows rapidly with r. The construction reveals the following general property: At any order, if the perturbation posesses only finitely many harmonics, the same will be true of the transformed Hamiltonian. This permits a kind of generalization of the preceding construction to perturbations of this type; in particular, χ will be given, at any order, by a finite sum (cf. [Gio] for details). Note that to apply the construction to a given order, the incommensurability relation $\omega.k \neq 0$ only has to be satisfied for $|k| \leq N$, where N depends on the requested order and the number of harmonics in the original perturbation.

(b) $\omega = \omega(I)$ satisfies Kolmogorov's (or Arnold's) condition; that is, ω parametrizes (locally) the action space or the energy surface (see Appendix 3). We note that cases falling between (a) and (b), i.e., $I \to \omega(I)$ degenerate and nonconstant have only been treated in particular problems such as in celestial mechanics.

In that case one works in a subdomain B of action space such that:

$$\exists\, \alpha,\ \forall\, I \in B,\ \forall\, k \in \mathbb{Z}^n - \{0\},\ f_k \neq 0 \Rightarrow |\,\omega(I).k\,| \geq \nu > 0.$$

Or, more simply, one defines $B = B(\nu, N)$, *independent* of the perturbation, by:

$$(12) \qquad \forall\, I \in B,\ \forall\, k \in \mathbb{Z}^n - \{0\},\ |\,k\,| \leq N \Rightarrow |\,\omega(I).k\,| \geq \nu > 0.$$

Here we recognize the definition of the nonresonant domain. N may arise naturally, if the perturbation posesses only a finite number of harmonics ($|\,k\,| \geq N \Rightarrow f_k = 0$), or N may be the *cutoff frequency*, or "ultraviolet cutoff," a term coined by Arnold for a procedure he introduced in this context. One then simply ignores harmonics of order higher than $N = N(\varepsilon)$ which have small amplitudes, so that the remainder comprises two terms that may be controlled by the choice of N. Using the notation in [Ben4], one writes:

$$(13) \qquad H = h + \varepsilon f^{\leq} + \varepsilon f^{>}$$

where $f^>$ is the "ultraviolet part" of the perturbation f:

$$f^> =_{def} \sum_{|k|>N} f_k e^{ik.\varphi};$$

and

$$\begin{aligned}
\exp(\varepsilon L_\chi) H &= h + \varepsilon\{\chi, H\} + H' \\
(14) \qquad &= h + \varepsilon(f^{\leq} + \omega.\partial\chi/\partial\varphi) + \varepsilon f^> + \varepsilon^2\{\chi, f\} + H' \\
&= h + \varepsilon f_0 + \varepsilon f^> + \varepsilon^2\{\chi, f\} + H',
\end{aligned}$$

H' being defined by this equation. The last two terms, which may be grouped together, are of order ε^2, according to formula 4; it is thus natural to choose $N(\varepsilon)$ such that $f^>$ is of order ε. With the domain of definition (2) for H, the coefficients f_k satisfy $|\,f_k\,| < c e^{-\sigma|k|}$ (see Appendix 1), leading to the choice $N(\varepsilon) = c\,|\log \varepsilon\,|$ as in Arnold's proof of the KAM theorem (cf. [Ar 2]).

ii) ω <u>is a constant Diophantine vector</u>:

Here we require more than simple incommensurability of the frequencies, namely the following inequalities, which prescribe a Diophantine condition (cf. Appendix 4):

$$(15) \qquad \exists\, \gamma > 0,\ \forall\, k \in \mathbb{Z}^n - \{0\},\ |\,\omega.k\,| > \gamma|\,k\,|^{-\tau},$$

where $\tau > n - 1$ (one often assumes $\tau = n$). The Hamiltonian is again of the form

$$(16) \qquad H(I, \varphi) = \omega.I + \varepsilon f(I, \varphi),$$

with an arbitrary perturbation. This case was closely examined by G. Benettin and

G. Galavotti in [Ben2] (see also [Gala2]) as a particularly simple case of Nekhoroshev's theorem. No geometry enters, since ω is independent of I, and the reduction to normal form may be performed uniformly in action space. G.Benettin and G.Galavotti proceed to show that the Hamiltonian may be put in the form:

$$H(I', \varphi') = \omega.I' + \varepsilon h_1(I', \varepsilon) + \varepsilon(\varepsilon/\varepsilon_0)^{(1/\varepsilon)^b} f_\infty(I', \varphi, \varepsilon), \quad (17)$$

where $h_1(I, 0) = f_0(I)$, ε_0 and b are positive constants, and f_∞ is comparable in norm to h and f, which may be assumed to have equal norms, by redefining ε if necessary. As in the general form of Nekhoroshev's theorem, (17) is obtained after a finite, ε-dependent number of canonical transformations.

iii) $\omega = \omega(I)$ satisfies Kolmogorov's condition (or Arnold's), but the reduction is not carried out on an open subset of action space:

Here we encounter Kolmogorov's remarkably insightful idea of *localization* in the neighborhood of tori with Diophantine frequencies; that is, the idea of working near points where $\omega(I) = \omega^* \in \mathbb{C}^n$ satisfies inequalities of the type (15) (see Section 7.2). A Newton's method is then required to accelerate the convergence, and this will be further discussed in §3. We note that the vector ω^* is of course *intrinsic* to the motion on the preserved torus (e.g., it can be read off the spectrum; cf. Appendix 2).

2. Resonant normal forms

This is a generalization of the preceding case; one does not seek to eliminate *all* the harmonics in the perturbation, but only those not belonging to the so-called *resonant module* $M \subset \mathbb{Z}^n$ (a discrete subgroup; cf. Appendix 3), which corresponds to vanishing or very small denominators $\omega(I).k$. The exact definition varies according to the properties of motion one wishes to display. The nonresonant case examined above corresponds to $M = \{0\}$.

Beginning with a Hamiltonian of the form (1), one seeks to transform it by means of a reduction of finite order r to the form:

$$\begin{aligned} H^{(r)}(I, \varphi) &= h(I) + \textstyle\sum_{j=1}^{r} \varepsilon^j Z_j(I, \varphi) + \varepsilon^{r+1} R(I, \varphi, \varepsilon) \\ &= Z^{(r)}(I, \varphi) + \varepsilon^{r+1} R(I, \varphi, \varepsilon), \end{aligned} \quad (18)$$

where, as one often does, we do not distinguish (notationally) between the "old" and "new" variables and R remains to be evaluated; Z_j is given as:

$$(19) \qquad Z_j = \sum_{k \in M} z_{j,k}(I) e^{ik.\varphi}.$$

To carry out this reduction, one solves at each step a linear equation of the form:

$$(20) \qquad f + \omega.\partial\chi/\partial\varphi = P_M f; \quad P_M f = \sum_{k \in M} f_k(I) e^{ik.\varphi}$$

which eliminates (to first order) the non resonant part of f.

In order to obtain convergent series, or even finite sums, one is led to distinguish an ultraviolet part (as in (13) and (14)), which leads to replacing the second member of (20) by $P_M f^{<} + f^{>}$, and thus to the solution:

$$(21) \qquad \chi(I, \varphi) = \sum_{k \notin M,\ |k| \leq N} i f_k(I)(\omega(I).k)^{-1} e^{ik.\varphi}.$$

We refer the reader to Chapter 7 for a further elaboration of this technique and its use in the proof of Nekhoroshev's theorem.

The most important information contained in the normal form (18) resides in the fact that:

$$(22) \qquad dI/dt = -\partial H/\partial\varphi = -\partial Z^{(r)}/\partial\varphi - \varepsilon^{r+1}\partial R/\partial\varphi,$$

where the term $\partial Z^{(r)}/\partial I$ belongs to the subspace generated by the vectors $k \in M$, in action space. In this way we see that Hamilton's equations give rise to a connection between *Fourier* space $\mathbb{Z}^n$ and *action* space $\mathbb{R}^n$.

Let us now briefly review in the resonant case the different possibilities considered above and see if and how these can be generalized.

i) (a) One can assume in this case that $\omega.k \neq 0$ only for $k \notin M$; one thus simply obtains, instead of (11), the resonant form (18), where $h(I) = \omega.I$.

i) (b) This is the case to be found in the proof of Nekhoroshev's theorem after defining a cutoff frequency at each step. One distinguishes (roughly speaking) the domains $B_M(\nu, N)$ such that (cf. Chapter 7):

$$(23) \qquad \forall\, I \in B_M,\ \forall\, k \in M,\ |k| \leq N \Rightarrow |\,\omega(I).k\,| > \nu = \nu(M, N) > 0.$$

On each domain B_M, one obtains a normal form (18); the geometric part of the proof mainly consists in optimally adjusting the various parameters, among which $N = N(\varepsilon)$ and $\nu(M, N) = \nu(r, k)$, where r is the dimension of M.

ii) Here one may assume the existence of two modules M and M', where $M \oplus M' = \mathbb{Z}^n$ and ω is Diophantine with respect to M'. The latter statement means that ω satisfies inequalities of type (15) for $k \in M' - \{0\}$. We may then eliminate, to any finite order, terms in the perturbation whose harmonics do not belong to M, thereby obtaining the form (18) again with $h(I) = \omega.I$. We see again that, in a perturbed system of noninteracting oscillators, it is the relationship of resonance among the frequencies that essentially governs the motion. To our knowledge, Nekhoroshev's limit $\varepsilon \to 0$, $r \to \infty$, $r = r(\varepsilon)$ has not been examined in details, although by simply repeating the proof of the nonresonant case one should obtain the fact that on exponentially long times (of order $\exp(1/\varepsilon)^b$, $b > 0$), the drift in action takes place in the plane generated by the vectors $k \in M$, this being a consequence of a normal form of the type (17), with h_1 now φ-dependent but containing only frequencies belonging to M.

iii) Here, hardly any generalization is possible. When one considers relative Diophantine conditions as above, that is, inequalities of type (15) satisfied for k in the complement of a certain module of $\mathbb{Z}^n$, the resulting drift in action space immediately nullifies the conditions. In addition, we know that the Poincaré-Fermi theorem denies the existence of invariant manifolds of dimension d which are regular in ε for $\varepsilon < \varepsilon_0$, whenever $n + 1 < d < 2n - 1$. Moreover, for $d = 2n - 1$, the only possible manifolds are the energy surfaces, and for $d = n + 1$ they are given by an equation of the form $\omega(I) = \lambda\omega^*$, where ω^* is a Diophantine vector and $\lambda \neq 0$ is a real parameter ($\lambda\omega^*$ is thus also Diophantine); in other words, they are smooth one parameter families of KAM tori (see [Ben1]).

3. Quadratic convergence; the method of Newton, Siegel and Kolmogorov

Since the appearance of Siegel's and Kolmogorov's theorems, the literature concerning these questions is so extensive that we can only refer the reader to other sources (see for example the large bibliography in [Bos]) and provide here a few simple ideas which are useful to keep in mind.

One is concerned with finding the best algorithms which give convergence of certain iterative methods (here, the elimination of angles in the Hamiltonian (1)). Three methods are generally used in this context: those of Picard, Newton, and Nash-Moser. We will not have much to say about the last method (cf. [Bos]), which is an improvement of the second, allowing for a treatment of nonanalytic problems (C^p, $p = 0, \ldots, \infty$) by regularizing the result at each step with an appropriate smoothing operator (generally a convolution). The same idea has been used in PDE's, but its usefulness there is not yet fully established.

Before detailing the other two methods in the cases of interest to us, we should point out that these algorithms provide proofs of fixed point, local inversion, and implicit function theorems (three families of closely connected results). In this way abstract theorems in certain Fréchet spaces (and not just Banach spaces) have been shown; the first of these are due to Nash and Hamilton (cf. [Bos]), and give KAM-type results in special cases. However, in a given situation it is often better to reconstruct the algorithm rather than apply the abstract theorem; in this way a more precise result is acheived by taking advantage of particular features of the problem.

We briefly recall the principle of Picard's method, which is both the oldest (at least formally) and simplest, and is also the one we used above. Among other things it gives a proof of the fixed point theorem for contraction maps in Banach spaces, and hence also of the ordinary implicit function theorem or of the local solubility of ODE's with Lipschitz coefficients. The model problem consists in solving the equation:

$$(24) \qquad f(x) = y, \quad (x, y) \in \mathbb{R}^2, \quad f \in C^1(\mathbb{R}, \mathbb{R})$$

in the neighborhood of y_0 (y close to y_0) for which $f(x_0) = y_0$ and $f'(y_0) \neq 0$. Starting with x_0, one constructs the sequence (x_n) such that:

$$(25) \qquad y = f(x_n) + f'(x_0).(x_{n+1}-x_n) \quad \text{or} \quad x_{n+1} = x_n + f'(x_0)^{-1}.(y-f(x_n)),$$

which amounts to the following classical procedure: take the intersection of the straight line passing through x_n with slope $f'(x_0)$ *independent of* n, with the horizontal ordinate y. If the initial approximation is of order ε ($|y - y_0| \sim \varepsilon$), it will be of order ε^n after n steps. It is easy to see that the formulas (25) still make sense if x and y are elements of a Banach space (or even a Fréchet space), and f is a C^1 map in the Gâteaux sense whose differential is invertible in x_0. The convergence proofs in the real case carry over word for word to Banach spaces (i.e. complete normed spaces).

In Newton's method, again following a classical construction, the straight line of fixed slope $f'(x_0)$ is replaced by the tangent to the curve at the point with abscissa z_n. In other words, the sequence (z_n) is constructed such that $z_0 = x_0$ and:

$$(26) \qquad y = f(z_n) + f'(z_n).(z_{n+1}-z_n) \quad \text{or} \quad z_{n+1} = z_n + f'(z_n)^{-1}(y - f(z_n)).$$

One then shows that the error at the nth step is only of order $\eta_n \sim \varepsilon^{2^n}$, which justifies the name of quadratic convergence ($\eta_{n+1} \sim (\eta_n)^2$). Here also, the formulas make sense in Banach or Fréchet spaces; in the latter case however, invertibility of the differential f' is required in an entire neighborhood of x_0 (this is no longer necessarily an "open" property).

It is natural at this point to ask how these constructions carry over to the case of Hamiltonian perturbations. Practically speaking, there are two differences between the algorithm used in §1, which we will see is essentially a Picard method, and the Newton's method used in proving KAM-type theorems:

1 - The unperturbed part of the Hamiltonian changes at each step.

2 - The perturbation must be treated as a whole; there is no "order-by-order" elimination.

One writes instead of (1):

$$(27) \qquad H(I, \varphi) = h(I) + f(I, \varphi); \quad \| f \|_D = \varepsilon.$$

A canonical transformation of the form $\exp(L_\chi)$ is then used, again with $\chi = \sum_k \chi_k e^{ik.\varphi}$ and $\chi_k = f_k(\omega.k)^{-1}$; in fact, this will be valid in a neighborhood $V^{(1)}$ of $\omega(I) = \omega^*$, where ω^* is Diophantine and the neighborhood will shrink at each step. After evaluating χ and $\exp(L_\chi)$, possibly using a cutoff frequency, one arrives at the expression:

$$(28) \qquad H^{(1)}(I, \varphi) = h^{(1)}(I) + f^{(1)}(I, \varphi); \quad \| f^{(1)} \|_{V^{(1)}} \sim \varepsilon^c .$$

$$H^{(1)} = \exp(\varepsilon L_\chi) H, \quad h^{(1)}(I) = h(I) + \langle f \rangle (I); \quad 1 < c < 2.$$

Beginning with (28), one then iterates the procedure with $h^{(1)}$ as the new unperturbed Hamiltonian, thus taking into account at each step the integrable part of the perturbation (its angular average). This, together with the Diophantine inequalities (15), allows one to control the effect of the small divisors. However, their existence implies that the convergence is never strictly quadratic: In the formulas (28) the constant c is always strictly less than 2, and the nth order remainder is of order ε^{c^n} on an open set $V^{(n)}$.

Let us briefly stress the differences between the classical (Picard like) algorithm, also used in the proof of Nekhoroshev theorem (in its resonant version) and the quadratic (Newton like) algorithm, from the practical viewpoint of Lie series and for Hamiltonians analytic in an (explicit) small parameter. Starting from:

$$(29) \qquad H(I, \varphi, \varepsilon) = h(I) + \varepsilon f(I, \varphi, \varepsilon)$$

n steps lead, in the P-algorithm, to:

$$(30) \qquad H^{(n)}(I_n, \varphi_n, \varepsilon) = h^{(n)}(I_n, \varepsilon) + \varepsilon^n f^{(n)}(I_n, \varphi_n, \varepsilon)$$

with $h^{(n)}(I_n, 0) = h$ and a remainder $f^{(n)}$ which has zero average in the angles. The generating function χ_n for the next transformation is found as the solution of the equation:

$$(31) \qquad \omega(I_n).\partial\chi_n/\partial\varphi_n = - f^{(n)<}(I_n, \varphi_n, 0)$$

where the original frequency vector $\omega = \nabla h$ is used together with the lowest order term on the r.h.s. and the cutoff is defined so that $f^{(n)>}(I_n, \varphi_n, 0)$ is of order ε^2, which for perturbations analytic in φ gives a logarithmic dependence (with respect to ε) of the cut-off frequency. The transformation $T^{(n)}: (I, \varphi) \to (I_n, \varphi_n)$ is thus given as a product:

$$(32) \qquad T^{(n)} = \prod_{k=1}^{n} \exp(\varepsilon^k L_{\chi_{k-1}}).$$

Instead, in the N-algorithm, n steps lead to:

$$(33) \qquad H^{(n)}(I_n, \varphi_n, \varepsilon) = h^{(n)}(I_n, \varepsilon) + \varepsilon^{2^n} f^{(n)}(I_n, \varphi_n, \varepsilon)$$

which satisfies $\| \varepsilon^{2^n} f^{(n)} \|_{V^{(n)}} \sim \varepsilon\, c^n$ $(1 < c < 2)$ in a neighborhood $V^{(n)}$ of the manifold $\omega_n(I_n, \varepsilon) = \omega^*$ with $\omega_n = \nabla h^{(n)}$ depending on ε. To proceed, one then has to solve:

$$(34) \qquad \omega_n(I_n, \varepsilon).\partial\chi_n / \partial\varphi_n = -\, g_n(I_n, \varphi_n, \varepsilon)$$

instead of (31) where g_n is constructed as follows: one first takes the terms in the ε–expansion of $f^{(n)}$ of order $0, \dots, \varepsilon^{2^{n+1}-1}$ and then one performs the usual ultraviolet cutoff. This of course serves to stress the fact that ε is not the natural ordering parameter since one should rather use $\varepsilon^{(n)} = \varepsilon^{2^n}$ at the nth step. χ_k is thus ε-dependent, contrary to what occurs in (32). Finally, the transformation $T^{(n)}$ is given as:

$$(35) \qquad T^{(n)} = \prod_{k=1}^{n} \exp(\varepsilon^{2^k} L_{\chi_{k-1}})$$

and the main content of the KAM theorem is that the sequence $T^{(n)}$ converges on what is seen to be a perturbed torus (this implies a careful choice of the size of the domains $V^{(n)}$).

In conclusion it is perhaps worthwhile to indicate why the quadratic convergence method does not improve the treatment of the resonant case. The solution of the linearized equation hinges on the operator L_h, i.e., the Poisson bracket with h, the unperturbed Hamiltonian, which reduces to $\nabla h.\partial/\partial\varphi$ when h is independent of the angles. In Newton's method, one replaces h by $h^{(n)}$, which is defined recursively as in (28); in the resonant case, one needs the analog of the " tangent to the point $Z^{(n)}$ " (see formula (18)), but the action of $L_{Z^{(n)}}(I,\varphi)$ is not known precisely enough in advance to carry out the analogy.

APPENDIX 9: STEEPNESS

1. Definition of steepness for a Hamiltonian

In this appendix, we define the concept of *steepness* of a function $H(I)$ of several variables ($I \in \mathbb{R}^m$), with a view towards applications to the integrable part of a Hamiltonian. We recall that steepness is the optimal hypothesis for the proof of Nekhoroshev's theorem (cf. Chapter 7), and we mainly follow the latter author's original articles ([Nek1] and [Nek2]) in our exposition here.

Although we do not prove Nekhoroshev's theorem in the general case of steep functions, the discussion in this appendix should serve to clarify the notion, so that the interested reader may then examine the few technical geometric lemmas which serve to pass from the (pseudo) convex to the steep case.

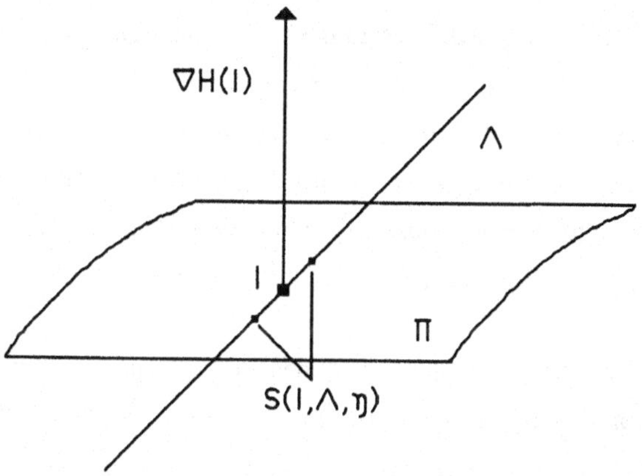

Figure A9.1: Λ is a plane passing trough the point I (which belongs to the energy surface Π) and perpendicular to $\nabla H(I)$. $S(I, \Lambda, \eta)$ is the intersection of Λ and the sphere with center I and radius η.

Consider a level surface of the Hamiltonian H, assumed for the moment to be *real analytic*, a point I on this surface, and a plane Λ of dimension r passing through I and perpendicular to $\nabla H(I)$ (Figure A9.1). The steepness condition amounts to requiring that the curvature of the intersection of the level surface and the plane Λ be sufficiently large in the neighborhood of I, independently of the choice of Λ.

To estimate the curvature of the level surface, we introduce the function:

$$m(I, \Lambda, \eta) = \text{Inf}_{I' \in S(I,\Lambda,\eta)} \| \nabla H_\Lambda(I') \|,$$

where H_Λ is the restriction of H to the plane Λ, ∇H_Λ is its gradient, i.e., the projection of $\nabla_I H$ on Λ, and $S(I, \Lambda, \eta)$ is the intersection of Λ and the sphere in $\mathbb{R}^n$ with center I and radius η.

The Hamiltonian H is said to be *steep at the point* I *on the plane* Λ if:

a) ∇H does not vanish as a function of I (by contrast, $\nabla H_\Lambda(I)$ is zero since Λ is tangent to the level surface).

b) There exist constants $C^* > 0$, $\delta > 0$ called the steepness coefficients, and $\alpha^* \geqslant 0$, the steepness index, such that for all η in $(0, \delta)$, $m(I, \lambda, \eta) > C^* \eta^{\alpha^*}$ (δ may perhaps be infinite).

The steepness condition therefore signifies that I is not a singular point of H and that in a δ-neighborhood of I, the variation of the gradient of H with the distance from I in *any* direction away from the plane Λ increases at least as a power of the distance. We note that $m(I, \Lambda, \eta)$ is defined as the lower bound of $\| \nabla H_\Lambda(I') \|$ on the sphere $S(I, \Lambda, \eta)$ in the plane Λ in order to give an isotropic formulation of the steepness condition. We further note that since H_Λ is a real analytic function, the steepness index is necessarily a positive integer.

If $\Lambda^r(I)$ denotes the set of planes of dimension r passing through the point I, we say that H is *steep at the point* I if:

a) H is steep at I on every plane of dimension 1 through $(n-1)$ passing through I.

b) The steepness coefficients C^*_r and δ_r and the steepness indices α^*_r depend (possibly) on the dimension r, but not on the choice of plane in $\Lambda^r(I)$ for fixed r.

We remark that H may possibly be steep for only certain dimensions of the plane Λ. For example, consider the three degree of freedom Hamiltonian

$$H(I) = (I_1 - I_2^2)^2 + I_3$$

(cf. [Nek2]). It is not steep at the origin; in fact, the gradient of H is equal to the constant vector $(0,0,1)$ at every point of the parabola $I_3 = 0$, $I_1 = I_2^2$, which passes through the origin. H is therefore not steep on the two dimensional plane $I_3 = 0$ containing the parabola (Figure A9.2). On the other hand, H is steep on every straight line passing through the origin in the plane $I_3 = 0$. On the straight line $I_1 = 0$ the restriction of the Hamiltonian is I_2^4, the norm of the gradient is $4|I_2^3|$, and the distance to the origin (in the Euclidean norm) is $|I_2|$. H is therefore steep with coefficients $C^* = 4$, $\delta = \infty$, and with index $\alpha^* = 3$. On all other straight lines through the origin, which may be written $I_2 = \mu I_1$, the restriction of H is $(1 - \mu^2)^2 I_1^2$, the norm of the gradient is:

$$2\mu|I_1 - \mu^2 I_1^2|.|1 + 2I_1|/(1 + \mu^2)^{1/2},$$

and the distance to the origin is $|I_1|(1 + \mu^2)^{1/2}$. H is therefore steep on such lines in a neighborhood of the origin, with steepness index $\alpha^* = 1$, and with steepness coefficients C^* and δ which depend on μ ($C^* \to 2$ and $\delta \to \infty$ as $|\mu| \to 0$).

This behavior is easily understood. The gradient does not vary on the parabola $I_1 = I_2^2$, and all straight lines in the plane other than $I_1 = 0$ or $I_2 = 0$ meet this curve at a finite distance from the origin. For this reason, on all such lines H is steep with index 1 in a neighborhood of the origin. The straight line $I_2 = 0$ does not intersect the parabola, so that the steepness coefficient C^* is maximal and the coefficient δ is infinity. As for the line $I_1 = 0$, it is tangent to the parabola at the origin, and this degeneracy causes the steepness index to be greater than 1.

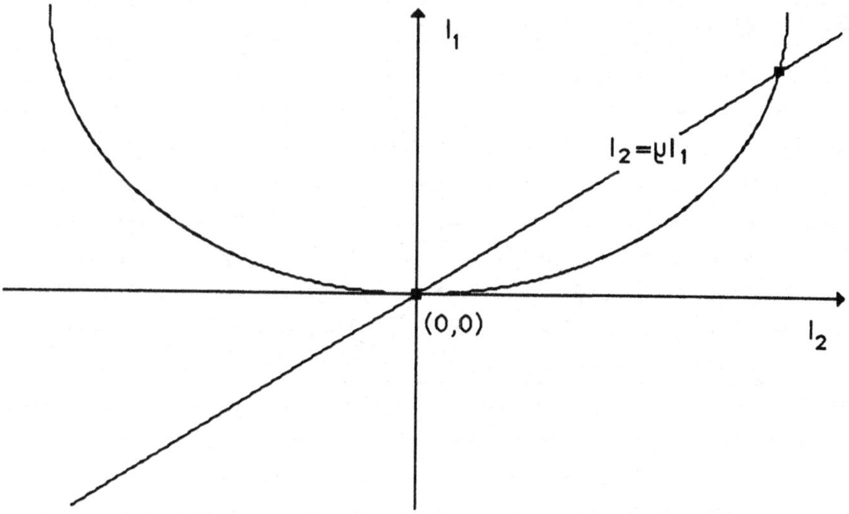

Figure A9.2: The Hamiltonian $H(I) = (I_1 - I_2^2)^2 + I_3$ is not steep at the origin since its gradient is constant along the parabola $I_3 = 0$, $I_1 = I_2^2$.

We remark that because of the presence of the parabola, the steepness is not uniform with respect to straight lines. In fact, on a line $I_1 = \lambda I_2$ with small slope λ, the norm of the gradient of H is equal to:

$$2|I_2|.|\lambda - I_2|.|\lambda - 2I_2|/(1 + \lambda^2)^{1/2}.$$

This norm vanishes for $I_2 = \lambda$, in other words at the point of intersection with the parabola (λ^2, λ) and at the point $(\lambda^2/2, \lambda/2)$. The steepness coefficient δ is less than $\lambda/2$, and it is therefore not possible to find an estimate of this coefficient which is uniform with respect to the slope λ.

We see here that the function H is not steep at the origin since its restriction to a curve passing through the origin is constant. For an *analytic* function, this is the *only* case where steepness conditions are not satisfied. For a function H to be steep at a point I (where ∇H does not vanish) on a plane Λ, it is necessary and sufficient that it be steep on every analytic curve γ in the plane which passes through I. In other words, at every point I' of γ with $\| I' - I \| = \xi \leqslant \delta$, we have the inequality:

$$\| \nabla H_\Lambda(I') \| > C^* \xi^\alpha.$$

It follows that a function steep at I on a plane Λ is steep on every plane of lower dimension passing through I and contained in Λ.

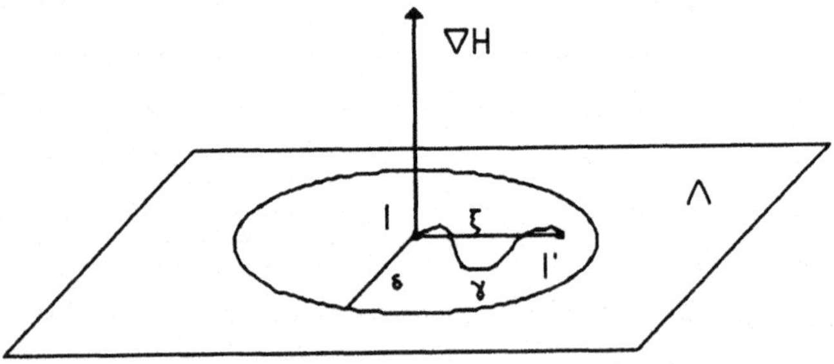

Figure A9.3: For H to be steep at point I on a plane Λ it is necessary and sufficient that it be steep on every analytic curve γ in the plane Λ which passes through I.

A function H is *steep on a domain* if the preceding inequalities are satisfied uniformly on the domain, that is, if:

a) $\| \nabla H(I) \| \geq c > 0$ at every point of the domain.

b) H is steep at every point in the domain with steepness coefficients and indices $\{C^*_r , \delta_r , \alpha^*_r\}$ independent of the point considered.

Suppose now that H is *differentiable* but not analytic. The steepness behavior in the neighborhood of a given point I no longer reduces to a study of the Taylor's series of H at I, and the variation of $\| \nabla H_\Lambda \|$ with distance in the plane Λ from I does not necessarily follow a power law. For such H, the preceding definition of steepness must be slightly modified:

H is steep at I on the plane Λ with coefficients C^* and δ and with index α^* if in every ball with center I and radius $\xi \leq \delta$, there exists a sphere of radius $\eta \leq \xi$ on which $\| \nabla H_\Lambda \| \geq C^* \xi^\alpha$ (the sphere guarantees the isotropic character of the estimate).

This definition of steepness ensures that the norm of the gradient takes on sufficiently large values with increasing distance from I, at least at certain points, independently of the direction in which that distance is measured. We note that α^* is not necessarily an integer as in the analytic case, but is always greater than or equal to one for H at least of class C^2 (by first order Taylor expansion).

Given this definition of steepness at a point I on a plane, the definitions of *steepness at a point* and *steepness on a domain* are the same as before.

Steepness is a very general property. It may be shown that among C^∞ functions, the only ones not steep at points where their gradient is nonzero (regular points) are those with Taylor coefficients satisfying an infinite number of independent algebraic equations. For each r, such functions define a subspace of the space of jets of order r with codimension tending to infinity with r. We shall see in the next section how such algebraic conditions arise.

2. Steepness and quasi-convexity; the three dimensional case

Quasiconvex Hamiltonians constitute an important class of steep Hamiltonians. A Hamiltonian H is quasiconvex at a given point I if its gradient does not vanish there and if the restriction of the quadratic form $\nabla^2 H$ (with matrix $(\partial^2 H/\partial I_j \partial I_k)$) to the hyperplane Λ normal to $\nabla H(I)$, and thus tangent to the level surface, is definite (positive or negative). In other words, H is quasiconvex if and only if the system:

$$\nabla H(I).\eta = 0; \quad \nabla^2 H(I)\eta.\eta = 0$$

admits no nontrivial solution $\eta = (\eta_1, \ldots, \eta_n)$. The quasiconvexity of H at the point I is equivalent to convexity of the level surface of H at the same point, and the class of functions quasiconvex at a point of course contains the class of functions convex at this point.

As usual we define functions to be *quasiconvex on a domain* when they are quasiconvex at every point of the domain. Quasiconvex functions are the "steepest" possible, since at each fixed point I, the steepness index is equal to 1 for each direction orthogonal to $\nabla H(I)$; the variation of the projection of the gradient of H on Λ with the distance to the reference point I is linear. In contrast, all steep non-quasiconvex functions necessarily possess at least one steepness index strictly greater than 1. It is also clear that quasiconvexity at a given point is a generic property of two degree of freedom Hamiltonians. The proof of Nekhoroshev's theorem is simplified by restricting consideration to convex or quasiconvex Hamiltonians, as we do in Chapter 7.

In the case of three degree of freedom Hamiltonians, there is an interesting relation between quasiconvexity and Arnold's nondegeneracy condition for the frequencies. We first recall the statement of Arnold's condition for an n-degree of freedom Hamiltonian (see Appendix 3):

H satisfies Arnold's condition at the point I if the map from $\mathbb{R}^{n+1}$ to itself which takes the point (I, λ) to the point $(\lambda\omega(I), H(I))$ is of maximal rank at $(I, 1)$. The Jacobian of this map at $(I, 1)$ is the determinant:

$$D_2(I) = \begin{vmatrix} \partial^2 H_0(I)/\partial I^2 & \partial H_0(I)/\partial I \\ \partial H_0(I)/\partial I & 0 \end{vmatrix}$$

Arnold's condition thus requires that this determinant be nonzero. $D_2(I)$ may also be considered as the Hessian of the map $(I, \lambda) \to \lambda H(I)$ evaluated at $(I, 1)$ and in this formulation, Arnold's condition appears quite close to Kolmogorov's, which requires that the Hessian of H itself be nonvanishing at point I.

For two degree of freedom Hamiltonians, the determinant $D_2(I)$ is simply the value of the quadratic form $\partial^2 H(I)/\partial I^2$ on the vector $[\partial H(I)/\partial I_1\ ,\ -\partial H(I)/\partial I_2]$, i.e., the direction vector of the line orthogonal to $\nabla H(I)$. Since $D_2(I)$ vanishes at the points where

∇H is zero (for any number of degrees of freedom), we see that Arnold's condition for $n = 2$ is equivalent to quasiconvexity of H. At points where Arnold's condition is not satisfied, the curve $H = C^{st}$ has an inflection point.

We will now show the following somewhat less trivial property: A function of three variables is quasiconvex at a point I if and only if the determinant $D_2(I)$ is strictly negative.
This assertion is a consequence of the following

<u>Proposition</u>:

Let Q be a quadratic form on $\mathbb{R}^n$ and let w be a nonzero vector. The sign of the discriminant of the restriction of Q to the hyperplane Λ orthogonal to w is opposite the sign of the determinant of the $(n+1) \times (n+1)$ matrix M, defined on an orthogonal basis $(v_1, \dots, v_n)$ of $\mathbb{R}^n$, given by:

$$M = \begin{bmatrix} Q(v_1,v_1) & & Q(v_1,v_n) & w_1 \\ Q(v_2,v_1) & & Q(v_2,v_n) & w_2 \\ & & & \\ Q(v_n,v_1) & & Q(v_n,v_n) & w_n \\ w_1 & & w_n & 0 \end{bmatrix}$$

Moreover, the discriminant and the determinant vanish simultaneously.

To prove this, we first note that neither sign depends on the basis chosen for $\mathbb{R}^n$. We may therefore choose a basis such that $w = v_n$ is the last vector of the basis and all other basis vectors are orthogonal to it. Writing the matrix M in this basis, we see that its determinant is the negative of the discriminant of the restriction of Q to Λ in the basis $(v_1, \dots, v_{n-1})$. ▪

Applying this proposition to a quadratic form in $\mathbb{R}^3$, we deduce that the restriction of Q to a 2-dimensional Λ is definite and thus has strictly positive discriminant if and only if the determinant of M is strictly negative. By taking the quadratic form to be $\nabla^2 H$ and $w = \nabla H$, this assures that the 3-degree of freedom Hamiltonian H of class C^2 is quasiconvex at I if and only if Arnold's determinant $D_2(I)$ is strictly negative.

It is in fact possible to show that the functions of class C^3 satisfying Arnold's condition belong to a particular class of functions which contains the quasiconvex functions of class C^3. These functions are defined by a condition on the jets of order three at the point I which generalizes quasiconvexity:

a) Their gradient at I is nonzero.

b) The restrictions of the bilinear form $\nabla^2 H$ and the trilinear form $\nabla^3 H$ (with components $\partial^3 H(I)/\partial I_i \partial I_j \partial I_k$) to the plane orthogonal to the gradient are such that the system:

$$\nabla H.\eta = 0; \quad \nabla^2 H.\eta.\eta = 0; \quad \nabla^3 H.\eta.\eta.\eta = 0$$

admits no nontrivial solution.

This condition ensures that, for every plane Λ tangent to the energy surface of H at I, the variation of $\| \nabla H_\Lambda \|$ with distance from I is at least quadratic in any fixed direction lying in the plane Λ. Functions satisfying conditions a) and b) are thus steep, and these conditions are generic for jets of functions of three variables. This demonstrates the genericity of the steepness condition in this particular case.

3. Higher dimensions; genericity

For $n \geqslant 3$, it does not appear possible to determine such explicit conditions guaranteeing steepness. Nevertheless, Nekhoroshev proved strong genericity results, and gave (somewhat implicit) algebraic conditions which test whether a given function is steep. Here we shall be content to state the main genericity result.

We denote by $J^r(n)$ the space of jets of order r of real-valued functions of n variables, i.e., the collection of of all derivatives of total order less than r. It can be shown that there exists a semi-algebraic set (defined by polynomial equalities and inequalities) $\Sigma^r(n) \subset J^r(n)$ such that every function H whose r^{th} order jet at I does not belong to $\Sigma^r(n)$ is either steep at I, or else is critical at I ($\nabla H(I) = 0$). In addition, the codimension of $\Sigma^r(n)$ in $J^r(n)$ is given by:

$$\text{codim } \Sigma^r(n) = \max(0, r - 1 - n(n-2)/4) \qquad n \text{ even}$$

$$= \max(0, r - 1 - (n-1)^2/4) \qquad n \text{ odd.}$$

Now let $r_m(n)$ be the smallest integer r such that a function of n variables with a generic jet of order r in $J^r(n)$ is steep at a noncritical point, which amounts to defining $r_m(n)$ by $\text{codim } \Sigma^{r_m}(n) = 1$, then:

$$r_m(n) = n(n-2)/4 + 2 \qquad \text{if } n \text{ is even,}$$

and

$$r_m(n) = (n-1)^2/4 + 2 \qquad \text{if } n \text{ is odd.}$$

This clarifies what is meant by genericity of the steepness condition. We make two additional remarks about the practical consequences of this result:

- The only *quadratic* Hamiltonians which are steep in dimension greater than or equal to 3 are those given by a *definite* quadratic form; they are thus convex and equal to the ordinary expression for kinetic energy. All other quadratic Hamiltonians are *nowhere* steep (because hyperboloids are ruled surfaces...).
- *Polynomial* Hamiltonians with n degrees of freedom are generically (with respect to their coefficients) steep only when they are at least of degree $r_m(n)$, a number which grows rapidly with n. Again, polynomial Hamiltonians of lower degree are generically nowhere steep.

BIBLIOGRAPHY

[Ab] Abraham R.H. and Marsden J.E., "Foundations of mechanics", second edition, Benjamin-Cummings, Reading, 1978.

[And] Andronov A.A. and Khaïkine C.E., "Theory of oscillations", Nauka, Moscow, 1937.
English translation: Princeton Univ. Press, Princeton N.J., 1949.

[Ano] Anosov, 1960, Izvestia Akad. Nauk SSSR Mathematics, **24 (5)**, 721-742, "Oscillations in systems of O.D.E. with rapidly oscillating solutions".

[Ar1] Arnold V.I.,1962, Doklady Akad. Nauk SSSR, **142**, 758-761, "On the behavior of an adiabatic invariant under slow periodic variation of the Hamiltonian".
English translation: 1962, Soviet Math. Doklady, **3**, 136-139.

[Ar2] Arnold V.I., 1963, Uspekhi Mat. Nauk, **18 (5)**, 13-39, "Proof of a theorem of A.N. Kolmogorov on the invariance of quasi-periodic motions under small perturbations of the Hamiltonian".
English translation: 1963 , Russian Math. Surveys, **18 (5)**, 9-36.

[Ar3] Arnold V.I., 1963, Uspekhi Mat. Nauk, **18 (6)**, 91, "Small denominators and problems of stability of motion in classical and celestial mechanics".
English translation: 1963, Russian Mat. Surveys, **18 (6)**, 85-191.

[Ar4] Arnold V.I., 1964, Doklady Akad Nauk SSSR, **156 (1)** ,"Instability of dynamical systems with several degrees of freedom".
English translation: 1964, Sov. Mat. Doklady , **5 (3)**, 581-585.

[Ar5] Arnold V.I., 1965, Doklady Akad. Nauk, **161 (1)**, 331-334, "Conditions for the applicability, and estimate of the error, of an averaging method for systems which pass trough states of resonance in the course of their evolution".
English translation: 1965, Sov. Mat. Doklady, **6 (2)**, p. 331.

[Ar6] Arnold V.I. and Avez A., "Ergodic problems of classical mechanics" Benjamin, New York, 1968.
Traduction française: "Problèmes ergodiques de la mécanique classique", Gauthier-Villars, Paris, 1967.

[Ar7] Arnold V.I.," Mathematical methods of classical mechanics", Nauka, Moscow, 1974.
English translation: Graduate Texts in Mathematics **60**, Springer Verlag, New York, 1978.

Traduction française: "Méthodes mathématiques de la mécanique classique", Editions Mir, Moscow, 1976.

[Ar8] Arnold V.I., "Geometrical methods in the theory of ordinary differential equations", Nauka, Moscow, 1978.
English translation: Springer Verlag, New York, 1983.
Traduction française: "Chapitres supplémentaires de la théorie des équations différentielles ordinaires", Editions Mir, Moscou,1980.

[Ar9] Arnold V.I., 1983, Uspekhi Mat. Nauk, **38** **(4)**, 189-203, "Remarks on the perturbation theory for problems of Mathieu type".
English translation: 1983, Russian Math. Surveys, **38** **(4)**, 215-233.

[Bac] Bachtin V.I., 1986, Functional Analysis and Applications, **20** **(2)**, 1-7, "Averaging in many frequency systems".
English translation: 1986, Functional Analysis and Applications, **20** **(2)**, p. 1.

[Bak] Bakaï A.C. and Stepanovski Y.P., "Adiabatic invariants", Naukova Dumka, Kiev, 1981.

[Ben1] Benettin G., Ferrari G., Galgani L. and Giorgilli A., 1982, Il Nuovo Cimento, **72** **B(2)**, 137-148, "An extension of the Poincaré Fermi-theorem on the non-existence of invariant manifolds in nearly integrable Hamiltonian systems".

[Ben2] Benettin G. and Galavotti G., 1986, J. Stat. Phys., **44** **(3-4)**, 293-338, "Stability of motion near resonances in quasi-integrable Hamiltonian systems".

[Ben3] Benettin G., Galgani L. and Giorgilli A., 1984, Nature, **311** **(5985)**, 444-445, "Botzmann's ultraviolet cutoff and Nekhorochev's theorem".

[Ben4] Benettin G., Galgani L. and Giorgilli A., "Poincaré non-existence theorem and classical perturbation theory for nearly integrable Hamiltonian systems" in "Advances in nonlinear dynamics and stochastic processes", R. Live and A. Politi Editors, World Scientific Publ. Company, Singapore, 1985.

[Ben 5] Benettin G., Galgani L. and Giorgilli A., 1985, Celestial Mechanics, **37**, 1-25, "A proof of Nekhoroshev's theorem for the stability times in nearly integrable Hamiltonian systems".

[Ben6] Benettin G., Galgani L., Giorgilli A. and Strelcyn J.M., 1984, Il Nuovo Cimento, **79B**, 201-223, "A proof of Kolmogorov's theorem on invariant tori using canonical transformation defined by the Lie method".

[Ber1] Berry M.V., "Regular and irregular motion" in "Topics in nonlinear dynamics", Jorna S. Editor, Am. Inst. Phys. Conf. Proc., **46**, 16-120,1978.

[Ber2] Berry M.V., 1984, Journal of Physics A: Math. Gen., **17**, 1225-1233.

[Ber3] Berry M.V., 1984, Proceedings of the Royal Society A, **392**, 45-57, "Quantal phase factors accompanying adiabatic changes".

[Ber4] Berry M.V., 1985, Journal of Physics A: Math. Gen., **18**, 15-27, "Classical adiabatic angles and quantal adiabatic phase".

[Ber5] Berry M.V. and Mount K.E., Report of Progress in Physics, 1972, **35**, 315-397, "Semiclassical approximations in wave mechanics".

[Bes] Besjes J.G., 1969, Journal de Mécanique, **8**, 357-373, "On the asymptotic methods for nonlinear differential equations".

[Bi] Birkhoff G.D., "Dynamical systems", Amer. Math. Soc. Coll. Publ. **9**, Amer. Math. Soc., New York, 1927.

[Bog] Bogoliubov N.N. and Mitropolski Y.A., "Asymptotic methods in the theory of nonlinear oscillations ", Fizmatgiz, Moscou, 1958.
English translation: Gordon and Breach, New York, 1961.
Traduction française: "Les méthodes asymptotiques dans la théorie des oscillations non linéaires", Gauthier-Villars, Paris, 1963.

[Boh] Bohr H., "Fastperiodische Funktionen", Springer Verlag, Berlin, 1932.

[Bor] Born M. and Fock V., 1928, Zeitschrift für Physik, **51**, p. 965.

[Bos] Bost J.-B., Séminaire Bourbaki **639** (1984-85), Astèrisque **133-134**, Paris, 1985.

[Bri] Brillouin L., "Les Tenseurs en Mécanique et en Elasticité" (Chapitre 8), Masson et Cie, Paris, 1960.

[Brj] Brjno A.D., 1971, Trudy Mosk. Mat. Obsc., **25**, "Analytical form of differential equations".
English translation: 1971, Trans. Moscow. Math. Soc., **25**.

[Car] Cary J.R., 1981, Physics reports, **79** **(2)**, 129-159, "Lie transform perturbation theory for Hamiltonian systems".

[Cas] Cassels J.W.S., "An introduction to the theory of Diophantine approximation", Cambridge University Press, Cambridge U.K., 1957.

[Chie1] Chierchia L. and Gallavotti G., 1982, Il Nuovo Cimento, **67B**, 277, "Smooth prime integrals for quasi-integrable Hamiltonian systems".

[Chie2] Chierchia L. and Gallavotti G., 1982, Il Nuovo Cimento, **72B**, 137.

[Chir] Chirikhov B.V., 1979, Physics Reports, **52** **(5)**, 263-379, "A universal instability of many-dimensional oscillator systems".

[Cho] Chow S.N. and Hale J.K., "Methods of bifurcation theory", Springer Verlag, New York, 1982.

[Co] Coddington E.A. and Levinson N., "Theory of ordinary differential equations", Mac Graw Hill, New York, 1955.

[Di] Dirac P.A.M.,1925, Proceedings of the royal society, **107**, 725-734.

[Dy] Dykhne A.M., 1960, J.E.T.P., **38** **(2)**, 570-578, "Quantum transitions in adiabatic approximation".

[Ei1] Einstein A., in "La théorie du rayonnement et les quanta", P. Langevin and M. De Broglie Editors, Gauthier-Villars, Paris, 1912.

[Ei2] Einstein A., 1917, Verhandlungen der Deutschen Physikalischen Gesselschaft, **19**, 82, "Zum Quantensatz von Sommerfeld und Epstein". Traduction française: 1986, Annales de la Fondation Louis de Broglie, **11**, 261.

[El] Ellison J.A., Saenz A.W. and Dumas H.S., 1988, "Improved N^{th} order averaging theory for periodic systems", submitted to Arch. Rat. Mech. Anal.

[Fed] Fedoriuk M.V., 1976, Differentsial'nye Uravneniya, **12** **(6)**, 1012-1018, "Adiabatic invariants for a system of harmonic oscillators and dispersion theory".

[Fer] Fer F., 1962, Le journal de physique et le radium, **23** **(12)**, 973-978, "Quelques remarques sur les systèmes thermodynamiques de Boltzmann et les invariants adiabatiques".

[Fi] Fink A.M., "Almost periodic differential equations", Lecture Notes in Mathematics **377**, Springer Verlag, Berlin, 1974.

[Fre] Freidlin M.I. and Wentzell A.D., "Random perturbations of dynamical systems", Nauka, Moscou, 1979.
English translation: Grundlehren der mathematischen Wissenschaften **260**, Springer Verlag, New York, 1984.

[Frö] Fröman N. and Fröman P.O., "JWKB approximation; contributions to the theory", North Holland, Amsterdam, 1965.

[Gala1] Galavotti G., " The elements of mechanics", Boringhieri, Torino, 1980.
English translation: Springer Verlag, Berlin, 1983.

[Gala2] Galavotti G., "Perturbation theory for classical Hamiltonian systems" in "Scaling and self-similarity in physics", Frölich J. Editor, 359-426, Birkhäuser, Boston, 1983.

[Galg] Galgani L., "Ordered and Chaotic motion in Hamiltonian Systems and the problem of Energy partition", in "Chaos in Astrophysics", Buchler J.R. et al. Editors, 245-257, Reidel, Dordrecht, 1985.

[Gia] Giacaglia G.E.O., "Perturbations methods in non linear systems", Appl. Math. Sc., **8**, Springer Verlag, New York, 1972.

[Gio] Giorgilli A. and Galgani L., 1985, Celestial Mechanics, **37**, 95-112, "Rigorous estimates for the series expansions of Hamiltonian perturbation theory".

[Go] Goldstein H., "Classical mechanics", second edition, Addison-Wesley, Reading (Massachusets), 1980.

[Gu] Guckenheimer and Holmes, "Nonlinear oscillations, dynamical systems and bifurcations of vector fields", Applied Mathematical Sciences **42**, Springer Verlag, New York, 1983.

[Hal] Hale J. K., "Ordinary differential equations, Wiley-Interscience Publishers, New York, 1969.

[Han] Hannay J.H., 1985, Journal of Physics A: Math. Gen., **18**, 221-230, "Angle variable holonomy in adiabatic excursion of an integrable Hamiltonian".

[Hea] Heading J., "An introduction to Phase-Integral methods", Methuen, London, 1962.

[Her] Herman M.R., "Sur les courbes invariantes par les difféomorphismes de l'anneau", Vol. 1, Astèrisque **103-104**, Paris, 1983 (volume 2 to appear).

[Hoc] Hochstadt H., 1963, Proc. Amer. Math. Soc., **14**, 930-932, "Estimates on the stability intervals of Hills equation".

[Hol1] Holmes C. and Holmes P.J., 1981, J. Sound and Vib., **78**, 161-174.

[Hol2] Holmes P.J. and Marsden J.E., 1982, Communications in mathematical physics, **82**, 523-544, "Horseshoes in perturbations of Hamiltonian systems with two degrees of freedom".

[Hol3] Holmes P.J. and Marsden J.E., 1982, J. M. P., **23** **(4)**, "Melnikov's method and Arnold diffusion for perturbation of integrable dynamical systems".

[Hol4] Holmes P.J. and Marsden J.E., 1983, Indiana Univ. Mathematics Journal, **32** **(2)**, "Horseshoes and Arnold diffusion for Hamiltonian systems on Lie groups".

[Hol5] Holmes P. J., Marsden J.E. and Scheurle J., 1987, preprint, "Exponentially small splitting of separatrices".

[Kas] Kasuga T., 1961, Proceedings of the Academy of Japan, **37**, 366-371, 372-376, 377-382, "On the adiabatic theorem for Hamiltonian system of differential equations in the classical mechanics".

[Kat] Kato T., 1950, Phys. Soc. of Japan, **5**, 435-439, "On the adiabatic theorem of quantum mechanics".

[Khan] Khanin K.M. and Sinaï Y.G., "The renormalisation group method and Kolmogorov-Arnold-Moser theorem" in "Nonlinear Phenomena in Plasma Physics and Hydrodynamics", Sagdeev R.Z. Editor, Advances in Science and Technology in the USSR, Mir, Moscow, 1986.
English translation: Mir Publishers, Moscow, 1986.

[Kn] Knoll J. and Schäffer, 1976, Annals of Physics, **97**, 307.

[Ko] Kolmogorov A.N., 1954, Dokl. Akad. Nauk. SSSR, **98** (**4**), 527-530.
English translation: in "Stochastic behavior in classical and quantum Hamiltonian systems", Casati G. and Ford J. Editors, Lecture Notes in Physics **93**, Springer Verlag, Berlin, 51-56, 1979.

[Kru] Kruskal M., 1962, Journal of mathematical physics, **3** (**4**), 806-829, "Asymptotic theory of Hamiltonian and other systems with all solutions nearly periodic".

[Kry] Krylov N.M. and Bogoliubov N.N., "Introduction to nonlinear mechanics", Moscow, 1937.
English translation: Princeton University Press, Princeton, 1947.

[Ku] Kubo, "Statistical mechanics", North Holland, Amsterdam, 1965.

[La1] Landau L. and Lifchitz E., "Course of Theoretical Physics, Volume 3: Quantum Mechanics", Nauka, Moscow, 1963.
English translation: Pergamon Press, Oxford, 1958 (first edition).
Traduction française: "Mécanique quantique (théorie non relativiste)", Editions Mir, Moscou, 1966.

[La2] Landau L. and Lifchitz E., "Course of Theoretical Physics, Volume 1: Mechanics", third edition, Nauka, Moscou, 1969.
English translation: Pergamon Press, Oxford, 1976.
Traduction française: "Mécanique", Editions Mir, Moscou, 1966.

[Lén] Lénard A., 1959, Ann. Physics, **6**, 261.

[Lew] Lewis H.R. Jr., 1968, Journal of Math. Physics, **9**, 1976.

[Lic] Lichtenberg A.J. and Liebermann M.A., "Regular and stochastic motion", Applied mathematical Sciences **38**, Springer Verlag, New York, 1983.

[Lin] Lindstedt M., 1882, Astron. Nach., **103**, 211.

[Lit1] Littlejohn R.G., 1979, J. Math. Phys., **20**, 2445-2458, "A guiding center Hamiltonian: A new approach".

[Lit2] Littlejohn R.G., 1981, Phys. Fluids, **24** (**9**), 1730-1749, "Hamiltonian formulation of guiding center motion".

[Lit3] Littlejohn R.G., 1982, J. Math. Phys., **23** (**5**), "Hamiltonian perturbation theory in noncanonical coordinates".

[Lit4] Littlejohn R.G., 1983, J. Plasma Physics, **29** **(1)**, 111-125, "Variational principles of guiding center motion".

[LoG1] Lochak G., 1972, C.R.A.S., **275B**, 49.

[LoG2] Lochak G., 1973, C.R.A.S., **276B**, 103.

[LoG3] Lochak G. and Alaoui A., 1977, Ann. Fond. Louis de Broglie, **2**, 87.

[LoG4] Lochak G. and Vassalo-Pereira J., 1976, C.R.A.S., **282B**, 1121.

[LoP] Lochak P., 1982, C.R.A.S., **295 Série 1**, 193, "Démonstration linéaire du théorème adiabatique en mécanique classique: équivalence avec le cas quantique".

[Mag] Magnus W. and Winkler S., "Hill's equation", Wiley-Interscience Publishers", New York, 1966.

[Mar] Markus L. and Meyer K.R., 1974, Memoirs of the Amer. Math. Soc., 144.

[Mas] Maslov V.P. and Fedoriuk M.V., "Semiclassical approximation in quantum mechanics", Nauka, Moscou, 1976.
English translation: Reidel, Dordrecht, 1981.

[McK1] Mac Kean H.P. and van Moerbeke P., 1975, Inventiones Mathematicae, **30**, 217-274, "The spectrum of Hill's equation".

[McK2] Mac Kean H.P. and Trubowitz E., 1976, C.P.A.M., **29**, 143-226, "Hill's operator and hyperelliptic function theory in the presence of infinitely many branch points".

[Mel1] Melnikov V.K., 1962, Doklady Akad. Nauk SSSR, **144** **(4)**, 747-750.
English translation: Soviet Physics Doklady, **7**, 502-504.

[Mel2] Melnikov V.K., 1963, Doklady Akad. Nauk SSSR, **148**, 1257.

[Mel3] Melnikov V.K., 1963, Trudy Moskov. Obsc., **12**, "On the stability of the center for time-periodic perturbations".
English translation: 1963, Trans. Moscow Math. Soc., **12**, 1-57.

[Mo1] Moser J., 1962, Nachr. Acad. Wiss. Göttingen Math. Phys. K1, **11a** **(1)**, 1-20, "On invariant curves of area preserving mappings of an annulus".

[Mo2] Moser J., 1966, Annali della Scuola Normale Superore di Pisa ser.3, **20** **(2)**, 265-315 and **20** **(3)**, 499-535, "A rapidly convergent method for nonlinear differential equations".

[Mo3] Moser J., 1968, Memoirs Am. Math. Soc., **81**, 1-60, "Lectures on Hamiltonian systems".

[Mo4] Moser J., "Stable and random motions in dynamical systems", Princeton University Press, Princeton, 1973.

[Mo5] Moser J., "Nearly integrable and integrable systems" in "Topics in nonlinear dynamics", Jorna S. Editor, Am. Inst. Phys. Conf. Proc., **46**, 1-15, 1978.

[Mu] Murdock J. and Robinson C., 1980, Journal of Diff. Equations, **36**, 425-441.

[Nei1] Neistadt A.I., "Some resonance problems in nonlinear systems", Thesis, Moscow University, 1975.

[Nei2] Neistadt A.I., 1975, Doklady Akad. Nauk. SSSR Mechanics, **221** (2), 301-304, "Passage through resonance in a two-frequency problem".
English translation: 1975, Soviet Phys. Doklady, **20** (3), 189-191.

[Nei3] Neistadt A.I., 1975, Doklady Akad. Nauk SSSR Mechanics, **223** (2), 314-317, "Averaging in multi-frquency systems".
English translation: 1975, Soviet Phys. Doklady, **20** (7), 492-494.

[Nei4] Neistadt A.I., 1976, Doklady Akad. Nauk. SSSR Mechanics, **226** (6), 1295-1298, "Averaging in multi-frequency systems II".
English translation: 1976, Soviet Phys. Doklady, **21** (2), 80-82.

[Nei5] Neistadt A.I., 1981, Prikl. Matem. Mekhan. SSSR, **45** (6), 1016-1025, "Estimates in the Kolmogorov theorem on conservation of conditionally periodic motions".
English translation: 1981, J. Appl. Math. Mech., **45**, 766-772.

[Nei6] Neistadt A.I., 1982, Prikl. Matem. Mekhan. SSSR, **45** (1), 58-63, "On the accuracy of conservation of the adiabatic invariant".
English translation: 1982, J. Appl. Math. Mech., **45**.

[Nei7] Neistadt A.I., 1984, Prikl. Matem. Mekhan. SSSR, **48** (2), 197-204, "On the separation of motions in systems with one rapidly rotating phase".
English translation: 1984, J. Appl. Math. Mech., **48**.

[Nek1] Nekhoroshev N.N., 1973, Math. Sbornik, **90** (3), 132.
English translation: 1973, Math. USSR Sbornik, **73** (3), 425-467.

[Nek2] Nekhoroshev N.N., 1977, Uspekhi Mat. Nauk, **32** (6), 5-66, "An exponential estimate of the time of stability of nearly-integrable Hamiltonian systems I".
English translation: 1977, Russian Math. Surveys, **32** (6), 1-65.

[Nek3] Nekhoroshev N.N., 1979, Trudy Sem. Petrovs., **5**, 5-50, "An exponential estimate of the time of stability of nearly-integrable Hamiltonian systems II".

[No1] Northrop T. G., "The Adiabatic Motion of Charged Particles", Wiley-Interscience Publishers, New York, 1963.

[No2] Northrop T.G. and Rome J.A., 1978, Physics of Fluids, **21** (3), 384-389, "Extensions of guiding center motion to higher order".

[Perk] Perko L.M., 1969, S.I.A.M. J. Applied Math., **17 (4)**, 698-724, "Higher order averaging and related methods for perturbed periodic and quasi-periodic systems".

[Pers] Persek S.C. and Hoppensteadt F.C., 1978, C.P.A.M., **31**, 133-156.

[Pö1] Pöschel J., 1982, Communications in pure and applied mathematics, **35**, 653-695, "Integrability of Hamiltonian systems on Cantor sets".

[Pö2] Pöschel J., 1982, Celestial Mechanics, **28**, 133.

[Poi] Poincaré H., Les méthodes nouvelles de la mécanique céleste (3 volumes), Gauthier-Villars, Paris, 1892, 1893, 1899.
English translation: N.A.S.A. Translation TT F-450/452, U.S. Fed. Clearinghouse, Springfield, 1967.

[Pok1] Pokrovskii V.L., Savvinykh S.K. and Ulinich F.R., 1958, Soviet Physics JETP, **34**, 879-882.

[Pok2] Pokrovskii V.L., Ulinich F.R. and Savvinykh S.K., 1958, Soviet Physics JETP, **34**, 1119-1120.

[Pu] Pugh C., 1969, Amer. J. Math., **91**, 363-367.

[Ra] Rapaport M., 1982, Celestial Mechanics, **28**, 291-293 .

[Re] Reed M. and Simon B., "Methods of modern mathematical physics 1: Functional analysis", Academic Press, New York, 1980.

[Ri] Rice J.R., The approximation of functions (2 volumes), Addison-Wesley Series in Computer Science and Information Processing, Addison-Wesley, Reading (Massachusetts), 1964.

[Ro] Robinson C., "Stability of periodic solutions from asymptotic expansions" in "Classical mechanics and dynamical systems", Devaney R.L. and Nitecki Z.H. Editors, Lecture Notes in Pure and Applied Mathematics **70**, Marcel Dekker Inc., New York and Basel, 1981.

[Rü] Rüssmann H., "On optimal estimates for the solutions of linear partial differential equations of first order with constant coefficients on the torus", Lecture Notes in Physics **38**, 598-624, Springer Verlag, Berlin, Heidelberg, New-York,1975.

[Sanc] Sanchez-Palencia E., 1976, Int. J. Nonlinear Mech., **11**, 251-263, "Méthode de centrage, estimation de l'erreur et comportement des trajectoires dans l'espace des phases".

[Sand1] Sanders J. A., 1980, S.I.A.M. J. Math. An., **11**, 758-770, "Asymptotic approximations and extension of time-scales".

[Sand2] Sanders J. A., 1982, Celestial Mechanics, **28**, 171-181, "Melnikov's method and averaging".

[Sand3] Sanders J. A. and Verhulst F., "Averaging methods in non linear dynamical systems", Applied Mathematical Sciences **59**, Springer Verlag, New York, 1985.

[Sc1] Schmidt W., 1969, J. Number Theory, **1**, 139-154, "Badly approximated systems of linear forms".

[Sc2] Schmidt W., "Diophantine approximations", Lecture Notes in Math. **785**, Springer Verlag, Berlin, Heidelberg, New-York, 1980.

[Sh] Shep T.J., 1974, Physica, **74**, 397-415, "Series solutions and the adiabatic invariant of the Helmholtz equation".

[Sie1] Siegel C.L., 1941, Ann. Math., **42**, 806-822.

[Sie2] Siegel C.L., 1942, Ann. Math., **43**, 607-612, "Iteration of analytic functions".

[Sie3] Siegel C.L., 1952, Nachr. Akad. Wissench. Göttingen Math.-Physik, **Kl. IIa**, 21-30.

[Sie4] Siegel C.L., 1954, Math. Ann., **128**, 144-170.

[Sie5] Siegel C. and Moser J., "Lectures on celestial mechanics", Springer Verlag, New York, 1971.

[Sim] Simon B., 1983, Physical Review Letters, **51 (24)**, 2167-2170, "Holonomy, the quantum adiabatic theorem, and Berry's phase".

[Sy] Symon K.R., 1970, Journal of Math. Phys., **11**, 1320.

[Vi1] Vittot M., "Théorie classique des perturbations et grand nombre de degrés de liberté", Thesis, Aix-Marseille, 1985.

[Vi2] Vittot M. and Bellissard J., 1985, preprint CNRS Marseille, "Invariant tori for an infinite lattice of coupled classical rotators".

[Vo] Voros A., 1983, Annales de l'Institut Henri Poincaré, Section A, **39 (3)**, 211-338.

[Was] Wasow W., "Linear turning point theory", Applied Mathematical Sciences **54**, Springer Verlag, New York, 1985.

[Way] Wayne E., 1984, Comm. Math. Phys., **96**, 311-329 and 331-344, "The KAM theory of systems with short range interactions".

[White] Whiteman K.J., 1977, Rep. Prog. Phys., **40**, 1033-1069.

[Whitn] Whitney H., 1934, Trans. A.M.S., **36**, 63-89, "Analytic extensions of differentiable functions defined on closed sets".

[Whitt] Whittaker E.T. and Watson G.N., "A course of Modern Analysis", Cambridge University Press, Cambridge U.K., fourth edition (1927) reprinted without changes, 1984.

[Ze] Von Zeipel H., 1916, Ark. Astron. Mat. Phys., **11** **(1)**.

[Zy] Zygmund A., "Trigonometric series", Volumes I and II combined, Cambridge University Press, Cambridge U.K., second edition (1959) reprinted 1977.

BIBLIOGRAPHICAL NOTES

1. On the book by N.N. Bogoliubov and Y.A. Mitropolski

The three books [And], [Kry] and [Bog] are undoubtedly among the important early works on nonlinear phenomena. Here we shall give a brief account of the latter, which to some extent has remained to these days an important, oft-quoted reference on the method of averaging, to which the authors - together with N.M. Krylov - very significantly contributed, although some of the ideas may be traced back at least to Lagrange. We refer to [Sand3] for a brief survey of the very early stages of the theory.

The book, which first appeared in Russian in 1958, is essentially written for engineers, so that, while making a relatively simple mathematical formalism widely accessible, its interest also lies in the treatment of numerous example problems in physics and electronics, many of which are still relevant despite their 1950's vintage.

This serves to stress that the terse résumé given below is very unfair to a remarkable book, in which purely mathematical considerations occupy only the last two chapters. We of course offer no justification for this, except to provide the reader with a convenient summary at a purely mathematical level before perusing the book.

i) The averaging method for one frequency systems (hence without resonance), taken to arbitrary order, is treated and illustrated with many examples. The general one frequency system in standard form is introduced in §25, and generalized averages, related to what we called KBM vector fields, are dealt with in §26 (for this material, see here Sections 3.1 and 3.2).

ii) n-frequency perturbations of linear systems, or in other words, sets of (isochronous) harmonic oscillators are also considered. In this very restricted treatment of n-frequency systems, the resonances are fixed independently of the values of the actions; there is thus no resonance *geometry* to consider (see here Section 3.2 and the final remarks of Chapter 6). This is the analog of Birkhoff's series construction in the neighborhood of elliptic fixed points of Hamiltonians (cf. Appendix 8). Specifically, Bogoliubov and Mitropolski first study the equation (§13):

$$x'' + \omega^2 x = \varepsilon f(\nu t, x, x'),$$

which may be expressed in the form of a system with two *fixed* frequencies ω and ν, in which case the resonance is given by a relation of the form $p\omega + q\nu = 0$, and its realization of course depends only on the system considered (and not on the values of x and x'). The approximations are then written down to arbitrarily high order, allowing an investigation of what happens as the pair (ω, ν) approaches a resonant value by treating ω and ν as parameters (§14). The generalization to n frequencies is next treated in the form of perturbations of a set of n harmonic oscillators, corresponding to normal modes of a linear system with constant coefficients (§20).

iii) The authors also investigate a system with two variable frequencies, with equations (§19):

$$d/dt\,[m(\tau)dx/dt] + k(\tau)x = \varepsilon F(\tau, \theta, x, dx/dt),$$

$$d\theta/dt = \nu(\tau), \quad \tau = \varepsilon t,$$

but this system is studied only in the neighborhood of a *given* resonance $\omega = p/q\ \nu$. One can recover a system identical to those considered in this book, but the restriction to a single resonance again makes geometry irrelevant.

To summarize, one may venture to state that the main contribution of the recent theories lies in the consideration of the *global geometry* of the resonances, although one should keep in mind that the content of Chapters 4 to 6 of the present book would have appeared quite elementary to Poincaré (or perhaps even to Gauss).

iv) In the last three paragraphs of the book ([Bog]), geometry enters, as it does here in Section 3.3, using results of Poincaré, Liapunov and others which opened the way to the modern theory of invariant manifolds. Unfortunately, though they are remarkably insightful, these sections are not easy to read today, except for the statement of the results, which are close to those presented or alluded to in Section 3.3 of the present book. In any case, these pertain less to the method of averaging itself than to the general theory of dynamical systems, many subsequent developments of which are adequately surveyed in [Gu].

2. On the book by M.I. Freidlin and A.D. Wentzell

The present work is limited to the deterministic case, but averaging results may also be established when the fast variables are stochastic processes. The fundamental reference in this case is surely the book by M.I. Freidlin and A.D. Wentzell, "Random Perturbations of Dynamical Systems" ([Fre]).

The authors consider systems of equations of the form:

$$(1) \qquad dX^{\varepsilon}(t)/dt = \varepsilon F(X^{\varepsilon}(t), \xi(t)),$$

where F is Lipschitz in all its arguments and $\xi(t)$ is a vector stochastic process, almost every trajectory of which is continuous. $\xi(t)$ is either given in advance or is the solution of a system of evolution equations; for simplicity, we will examine only the first case.

In contrast with the deterministic case, one can consider various kinds of convergence of the exact system to the averaged system: almost sure convergence, convergence in probability, and so on. An averaging theorem may be proved in the sense of almost sure convergence, provided the limit:

$$(2) \qquad \langle F\rangle(x, \omega) = \lim_{T\to\infty} 1/T \,.\, \int_0^{\infty} F(x, \xi(s), \omega)\, ds$$

exists uniformly in x for almost every realization of the process ξ. This theorem ensures that, on any interval $[0, T/\varepsilon]$, the solution of (1) with initial condition $X^{\varepsilon}(0) = x$ converges uniformly as ε tends to 0 to the solution of the averaged system:

$$(3) \qquad dX^{\omega}(t)/dt = \varepsilon\langle F\rangle(X^{\omega}, \omega)$$

with the same initial condition. Because of the uniform convergence, this averaging result is in some sense the strongest that can be proved. However, more interesting results are possible for random perurbations. The above theorem establishes a result similar to the deterministic case for almost every realization, but the averaged system *depends* on the realization ω considered.

More interesting averaging results are obtained by weakening the notion of convergence to convergence in probability. The averaged system is then deterministic and the theorem may be stated as:

Theorem:

If

a) $\sup_t E(F^2) < \infty$,

b) there exists an averaged vectorfield $\langle F\rangle(x)$, expressed as the probability limit of the time average of $F(x, \xi(t))$:

$$(3) \qquad \forall x, \forall \delta > 0 \;\; \lim_{T\to\infty} P\{\, |1/T.\int_t^{t+T} F(x, \xi(s))\, ds - \langle F\rangle(x)| > \delta \,\} = 0$$

which in some sense expresses the average size of the random perturbation,

then the solution $X^{\varepsilon}(t)$ of the exact system converges in probability to the solution $Y^{\varepsilon}(t)$ of the averaged system with the same initial condition:

$$(4) \qquad \forall T > 0, \forall \delta > 0 \quad \lim_{\varepsilon\to 0} P\{\mathrm{Sup}_{0\leq T\leq 1/\varepsilon} |X^{\varepsilon}(t) - Y^{\varepsilon}(t)| > \delta\} = 0.$$

This "law of large numbers" type result assures that the probability of a significant deviation between the two systems tends to 0 with ε; with nearly unit probability, the trajectory of the exact system thus remains in a small neighborhood of the averaged trajectory on the time interval $[0, 1/\varepsilon]$.

It is possible to go further and to analyze the distribution of the deviations $X^\varepsilon(t) - Y^\varepsilon(t)$ with respect to the averaged system. One shows that the average quadratic separation is of order $\sqrt{\varepsilon}$ (cf. the estimates in the deterministic case), and that the normalized deviation $\zeta^\varepsilon(t) = (X^\varepsilon(t) - Y^\varepsilon(t))/\sqrt{\varepsilon}$ has a normal distribution for every fixed t in the limit $\varepsilon \to 0$.

This "central limit" type result may still be improved by showing that the stochastic process $\zeta^\varepsilon(t)$, $t \in [0, 1/\varepsilon]$, converges weakly to a Gaussian process. The proof requires sufficient regularity of F along with sufficiently rapid decrease of the correlations of the stochastic process $\xi(t)$ (a property called strong mixing). It is carried out in two steps:

1) As $\zeta^\varepsilon(t)$ satisfies a system of evolution equations, one first shows the result for normalized deviations of the corresponding linearized system.

2) One next shows that, in the limit $\varepsilon \to 0$, the exact trajectory and the solution corresponding to the linearized system differ by an infinitesimal of order greater than $\sqrt{\varepsilon}$.

The deviations with respect to the averaged system may thus be approximated by a diffusion process. In the particular case where the averaged vectorfield is identically zero, it is possible to establish an analogous result on the longer time interval $[0, 1/\varepsilon^2]$: the stochastic process $X^\varepsilon(\varepsilon^2 t)$ converges weakly to a Gaussian process. In this way it may be shown that deviations of order one arise with high probability on such time intervals.

Averaging results are thus only generally valid on a timescale of order $1/\varepsilon$. Let us consider the case of an averaged system possessing an asymptotically stable equilibrium position. The averaging results show that, for small enough ε, and with nearly unit probability, the trajectories of the exact system remain in a neighborhood of this equilibrium position on the time interval $[0, T/\varepsilon]$, where $T>0$ may be chosen arbitrarily. However, they do not permit a precise estimate of the probability of leaving such a neighborhood on this interval, nor do they describe events determined by the behavior of the system on time intervals of order $1/\varepsilon^2$ or longer. Such events result from *large deviations* (of order one), the probability of which is exponentially small. Their study requires the introduction of a more suitable formalism, based on the notion of *action functionals*. The normalized action S (with normalization constant $1/\varepsilon$) is defined as follows: in the limit $\varepsilon \to 0$ the "probability" of a fluctuation $\varphi(t)$ is $C[\varphi] \exp\{-S[\varphi]/\varepsilon\}$. Using this concept, one obtains estimates *up to logarithmic equivalence* of the probability of rare events, meaning that the coefficients appearing in front of the exponential are not estimated.

It is also possible to study *intermediate deviations* $(X^\varepsilon(t) - Y^\varepsilon(t))/\varepsilon^\chi$, where

$0 < \chi < 1/2$, and this possibility is not limited to the case of an asymptotically stable equilibrium position. These deviations are related to both ordinary and large deviations: their probability tends to 0 with ε, like the probability for large deviations, while their characteristics, like those of ordinary deviations of order $\sqrt{\varepsilon}$, are determined by the behavior of the linearized exact system in the neighborhood of the averaged trajectory (which is not necessarily a stable equilibrium position). These intermediate deviations may also be studied with the aid of a normalized action functional (the normalization coefficient is then $\varepsilon^{2\chi-1}$). For a noise of order one, they have the same asymptotic behavior as do order $\sqrt{\varepsilon}$ deviations for a noise of order $\varepsilon^{1/2-\chi}$.

INDEX

Primary references are *italicized*

Applied Mathematical Sciences

cont. from page ii